New Trends in
GREEN CHEMISTRY

New Trends in
GREEN CHEMISTRY

V. K. Ahluwalia

Visiting Professor
Dr. B. R. Ambedkar Centre for Biomedical Research
University of Delhi, Delhi - 110 007, India

M. Kidwai

Coordinator
International Chapter of Green Chemistry in India
Department of Chemistry, University of Delhi,
Delhi - 110 007, India

Kluwer Academic Publishers
BOSTON   DORDRECHT   LONDON

Anamaya Publishers
NEW DELHI

A C.I.P. catalogue record for the book is available from the Library of Congress

ISBN 1-4020-1872-X

Copublished by Kluwer Academic Publishers,
P.O. Box 17, 3300 AA Dordrecht, The Netherlands
with Anamaya Publishers, New Delhi, India

Sold and distributed in all the countries, except India, by
Kluwer Academic Publishers, 101 Philip Drive,
Norwell, MA 02061, U.S.A. and in Europe by
Kluwer Academic Publishers, P.O. Box 322,
3300 AH Dordrecht, The Netherlands

In India by Anamaya Publishers
F-230, Lado Sarai, New Delhi - 110 030, India

Printed in India.

To
Professor Sukhdev, FNA
on his 80th birthday

Foreword

Organic chemistry has played a vital role in the development of diverse molecules which are used in medicines, agrochemicals and polymers. Most of the chemicals are produced on an industrial scale. The industrial houses adopt a synthesis for a particular molecule which should be cost-effective. No attention is paid to avoid the release of harmful chemicals in the atmosphere, land and sea. During the past decade special emphasis has been made towards green synthesis which circumvents the above problems. Prof. V. K. Ahluwalia and Dr. M. Kidwai have made a sincere effort in this direction.

This book discusses the basic principles of green chemistry incorporating the use of green reagents, green catalysts, phase transfer catalysis, green synthesis using microwaves, ultrasound and biocatalysis in detail. Special emphasis is given to liquid phase reactions and organic synthesis in the solid phase.

I must congratulate both the authors for their pioneering efforts to write this book. Careful selection of various topics in the book will serve the rightful purpose for the chemistry community and the industrial houses at all levels.

PROF. JAVED IQBAL, PhD, FNA
Distinguished Research Scientist & Head
Discovery Research
Dr. Reddy's Laboratories Ltd.
Miyapur, Hyderabad - 500 050, India

Preface

Organic chemistry deals with the synthesis of molecules having diverse uses in medicines, agrochemicals and biomolecules. The industries producing such chemicals are basically concerned with the type of reaction involved, the percentage yield etc. so that synthesis becomes cost effective. Special attention is given to ensure that there is no environmental pollution. All these form the basis of Green Chemistry—the pressing need of all nations. This book discusses designing green synthesis, basic principles of green chemistry with its role in day-to-day life, environmental pollution, green reagents and catalysts, phase transfer catalysis in green synthesis, microwave induced green synthesis, ultrasound assisted green synthesis, biocatalysts in organic synthesis, aqueous phase reactions, organic synthesis in the solid state, versatile ionic liquids as green solvents and finally some examples of synthesis involving basic principles of green chemistry.

We are grateful to Dr. Dennis Hjersen, Director, Green Chemistry Institute, USA; Professor Sukhdev, FNA, INSA Professor, New Delhi; Professor Nityanand, FNA, Former Director, CDRI Lucknow; Prof. Harjeet Singh, FNA, Professor Emeritus, Guru Nank Dev University, Amritsar; Prof. Mihir K. Chaudhury, FNA, Head, Department of Chemistry, Indian Institute of Technology, Guwahati; Prof. S. Chandrasekharan, FNA, Head, Department of Chemistry, Indian Institute of Science (Bangalore); Professor Javed Iqbal, FNA, Distinguished Research Scientist and Head, Discovery Research, Dr. Reddy's Laboratories Ltd., Hyderabad, for their valuable discussions.

V. K. AHLUWALIA

M. KIDWAI

Contents

New Trends in
GREEN CHEMISTRY

1. Introduction

Chemistry brought about medical revolution till about the middle of twentieth century in which drugs and antibiotics were discovered. These advances resulted in the average life expectancy rising from 47 years in 1900 to 75 years in 1990's. The world's food supply also increased enormously due to the discovery of hybrid varieties, improved methods of farming, better seeds, use of insecticides, herbicides and fertilizers. The quality of life on earth became much better due to the discovery of dyes, plastics, cosmetics and other materials. Soon, the ill effects of chemistry also became pronounced, main among them being the pollution of land, water and atmosphere. This is caused mainly due to the effects of by-products of chemical industries, which are being discharged into the air, rivers/oceans and the land. The hazardous waste released adds to the problem. The use of toxic reactants and reagents also make the situation worse. The pollution reached such levels that different governments made laws to minimise it. This marked the beginning of *Green Chemistry* by the middle of 20th century.

Green chemistry is defined as environmentally benign chemical synthesis. The synthetic schemes are designed in such a way that there is least pollution to the environment. As on today, maximum pollution to the environment is caused by numerous chemical industries. The cost involved in disposal of the waste products is also enormous. Therefore, attempts have been made to design synthesis for manufacturing processes in such a way that the waste products are minimum, they have no effect on the environment and their disposal is convenient. For carrying out reactions it is necessary that the starting materials, solvents and catalysts should be carefully chosen. For example, use of benzene as a solvent must be avoided at any cost since it is carcinogenic in nature. If possible, it is best to carry out reactions in the aqueous phase. With this view in mind, synthetic methods should be designed in such a way that the starting materials are consumed to the maximum extent in the final product. The reaction should also not generate any toxic by-products.

2. Designing a Green Synthesis

In any synthesis of a target molecule, the starting materials that are made to react with a reagent under appropriate conditions. Before coming to a final decision, consider all the possible methods that can give the desired product. The same product can also be obtained by modifying the conditions. The method of choice should not use toxic starting materials and should eliminate by-products and wastes. Following are some of the important considerations.

2.1 Choice of Starting Materials
It is very important to choose the appropriate starting materials. The synthetic pathway will depend on this. Also consider the hazards that may be faced by the workers (chemists carrying out the reaction and also the shippers who transport these) handling the starting materials.

Till now, most syntheses make use of petrochemicals (made from petroleum), which are non-renewable. Petroleum refining also requires considerable amounts of energy. It is therefore important to reduce the use of petrochemicals by using alternative starting materials, which may be of agricultural/biological origin. For example, some of the agricultural products such as corn, potatoes, soya and molasses are transformed through a variety of processes into products like textiles, nylon etc. Some of the materials that have biological origin (obtained from biomass) are: butadiene, pentane, pentene, benzene, toluene, xylene, phenolics, aldehydes, resorcinol, acetic acid, peracetic acid, acrylic acid, methyl aryl ethers, sorbitol, mannitol, glucose, gluconic acid, 5-hydroxymethyl furfural, furfural, levulinic acid, furan, tetrahydrofuran, furfuryl alcohol etc.

2.2 Choice of Reagents
Selection of the right reagent for a reaction is made on the basis of efficiency, availability and its effect on environment. The selection of a particular reagent versus another reagent for the same transformation can effect the nature of by-products, percentage yield etc. (see Chapter 6 for details).

2.3 Choice of Catalysts
Certain reactions proceed much faster and at a lower temperature with the use of catalysts. Heavy metal catalysts should be avoided as they cause environmental problems and are toxic in nature. Use of visible light to carry

out the required chemical transformation should be explored. Certain biocatalysts (enzymes) can also be used for various steps. This aspect will be discussed separately in Chapters 7, 11 and 12.

2.4 Choice of Solvents

Most of the common solvents generally used cause severe hazards. One of the commonly used solvents, benzene is now known to cause or promote cancer in humans and other animals. Some of the other aromatic hydrocarbons, for example toluene could cause brain damage, have adverse effect on speech, vision and balance, or cause liver and kidney problems. All these solvents are widely used because of their excellent solvency properties. These benefits nevertheless, are coupled with health risks.

Commonly used halogenated solvents, like methylene chloride, chloroform, perchloroethylene and carbon tetrachloride have long been identified as suspected human carcinogens.

The chlorofluorocarbons (CFCs) were used up to 20th century as cleaning solvents, blowing agents for molded plastics and for refrigeration. Despite the fact that CFCs have very low direct toxicity to humans and wild life (being non-inflammable and non-explosive, they have low accident potential), the single effect of CFCs in causing depletion of the ozone layer is a sufficient reason for not using them.

A versatile solvent, carbon dioxide[1] is used as liquid CO_2 or supercritical CO_2 fluid (the states of CO_2 most commonly used for solvent use). A gas is normally converted to a liquid state by increasing the pressure exerted upon it. However, if the substance is placed at a temperature above its critical temperature T_c (31 °C for CO_2) and critical pressure P_c (72.8 atm for CO_2), a supercritical fluid is obtained. The T_c of a substance is the temperature above which a distinct liquid phase of the substance cannot exist, regardless of the pressure applied. P_c is the pressure at which a substance can no longer exist in gaseous state. In a supercritical liquid, the individual molecules are pressed so close together (due to high pressure) that they are almost in liquid state. Supercritical liquids have density close to that of the liquid state and viscosity close to that of gaseous state.

It is ideal to carry out the reaction in aqueous phase if possible. The use of water as a solvent has distinct advantages. This aspect has been discussed separately in Chapter 12.

Another way to carry out the reaction is without the use of solvent (solvent less reactions). One such reaction comprises those reactions in which the starting materials and the reagents serve as solvents. Alternatively, the reactions can be performed in the molten state to ensure proper mixing. There is still another reaction that can be carried out on solid surfaces such as specialized clays. All such methods obviate the need of the solvent and will be discussed subsequently in Chapter 13.

A versatile solvent, ionic-liquid for carrying out reactions is discussed in Chapter 14.

Reference

1. J.A. Hyatt, Liquid and Supercooled Carbon Dioxide as Organic Solvents, *J. Am. Chem. Soc.*, 1986, **49**, 5097-5101.

3. Basic Principles of Green Chemistry

Green chemistry is defined as environmentally benign chemical synthesis. Any synthesis, whether performed in teaching laboratories or industries should create none or minimum by-products which pollute the atmosphere. According to the work carried out by Paul T. Anastas, the following basic principles of green chemistry have been formulated[1]:

- Prevention of waste/by-products.
- Maximum incorporation of the reactants (starting materials and reagents) into the final product.
- Prevention or minimization of hazardous products.
- Designing of safer chemicals.
- Energy requirement for any synthesis should be minimum.
- Selecting the most appropriate solvent.
- Selecting the appropriate starting materials.
- Use of the protecting group should be avoided whenever possible.
- Use of catalysts should be preferred wherever possible.
- Products obtained should be biodegradable.
- The manufacturing plants should be so designed as to eliminate the possibility of accidents during operations.
- Strengthening of analytical techniques to control hazardous compounds.

3.1 Prevention of Waste/By-Products

It is most advantageous to carry out a synthesis in such a way so that formation of waste (by-products) is minimum or absent. It is especially important because in most of the cases the cost involved in the treatment and disposal of waste adds to the overall production cost. Even the unreacted starting materials (which may or may not be hazardous) form part of the waste. Hence, the next basic principle is important and should carefully be considered as "prevention is better than cure" applies in this case also. In other words, the formation of the waste (or by-products) should be avoided as far as possible. The waste (or by-products) if discharged (or disposed off) in the atmosphere, sea or land not only causes pollution but also requires expenditure for cleaning-up.

3.2 Maximum Incorporation of the Reactants (Starting Materials and Reagents) into the Final Product

Chemists globally consider that if the yield of a reaction is about 90%, the reaction is good. The percentage yield is calculated by

$$\% \text{ yield} = \frac{\text{Actual yield of the product}}{\text{Theoretical yield of the product}} \times 100$$

In other words, if one mole of a starting material produces one mole of the product, the yield is 100%. Such synthesis is deemed perfectly efficient by this calculation. A perfectly efficient synthesis according to the percentage yield calculations may, however, generate significant amount of waste (or by-products) which is not visible in the above calculation. Such synthesis is not green synthesis. Typical examples like Wittig reaction and the Grignard reactions illustrate the above contention. Both these reactions may proceed with 100% yield but do not take into account the large amounts of by-products obtained.

The reaction or the synthesis is considered to be green if there is maximum incorporation of the starting materials and reagents in the final product. We should take into account the percentage atom utilisation, which is determined by the equation

$$\% \text{ atom utilization} = \frac{\text{MW of desired product}}{\text{MW of (desired product + waste products)}} \times 100$$

The concept of atom economy developed by B.M. Trost[2] is a consideration of 'how much of the reactants end up in the final product'. The same concept determined by R.A. Sheldon[3] is given as

$$\% \text{ atom economy} = \frac{\text{FW of atoms utilized}}{\text{FW of the reactants used in the reaction}} \times 100$$

Let us consider some of the common reactions viz., rearrangement, addition, substitution and elimination to find out which is more atom economical.

3.2.1 Rearrangement Reactions

These reactions involve rearrangement of the atoms that make up a molecule. For example, allyl phenyl ether on heating at 200 °C gives o-allyl phenol.

Allyl phenyl ether   o-Allyl phenol

This is a 100% atom economical reaction, since all the reactants are incorporated into the product.

3.2.2 Addition Reactions
Consider the addition of hydrogen to an olefin.

$$H_3C–CH = CH_2 + H_2 \xrightarrow{\text{Ni}} H_3C–CH_2–CH_3$$

Propene   Propane

Here also, all elements of the reactants (propene and hydrogen) are incorporated in the final product (propane). The reaction is a 100% atom economical reaction.

Similarly, cycloaddition reactions and bromination of olefins are 100% atom economical reactions.

Butadiene   Cyclohexene

$$H_3C–CH=CH_2 + Br_2 \longrightarrow H_3C–CH–CH_2Br$$

Propene $\underset{Br}{|}$

1,2-Dibromopropane

3.2.3 Substitution Reactions
In substitution reactions, one atom (or group of atoms) is replaced by another atom (or group of atoms). The atom or group that is replaced is not utilised in the final product. So the substitution reaction is less atom-economical than rearrangement or addition reactions. Consider the substitution reaction of ethyl propionate with methyl amine

$$CH_3CH_2\overset{O}{\overset{||}{C}}–OC_2H_5 + H_3C–NH_2 \longrightarrow CH_3CH_2\overset{O}{\overset{||}{C}}–NHCH_3 + CH_3CH_2OH$$

Ethyl propionate   Methyl amine   Propionamide   Ethyl alcohol

In this reaction, the leaving group (OC_2H_5) is not utilised in the formed amide. Also, one hydrogen atom of the amine is not utilised. The remaining atoms of the reactants are incorporated into the final product.

The total of atomic weights of the atoms in reactants that are utilised is

87.106 g/mole, while the total molecular weight including the reagent used is 133.189 g/mole (see table). Thus, a molecular weight of 46.069 g/mole remains unutilised in the reaction.

	Reactants		Utilised		Unutilised	
	Formula	FW	Formula	FW	Formula	FW
	$C_5H_{10}O_2$	102.132	C_3H_5O	57.057	C_2H_5O	45.061
	CH_5N	31.057	CH_4N	30.049	H	1.008
Total	$C_6H_{15}NO_2$	133.189	C_4H_9NO	87.106	C_2H_5OH	46.069

Therefore, the % atom economy = $\dfrac{87.106}{133.189} \times 100 = 65.40$ %.

3.2.4 Elimination Reactions

In an elimination reaction, two atoms or group of atoms are lost from the reactant to form a π-bond. Consider the following Hofmann elimination

$$H_3C-HC\underset{H}{\overset{CH_2-N^+(CH_3)_2CH_3}{<}} \ \xrightarrow{OH^- \ \Delta} \ CH_3-CH=CH_2 \ + \ (H_3C)_2N-CH_3 \ + \ H_2O$$

The elimination reaction is not very atom-economical. The percentage atom economy is 35.30%. In fact this is least atom-economical of all the above reactions.

Consider another elimination reaction involving the dehydrohalogenation of 2-bromo-2-methylpropane with sodium ethoxide to give 2-methylpropene.

$$\underset{\text{2-Bromo-2-methylpropane}}{H_3C-\overset{CH_3}{\underset{\overset{|}{Br}\ \overset{|}{H}}{C}}-CH_2} \ \xrightarrow{NaOC_2H_5} \ \underset{\text{2-Methylpropane}}{H_3C-\overset{CH_3}{\overset{|}{C}}=CH_2} \ + \ C_2H_5OH \ + \ NaBr$$

This dehydrohalogenation (an elimination reaction) is also not very atom-economical. The percentage atom economy or utilisation is 27% which is even less atom-economical than the Hofmann elimination reaction.

3.3 Prevention or Minimization of Hazardous Products

The most important principle of green chemistry is to prevent or at least minimize the formation of hazardous products, which may be toxic or environmentally harmful. The effect of hazardous substances if formed may be minimised for the workers by the use of protective clothing, engineering controls, respirator etc. This, however, adds to the cost of production. It is found that sometimes the controls can fail and so there is much more risk involved. Green chemistry, in fact, offers a scientific option to deal with such situations.

3.4 Designing Safer Chemicals

It is of paramount importance that the chemicals synthesised or developed (e.g. dyes, paints, adhesives, cosmetics, pharmaceuticals etc.) should be safe to use. A typical example of an unsafe drug is thalidomide (introduced in 1961) for lessening the effects of nausea and vomiting during pregnancy (morning sickness). The children born to women taking the drug suffered birth defects (including missing or deformed limbs). Subsequently, the use of thalidomide was banned, the drug withdrawn and strict regulations passed for testing of new drugs, particularly for malformation-inducing hazards.

With the advancement of technology, the designing and production of safer chemicals has become possible. Chemists can now manipulate the molecular structure to achieve this goal.

3.5 Energy Requirements for Synthesis

In any chemical synthesis, the energy requirements should be kept to a minimum. For example, if the starting material and the reagents are soluble in a particular solvent, the reaction mixture has to be heated to reflux for the required time or until the reaction is complete. In such a case, time required for completion should be minimum, so that bare minimum amount of energy is required. Use of a catalyst has the great advantage of lowering the energy requirement of a reaction.

In case the reaction is exothermic, sometimes extensive cooling is required. This adds to the overall cost. If the final product is impure, it has to be purified by distillation, recrystallisation or ultrafiltration. All these steps involve the use of energy. By designing the process such that there is no need for separation or purification, the final energy requirements can be kept at the bare minimum.

Energy to a reaction can be supplied by photochemical means, microwave or sonication and will be discussed in subsequent chapters.

3.6 Selection of Appropriate Solvent

The solvent selected for a particular reaction should not cause any environmental pollution and health hazard. The use of liquid or supercritical liquid CO_2 should be explored. If possible, the reaction should be carried out in aqueous phase or without the use a of solvent (solventless reactions). A better method is to carry out reactions in the solid phase (for details see Chapter 13).

One major problem with many solvents is their volatility that may damage human health and the environment. To avoid this, a lot of work has been carried out on the use of immobilised solvents. The immobilised solvent maintains the solvency of the material, but it is non-volatile and does not expose humans or the environment to the hazards of that substance. This can be done by tethering the solid molecule to a solid support or by binding the solvent molecule directly on to the backbone of a polymer. Some new polymer substances having solvent properties that are non-hazardous are also being discovered.

3.7 Selection of Starting Materials

Starting materials are those obtained from renewable or non-renewable material. Petrochemicals are mostly obtained from petroleum, which is a non-renewable source in the sense that its formation take millions of years from vegetable and animal remains. The starting materials which can be obtained from agricultural or biological products are referred to as renewable starting materials (sec. 2.1). The main concern about biological or agricultural products however, is that these cannot be obtained in continuous supply due to factors like crop failure etc.

Substances like carbon dioxide (generated from natural sources or synthetic routes) and methane gas (obtained from natural sources such as marsh gas) are available in abundance. These are considered as renewable starting materials.

3.8 Use of Protecting Groups

In case an organic molecule contains two reactive groups and you want to use only one of these groups, the other group has to be protected, the desired reaction completed and the protecting group removed. For example

Reactions of this type are common in the synthesis of fine chemicals, pharmaceuticals, pesticides etc. In the above protection, benzyl chloride (a known hazard) and the waste generated after deprotection should be handled carefully.

Another reaction involving protection of a keto function by using 1,2-ethanediol is as follows:

Thus, we see that the protecting groups that are needed to solve a chemoselectivity problem should be added to the reaction in stoichiometric amounts only and removed after the reaction is complete. Since these protecting groups are not incorporated into the final product, their use makes a reaction less atom-economical. In other words the use of protective group should be avoided whenever possible. Though atom-economy is a valuable criteria in evaluating a particular synthesis as 'green', other aspects of efficiency must also be considered.

3.9 Use of Catalyst

It is well known that use of a catalyst facilitates transformation without the catalyst being consumed in the reaction and without being incorporated in the final product. Therefore, use of catalyst should be preferred whenever possible. Some of the advantages are:

(i) Better yields. Hydrogenation or reduction of olefins in presence of nickel catalyst

$$H_3C\text{–}CH\text{=}CH_2 + H_2 \xrightarrow{\text{Ni}} H_3C\text{–}CH_2\text{–}CH_3$$
$$\text{Propene} \qquad\qquad\qquad\qquad \text{Propane}$$

$$C_6H_5CH_2Cl + \text{aqueous } KCN \xrightarrow[\text{Catalyst}]{\text{Phase transfer}} C_6H_5CH_2CN$$
Benzyl chloride   90% yield
                                                        Benzyl cyanide

Toluene   Benzoic acid
                                                       85% yield

(ii) The reaction becomes feasible in those cases where no reaction is normally possible

$$HC\equiv CH + H_2O \xrightarrow[H_2SO_4]{HgSO_4} H_3C-CHO$$

$$H_3C-C\equiv CH + CO + H_3C-OH \xrightarrow{Pd} CH_3-\overset{\overset{O}{\|}}{\underset{\underset{CH_2}{\|}}{C}}-C-OCH_3$$

Methyl methacrylate
(shell corporation)

(iii) Selectivity enhancement

$$\underset{\text{Propyne}}{H_3C-C\equiv CH + H_2} \xrightarrow[\text{mono addition}]{Pd-BaSO_4} \underset{\text{Propene}}{H_3C-CH=CH_2}$$

Selectivity in C-methylation versus O-methylation

$$\underset{\underset{O}{\|}}{C_6H_5-C-CH_2COCH_3} \xrightarrow[CH_3I]{NaOEt} C_6H_5\overset{\overset{CH_3}{|}}{\underset{\underset{O}{\|}}{C}}-CH-COCH_3 \quad (R\ \&\ S)$$

In addition to the above mentioned beneficial use of catalysts, there is significant advantage in the energy requirement. With advances in the selectivity of catalysts, certain reactions in green synthesis have become very convenient. A special advantage of the use of catalysts is better utilisation of starting materials and minimum waste product formation.

3.10 Products Designed Should be Biodegradable

The problem of products not being biodegradable is encountered particularly in insecticides and polymers.

Insecticides

It is well known that farmers use different types of insecticides to protect crops from insects. The more widely used insecticides are organophosphates, carbamates and organochlorides. Of these, organophosphates and carbamates are less persistent in the environment compared to the organochlorides (for example aldrin, dieldrin and DDT). Though the latter are definitely effective but they tend to bioaccumulate in many plant and animal species and incorporate into the food chain. Some of the insecticides are also responsible for the population decline[4] of beneficial insects and animals, such as honeybees, lacewings, mites, bald eagles etc.

DDT

Organophosphate
X = O or S
R' = CH$_3$ or CH$_2$CH$_3$

Carbamate

Considering the above, it is of utmost importance that any product (e.g. insecticides) synthesised must be biodegradable. It is also equally important that during degradation the products themselves should not possess any toxic effects or be harmful to human health. It is possible to have a molecule (e.g. insecticide) which may possess functional groups that facilitate its biodegradation. The functional groups should be susceptible to hydrolysis, photolysis or other cleavage.

Some of the diacylhydrazines (developed by Rohm and Hass Company) which have been found to be useful as insecticides are:

Tebufenozide

Halofenozide

Methoxyfenozide

3.11 Designing of Manufacturing Plants

The importance of prevention of accidents in manufacturing units cannot be over emphasised. A number of accidents have been found to occur in industrial units. The gas tragedy in Bhopal (December 1984) and several other places has resulted not only in loss of thousands of human lives but also rendered many persons disabled for the rest of their lives. The hazards posed by toxicity, explosions, fire etc. must be looked into and the manufacturing plants should be so designed to eliminate the possibility of accidents during operation.

3.12 Strengthening of Analytical Techniques

Analytical techniques should be so designed that they require minimum usage of chemicals, like recycling of some unreacted reagent (chemical) for the completion of a particular reaction. Further, placement of accurate sensors to monitor the generation of hazardous by-products during chemical reaction is also advantageous.

References

1. Paul T. Anastas and John C. Warner, Green Chemistry, Theory and Practice, Oxford University Press, New York, 1998.
2. Barry M. Trost, *Science*, 1991, **254**, 1471-1477.
3. Roser A. Sheldon, Chem. Ind. (London), 1992, 903-906.
4. Colin Baird, Environmental Chemistry, W.H. Freeman, New York, 1999.

4. Green Chemistry in Day-to-Day Life

With the advancement of science, green chemistry has changed our life style. Some of its important applications are described.

4.1 Dry Cleaning of Clothes

Perchloroethylene (PERC), $Cl_2C=CCl_2$ is commonly being used as a solvent for dry cleaning. It is now known that PERC contaminates ground water and is a suspected human carcinogen.

A technology, known as Micell Technology developed by Joseph De Simons, Timothy Romack, and James McClain made use of liquid carbon dioxide and a surfactant for dry cleaning clothes, thereby replacing PERC. Dry cleaning machines have now been developed using this technique.

Micell technology has also evolved a metal-cleaning system that uses carbon dioxide and a surfactant, thereby eliminating the need of halogenated solvents[1].

4.2 Versatile Bleaching Agent

It is common knowledge that paper is manufactured from wood (which contains about 70% polysaccharides and about 30% lignin). For good quality paper, the lignin must be completely removed. Initially, lignin is removed by placing small chipped pieces of wood into a bath of sodium hydroxide and sodium sulfide (that is how pulp is formed). By this process about 80-90% of lignin is decomposed. The remaining lignin was so far removed through reaction with chlorine gas. The use of chlorine removes all the lignin (to give good quality white paper) but causes environmental problems. Chlorine also reacts with aromatic rings of the lignin (by aromatic substitution) to produce dioxins, such as 2,3,7,8-tetrachloro-p-dioxin and chlorinated furans. These compounds are potential carcinogens and cause other health problems.

2,3,7,8-Tetrachloro-
dibenzo-p-dioxin

Chlorinated
furans

These halogenated products find their way into the food chain and finally into products like dairy products, pork, beef and fish. In view of this, use of chlorine has been discouraged. Subsequently, chlorine dioxide was used. Other bleaching agents like hydrogen peroxide, ozone or oxygen also did not give the desired results.

A versatile bleaching agent has been developed by Terrence Collins of Carnegie Mellon University. It involves the use of hydrogen peroxide as a bleaching agent in the presence of some activators known as TAML activators[2] that act as catalysts which promote the conversion of H_2O_2 into hydroxyl radicals that are involved in oxidation/bleaching. The catalytic activity of TAML activators allows H_2O_2 to break down more lignin in a shorter time and at much lower temperature.

These bleaching agents also find use in laundry and result in lesser use of water.

References

1. Micell Technologies, Website: *www.micell.com*, accessed Dec. 1999.
2. J.A. Hall, L.D. Vuocolo, I.D. Suckling, C.P. Horwitz, R.W. Allison, L.J. Wright and T. Collins, A new catalysed hydrogen peroxide bleaching process, Proceedings of the 53rd APPITA Annual Conference April 19-22, 1999, Rotorua, New Zealand.

5. Environmental Pollution

The term pollution is used to describe the introduction of harmful substances into the environment as a result of domestic, agricultural or industrial activities. Substances which pollute the air include gases like oxides of carbon, sulphur and nitrogen and the hydrocarbons emitted by thermal plants, motor vehicles and chemical industries. The increase in the level of carbon dioxide produces greenhouse effect, while nitrogen and sulphur oxides come down as acid rain causing destruction of vegetation and aquatic life. Water is normally polluted by effluents from industries, pesticides and fertilizers washed down from agricultural fields and city sewage. The greatest release of hazardous waste to the environment is from industries. Toxic chemicals flowing into rivers often kill fish. People who eat the contaminated fish have been known to develop health related problems, at times even fatal. Oil spills are also a major cause of water pollution which, if happens near the shore may seriously affect coastal flora and fauna. The increasing level of noise is also another cause of pollution which effects the environment. Pollution in any form is harmful to human life, and unless checked in time may threaten our very survival.

As already stated, environmental pollution results in a variety of ways. Presently, only the environmental pollution caused by the use of solvents, reagents and products will be dealt with.

The solvents that are mostly employed include volatile organic solvents (VOCs) like methylene chloride, chloroform, perchloroethylene (PERC) and carbon tetrachloride. Some VOCs like, isopropyl alcohols, xylenes, toluenes and ethylenes have been used as cleaning fluids because of their ability to dissolve oils, waxes and greases. Also, they readily evaporate from the items that they are being used to clean (VOCs readily evaporate or vaporise at room temperature). When VOCs come in contact with sun light and nitrogen oxides (by-products from the combustion of fossil fuels), these are transformed into ozone, nitric acid and partially oxidised organic compounds.

$$\text{VOCs} + \text{NO} + \text{Sunlight} \longrightarrow \longrightarrow \longrightarrow O_3 + HNO_3 + \begin{array}{l}\text{Organic compounds} \\ \text{(partially oxidised)}\end{array}$$

This product mixture (consisting of ozone, nitric acid and partially oxidised organic compound) is formed at the ground level and is commonly called smog. Continued exposure to smog may result in aggravating asthma and may induce respiratory problems[1] and can also cause lung cancer.[2]

This phenomenon leads to elevated levels of tropospheric ozone (one of the components of smog) and may cause damage to crops, discolour fabrics and harden rubber.

The use of chlorofluorocarbons (CFCs) and hydrochlorofluorocarbons (HCFCs) as solvents creates environmental problems. When CFCs are released into the atmosphere, these being inert, rise through the troposphere into the stratosphere (the CFCs are also released to the atmosphere from refrigeration industries), where they are photochemically decomposed by high-energy ultraviolet radiation from the sun. In fact the ozone layer present in the stratosphere prevents the harmful ultraviolet radiations from reaching the earth. On photochemical decomposition, the CFCs result in the formation of atomic chlorine (chlorine radical), which destroys the ozone by abstracting an oxygen atom from an ozone molecule and converting it into oxygen. The ClO radical formed, can react with an oxygen atom to form an oxygen molecule and regenerate a chlorine radical.

$$CF_2Cl_2 + UV \longrightarrow CF_2Cl + Cl^{\bullet}$$

$$O_3 + Cl^{\bullet} \longrightarrow O_2 + ClO^{\bullet}$$

$$ClO^{\bullet} + O \longrightarrow Cl^{\bullet} + O_2$$

The overall reaction is the conversion of ozone into oxygen.

$$O_3 + O \longrightarrow 2O_2$$

This results in depletion of the ozone layer and so the harmful ultraviolet radiations reach the surface of the earth. These ultraviolet radiations are responsible for causing skin cancer and cataract among humans.

Due to above reasons, the governments around the world have forbidden the use of CFCs. The Montreal Protocol, signed in Canada in 1987 and 1989, and subsequently in London in 1990 decided to speed up the phasing out of CFCs. It has already been stated (sec 2.4) that liquid and supercritical carbon dioxide has replaced CFCs which were also used as blowing agents for polystyrene.[3] Subsequently in place of CFCs, aliphatic hydrocarbons (e.g., pentane) were used as blowing agents for polystyrene. Though aliphatic hydrocarbons are not ozone depleting, they can lead to the formation of ground level smog (a mixture of ozone, nitric acid and partially oxidised organic compounds) if their emissions are not captured. A harmless blowing agent viz. carbon dioxide has been developed. The advantage of CO_2 is that it does not deplete the ozone layer, does not form smog, is economical, handling is easier and is nonflammable.

However, carbon dioxide is a greenhouse gas. It is commonly known that in the atmosphere, CO_2 allows UV and visible radiations to reach the earth's surface but reflects the IR radiations (heat) coming from the earth's surface and directing it back to the earth. Thus, excess levels of CO_2 in the atmosphere can significantly raise the temperature of the earth's atmosphere, which is responsible for global warming.

Nitrous oxide (N_2O), commonly known as laughing gas is also known to cause environment pollution. Nitrous oxide is obtained as a by-product during the manufacture of adipic acid as follows:

Benzene   Cyclohexane   Cyclohexanone   Cyclohexanol

$$\xrightarrow[\text{or } NH_4NO_3]{HNO_3} \quad HOOC-(CH_2)_4-COOH + N_2O$$

Adipic acid

The production of N_2O as a by-product can cause 10% annual increase of N_2O levels.[4] The N_2O is obtained as a by-product in a number of reactions involving oxidation with HNO_3. The N_2O formed rises into the stratosphere and plays a role in the destruction of the ozone layer. The first step in this process is the reaction of N_2O with atmospheric oxygen to produce $NO^{\bullet}$, which acts as a catalyst to deplete ozone[1] (steps 2 and 3). Nitric oxide abstracts an oxygen atom from an ozone molecule, giving O_2 and $NO_2^{\bullet}$ (step 2). The resulting $NO_2^{\bullet}$ then reacts with an additional oxygen atom regenerating $NO^{\bullet}$ and forming another molecule of O_2. The regenerated $NO^{\bullet}$ (step 3) can react again (steps 2 and 3) resulting in significant loss of ozone for every molecule of N_2O.

$$N_2O + O^{\bullet} \longrightarrow 2NO^{\bullet} \tag{1}$$

$$NO^{\bullet} + O_3 \longrightarrow NO_2^{\bullet} + O_2 \tag{2}$$

$$NO_2^{\bullet} + O^{\bullet} \longrightarrow NO^{\bullet} + O_2 \tag{3}$$

The overall reaction is

$$O_3 + O \longrightarrow 2O_2$$

N_2O also acts as a greenhouse gas and causes global warming (as in the case of CO_2). Also, air pollution is a result of exhaust emissions particularly CO_2 (already discussed) and nitrogen oxides. The nitrogen oxide part of the exhaust mixes with moisture in the atmosphere and then comes down in the form of acid rain. During lightening and thunderstorm, the nitrogen in the atmosphere combines with oxygen to give nitric oxide, which combines with oxygen to give NO_2. The NO_2 dissolves in water to give HNO_2 and HNO_3, which finally come to earth in the form of acid rain.

$$N + O \xrightarrow{\text{Electric discharge}} NO$$

$$NO + O \longrightarrow NO_2$$

$$2NO_2 + H_2O \longrightarrow HNO_2 + HNO_3$$

Acid rain is detrimental to plants and aquatic life. In some extreme cases, when the pH falls below 4.5, acid rain is responsible for making lakes devoid of aquatic life.[5]

References

1. C. Baird, Environmental Chemistry, W.H. Freeman, New York, 1999.
2. Smog linked to lung cancer in men. *www.web.net/men/air.htm*, accessed Dec. 1999.
3. Ozone depleting substances, *www.epa.gov/spdpublic/ods.html*, accessed Dec. 1999.
4. K.M. Draths and J.W. Frost, Environmentally Compatible Synthetic of Adipic Acid from D-glucose, *J. Am. Chem. Soc.*, 1994, **116**, 399-400.
5. P. Buell and J. Girard, Chemistry: An Environmental Perspective, Prentice Hall, Englewood Cliffs, NJ, 1994.

6. Green Reagent

In order to carry out the transformation of selected feedstock into the target molecule the criteria of efficiency, availability and effect of the reagent used must be kept in mind. Some of the green reagents are as follows.

6.1 Dimethylcarbonate

Conventional methylation reactions employ methyl halides or methyl sulfate. The toxicity of these compounds and their environmental consequences render these syntheses somewhat undesirable. Tundo[1,2] developed a method to methylate active methylene compounds selectively using dimethylcarbonate (DMC) (Scheme 1) in which no inorganic salts are produced.

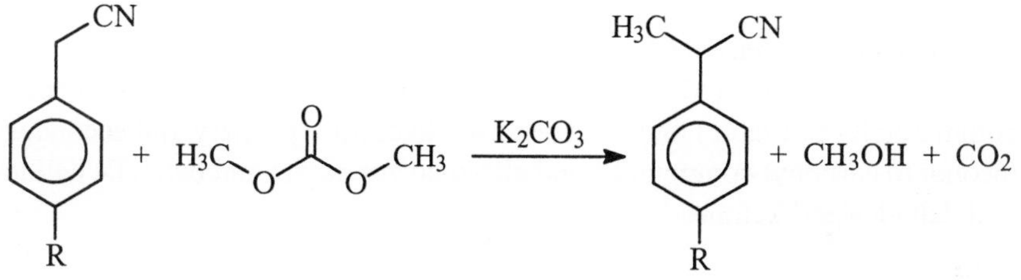

Scheme 1

Dimethylcarbonate (DMC) can also act as an organic oxidant. Cyclopentanone and cyclohexanone react with DMC and a base K_2CO_3 to yield adipic and pimelic methyl (or ethyl) esters, respectively[3] (Scheme 2).

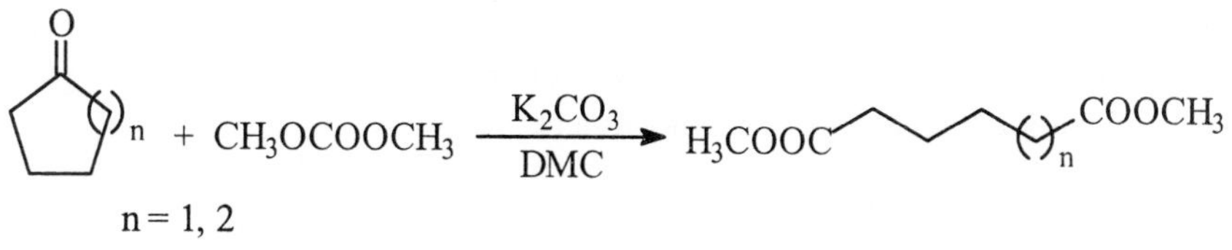

Scheme 2

Many oxidative processes have negative environmental consequences. By creating long-lived catalytic and recyclable oxidants, metal ion contamination in the environment can be minimized by using molecular oxygen as the primary oxidant. Several ligand systems that are stable towards oxidative decomposition in oxidizing environments are being developed by Collins.[4]

6.2 Polymer Supported Reagents

Besides DMC, there are a group of reagents which though are ordinary reagents, are bound to polymer support. The main advantage of using these reagents is that any excess of the reagent can be recovered by filtration and used again. Also, the isolation of the product is very easy. Some of such reagents are given as follows:

6.2.1 Polymer Supported Peracids

These are used for epoxidations of alkenes in good yields[5] (Scheme 3).

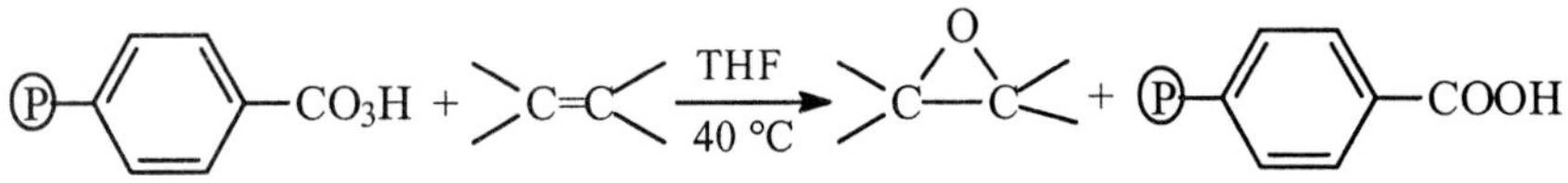

Scheme 3

6.2.2 Polymer Supported Chromic Acid

The polymer supported chromic acid (Amberlyst A-26, $HCrO_4^-$ form is commercially available) and has been used to oxidise primary and secondary alcohol to carbonyl compounds[6] and also oxidizes allylic and benzylic halides to aldehydes and ketones.[7]

6.2.3 Polymeric Thioanisolyl Resin

Polystyrene methyl sulphide on reaction with chlorine in the presence of triethylamine forms s-chloro sulfonium chloride resin, which acts as a selective oxidant for the alcohols[8] (Scheme 4).

6.2.4 Poly-N-Bromosuccinimide (PNBS)

It is an efficient polymer based brominating agent and is used as benzylic and allylic brominating agent. Thus, cumene on bromination[9] yield α,β,β'-tribromocumene (Scheme 5). However, bromination with NBS gives α,β-dibromocumene and α-bromocumene.

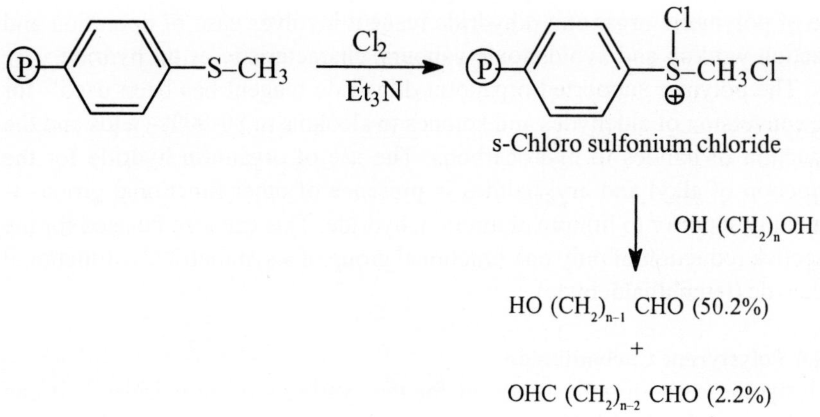

Scheme 4

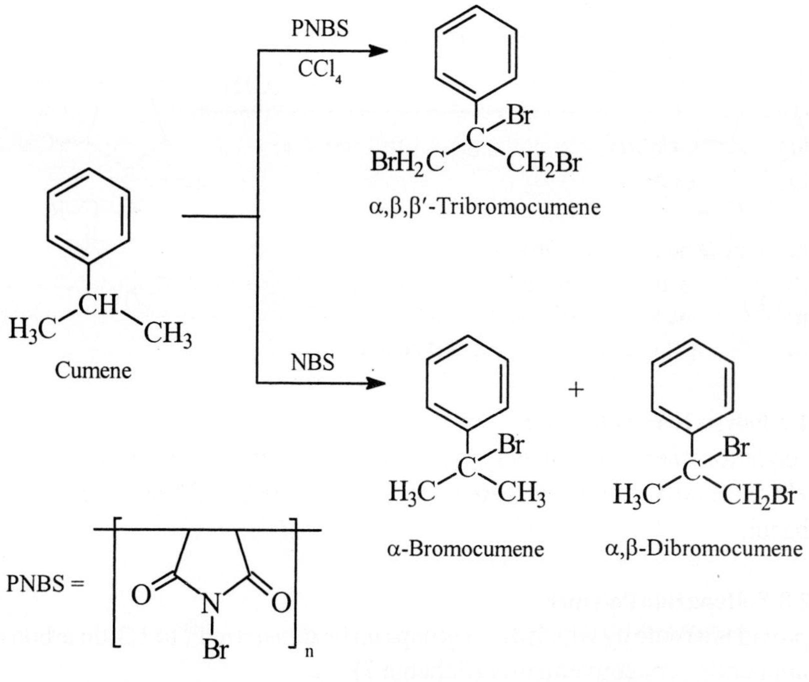

Scheme 5

6.2.5 Polymeric Organotin Dihydride Reagent as a Reducing Agent

The versatility and selectivity of organotin hydride is well documented. The use of polymeric organotin dihydride reagent involves ease of operation and reaction workup and avoids toxic vapours, characteristic of tin hydride.

The polymer supported organotin dihydride reagent has been used[10] for the conversion of aldehydes and ketones to alcohols in 80-90% yields and the reduction of halides to hydrocarbons. The use of organotin hydride for the reduction of alkyl and aryl halides in presence of other functional groups is generally superior to lithium aluminium hydride. This can also be used for the selective reduction of only one functional group of a symmetrical difunctional aldehyde (terephthaldehyde).

6.2.6 Polystyrene Carbodiimide

Polystyrene carbodiimide is useful for the synthesis of anhydrides[11]. It can also be used for the Moffatt oxidation of alcohols to aldehydes and ketones. Even the labile prostaglandin intermediate (A) is readily converted to the desired aldehyde (B) (Scheme 6).

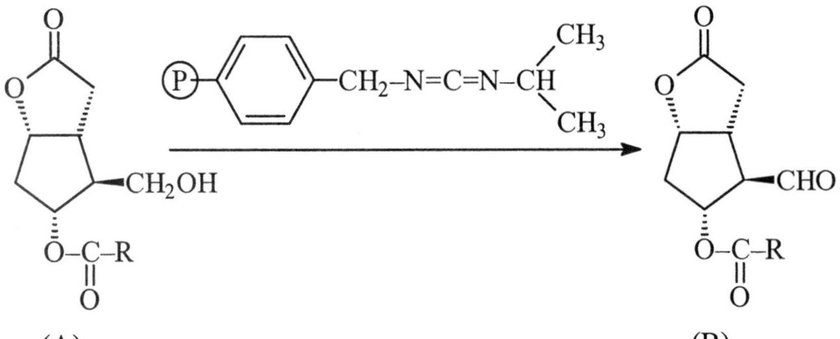

Scheme 6

6.2.7 Polystyrene Anhydride

Acetylation of aniline with polystyrene anhydride gives[12] benzanilide in 90% yield. Similarly, ethyl benzoate is obtained in 90% yield by acylation of ethanol.

6.2.8 Sulfonazide Polymer

It provides a route by which diazo group can be transferred[13] to 1,3-dicarbonyl compounds very conveniently (Scheme 7).

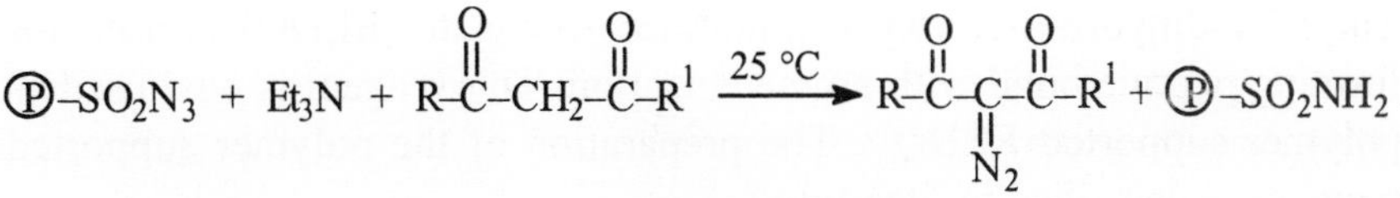

$$\text{P}-SO_2N_3 + Et_3N + R-\overset{O}{\overset{\|}{C}}-CH_2-\overset{O}{\overset{\|}{C}}-R^1 \xrightarrow{25\ °C} R-\overset{O}{\overset{\|}{C}}-\overset{}{\underset{\overset{\|}{N_2}}{C}}-\overset{O}{\overset{\|}{C}}-R^1 + \text{P}-SO_2NH_2$$

Scheme 7

6.2.9 Polystyrene Wittig Reagent

The polymeric Wittig reagent[14] (prepared as given in Scheme 8) reacts with carbonyl compounds (e.g. $C_6H_5COCH_3$, p-ClC_6H_4CHO, C_6H_5CHO etc.) to give the usual products (Scheme 8).

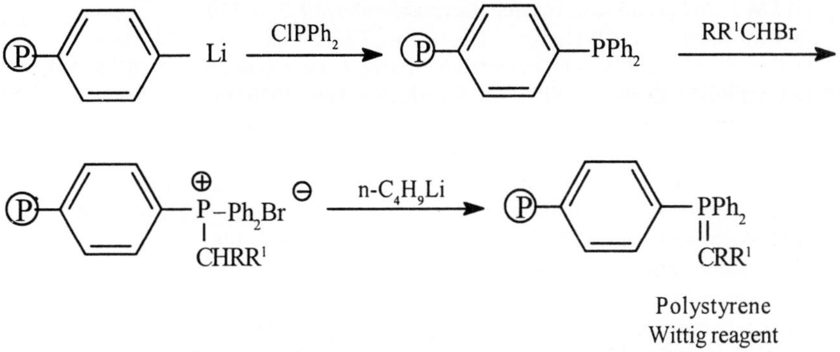

Polystyrene Wittig reagent

Scheme 8

6.2.10 Polymeric Phenylthiomethyl Lithium Reagent

The polymeric phenylthiomethyl lithium reagent[15] is useful for lengthening of side chain of alkyl iodide in good yield (Scheme 9).

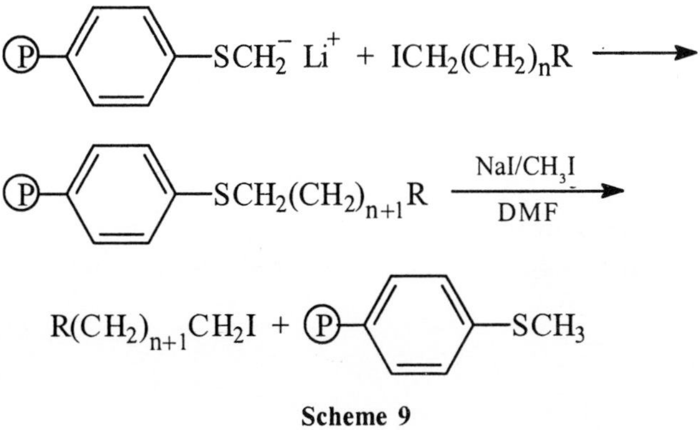

Scheme 9

6.2.11 Polymer Supported Peptide Coupling Agent

Ethyl-1,2-dihydro-2-ethoxy-1-quinolinecarboxylate (EEDQ) is used for forming peptide bond with no racemization. This reagent is now used as polymer supported EEDQ[16]. The preparation of the polymer supported reagents is described in literature[17].

References

1. P. Tundo, F. Trotta and G. Moraglio, *J. Org. Chem.*, 1987, **52**, 1300.
2. P. Tundo, F. Trotta and G. Moraglio, *J. Chem. Soc., Perkin Trans. 1*, 1989, 1070.
3. M. Selva, C.A. Marques and P. Tundo, *Gazz. Chim. Ital.*, 1993, **123**, 515.
4. T.J. Collins, *Acc. Chem. Res.*, 1994, **27**, 279.
5. (a) J.M.J. Frehet and K.E. Haque, *Macromolecules*, 1975, **8**, 130.
   (b) C.R. Harrison and P. Hodge, *J. Chem. Soc., Chem. Commun.*, 1974, 1009.
6. G. Cainelli, G. Cardillo, M. Orena and S. Sandri, *J. Am. Chem. Soc.*, 1976, **98**, 6737.
7. G. Cardillo, M. Orena and S. Sandri, *Tetrahedron Lett.*, 1976, 3985.
8. G.A. Crossby, N.M. Weinshenkar and H.S. Lin, *J. Am. Chem. Soc.*, 1975, **97**, 2232.
9. (a) C. Yaroslavsky, A. Patchorink and E. Katehalski, *Tetrahedron Lett.*, 1970, 3629.
   (b) C. Yaroslavsky, E. Patchonik and E. Katchalski, *Israel J. Chem.*, 1970, 37.
10. N.M. Weinshenkar, G.P. Crosby and J.Y. Wong, *J. Org. Chem.*, 1975, **40**, 1966.
11. N.M. Weinshenker and C.M. Shen, *Tetrahedron Lett.*, 1972, 3281.
12. M.B. Shambhu and G.A. Digens, *Tetrahedron Lett.*, 1973, 1627.
13. W.R. Roush, D. Feitler and J. Rebek, *Tetrahedron Lett.*, 1974, 1391.
14. (a) M.J. Farrall and J.M.J. Frechet, *J. Org. Chem.*, 1976, **41**, 3877.
    (b) W. Heitz and R. Michels, *Angew. Chem. Int. Ed.*, 1972, **11**, 298.
15. G.A. Crossby and M. Katu, *J. Am. Chem. Soc.*, 1977, **99**, 278.
16. J. Brown and R.E. Williams, *Can. J. Chem.*, 1971, **49**, 3764.
17. V.K. Ahluwalia and Renu Aggarwal, Organic Synthesis: Special Techniques, Narosa Publishing House, New Delhi, 2001, pp. 150-190 and the references cited therein.

7. Green Catalysts

Some of the major advances in chemistry especially industrial chemistry, over the past decade have been in the area of catalysts. Through the use of catalyst, chemists have found ways of removing the need for large quantities of reagents that would otherwise have been needed to carry out the transformations and ultimately would have contributed to the waste stream.

Catalysts play a major role in establishing the economic strength of the chemical industry and the clean technology revolution in the industry will provide new opportunities for catalysis and catalytic processes. Following are some different types of catalysts used.

7.1 Acid Catalysts

The traditional catalyst hydrogen fluoride, an extremely corrosive, hazardous and toxic chemical used in the production of linear alkylbenzenes (LAB's), has been successfully replaced by a solid acid catalyst, viz. fluorided silica-alumina catalyst, which does not require special material of construction (of the container), involves lower operating costs and obviates the need for an acid scrubbing system and waste disposal of calcium fluoride.[1]

Microencapsulated Lewis acids have replaced traditional corrosive monomeric Lewis acids in the reactions like Michael[2], Friedel Crafts[3], Mannich[4], Iminoaldol[5] reactions (Schemes 1 to 4).

Scheme 1

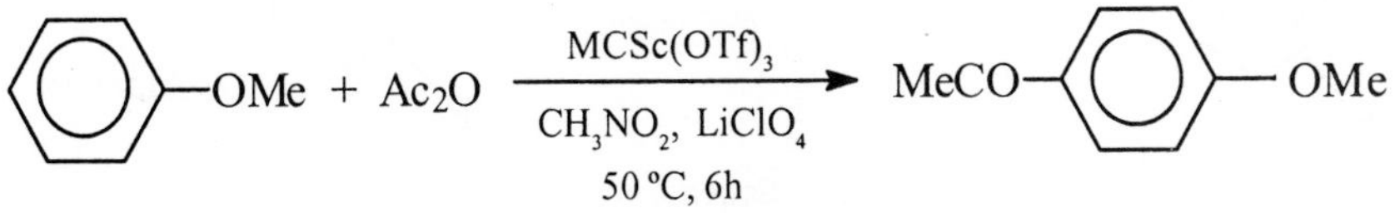

Scheme 2

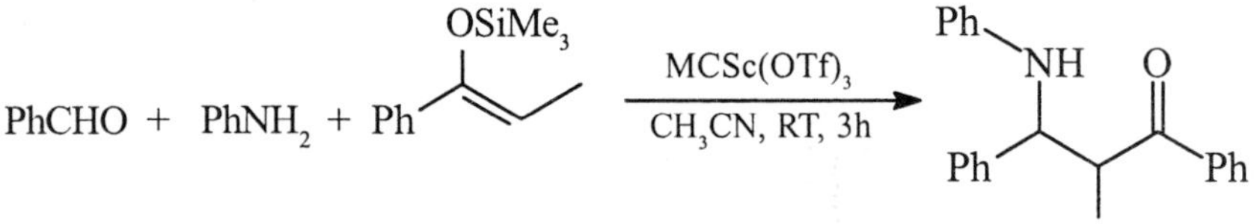

PhCHO + PhNH$_2$ + Ph

Scheme 3

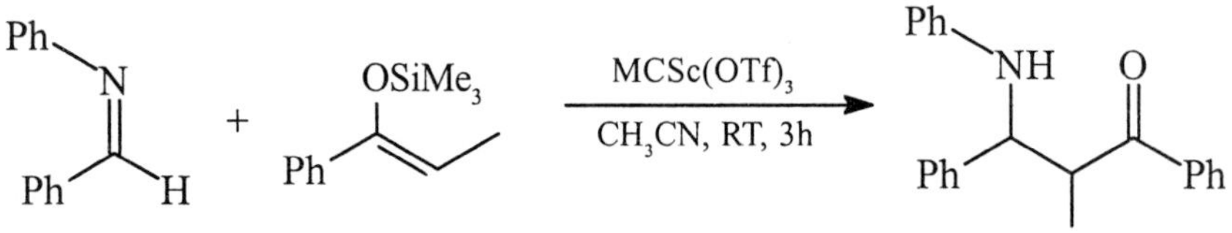

[MCSc(OTf)$_3$] = Microencapsulated scandium trifluoro methane sulfonate

Scheme 4

Several large-scale industrial processes utilizing heteropolyacids (HPA) catalysts exist. The two examples are hydration of the isobutylene and the polymerization of the tetrahydrofuran.[6-8]

Zeolites are widely used in the petrochemical industry in acid catalyzed processes and there are several reviews concerning recent developments in their use in the synthesis of fine and speciality chemicals[9,10] (Schemes 5 and 6).

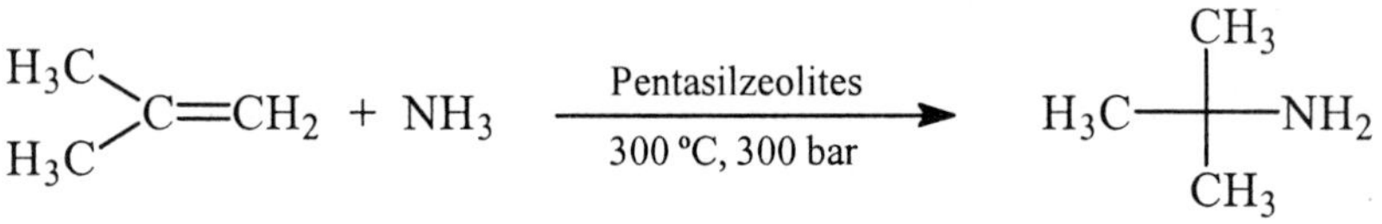

Scheme 5

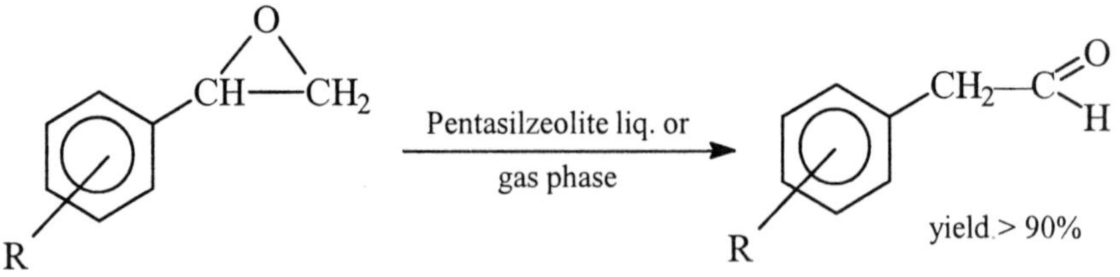

Scheme 6

Clayzic has been used to catalyze various Friedel-Crafts reactions including those of aromatic substrates with alkyl halides, aldehydes and alcohols. Other applications include the preparation of benzothiophenes by cyclisations of phenylthioacetals (normal catalysts can cause extensive polymerisation of the thiophene). The pores in clayzic are believed to favour the desired intramolecular cyclisation at the expense of polymerisation (Scheme 7)[11] and the olefination of benzaldehyde involving a previously unknown reaction mechanism (Scheme 8).[12]

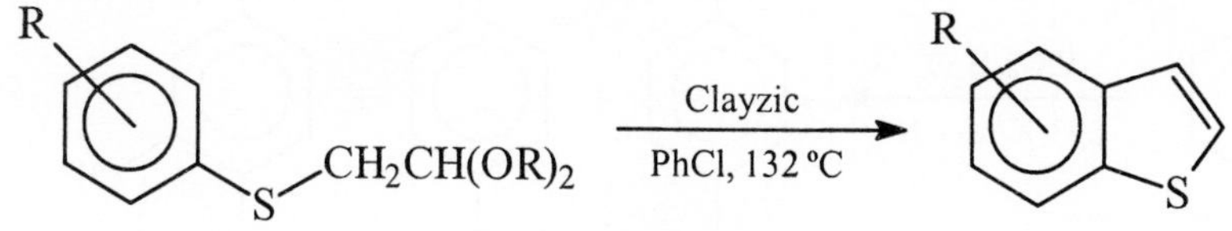

Scheme 7

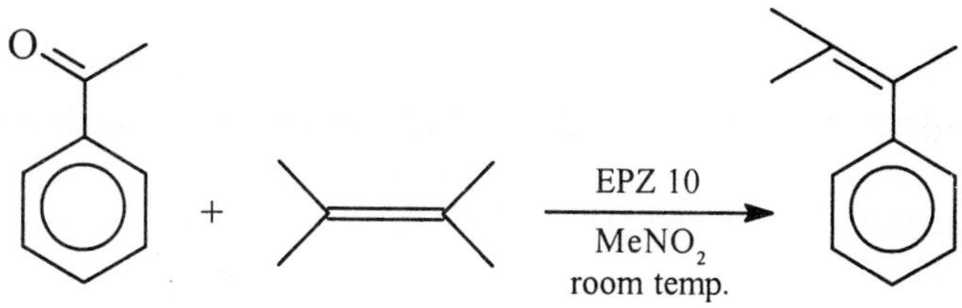

Scheme 8

7.2 Oxidation Catalysts

A large number of supported reagents have been used in the liquid phase partial oxidation of organic substrates.[13,14] There has been considerable success in the use of molecular sieves (titanium and vanadium) in commercial units.[15] The most important application of titanium silicates (TS-1) is the hydroxylation of phenol, giving mixtures of hydroquinone and catechol (Scheme 9). The process is clean, giving excellent conversion to product with very little waste.

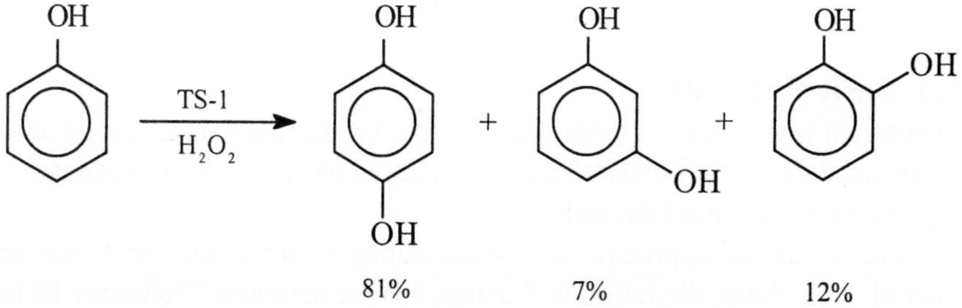

Scheme 9

Vanadium silicate molecular sieves are capable of selectively oxidising 4-chlorotoluene to 4-chlorobenzaldehyde using hydrogen peroxide as the source of oxygen in acetonitrile solvent (Scheme 10).[16]

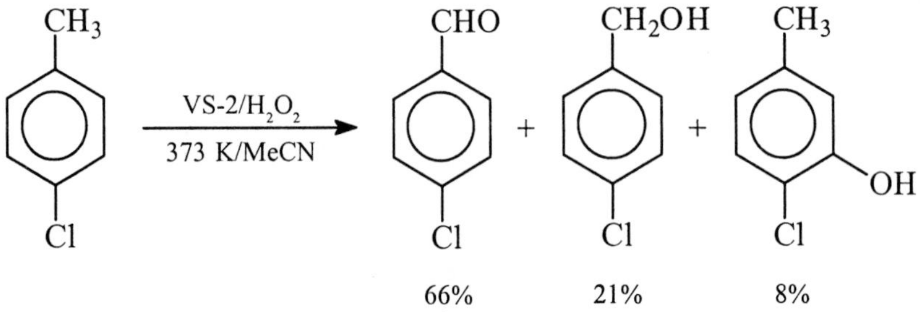

Scheme 10

Recent reports of oxidation catalysts based on chemically modified support materials include cobalt, copper and iron. Effective catalysts include cobalt immobilised on silica which has been derivatised with carboxylic acid functions (Scheme 11).[17]

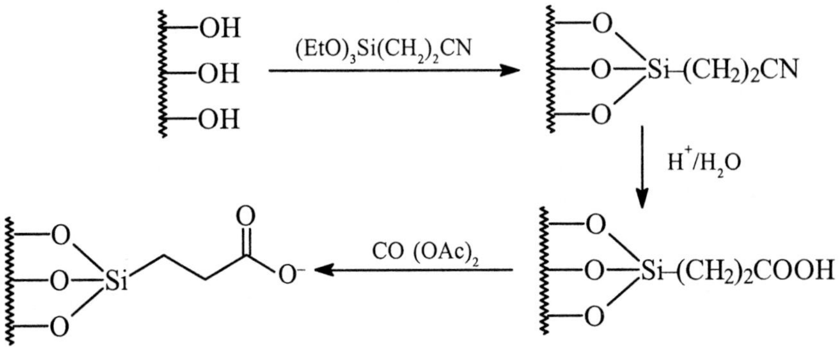

Scheme 11

7.3 Basic Catalysts

In contrast to the areas of heterogeneous oxidation catalysis and solid acid catalysis, the use of solid base catalysis in liquid phase reactions has not met the same level of breakthrough.

The industrial applications of basic catalysts are in the alkylation of phenol, side chain alkylation and isomerisation reactions (Schemes 12 to 14).[18,19]

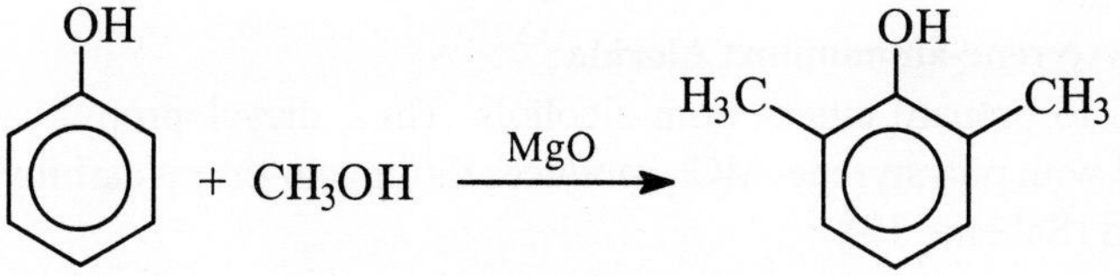

Scheme 12

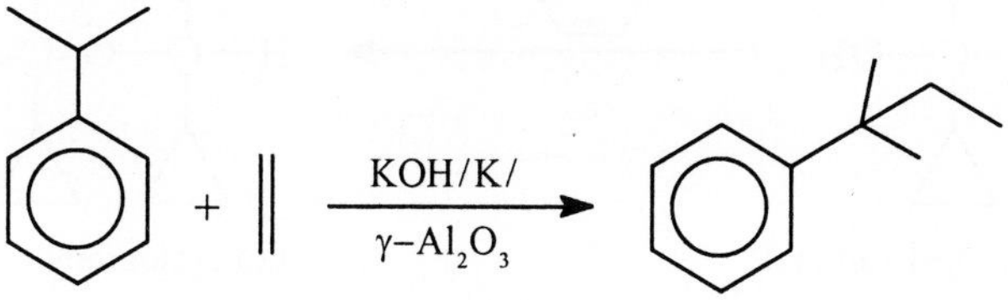

Scheme 13

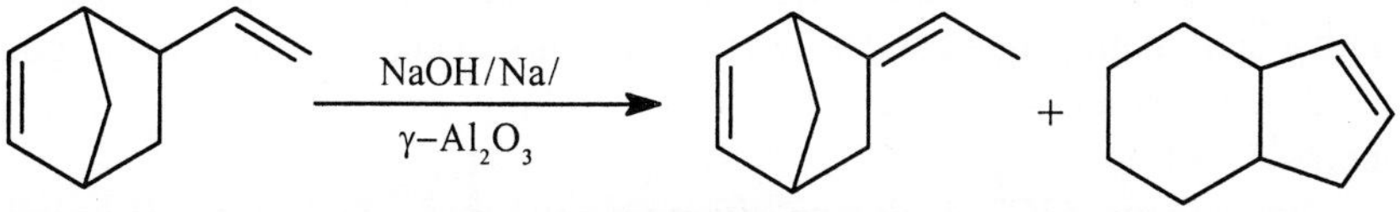

Scheme 14

Besides what has been stated above, the following catalysts/catalytic processes find wide applications in industry.

1. **Biocatalysis:** The most important conversions in the context of green chemistry is with the help of enzymes or biocatalysts[20] (for details see Chapter 11).

2. **Phase transfer catalysis and crown ethers:** These find numerous applications in organic synthesis in industry and in the laboratories[21] (for details see Chapter 8).

3. **Photocatalysis:** A large number of conversions/syntheses have been carried out photolytically (for details see Chapter 12).

7.4 Polymer Supported Catalysts

The conventional catalyst which is normally used in the homogenous phase, is linked to a polymer backbone and is used in this form to catalyse different reactions. Following are some of the polymer supported catalysts and their applications.

7.4.1 Polystyrene-aluminium Chloride

It is used to prepare ethers from alcohols. Thus, dicyclopropyl carbinol on treatment with polystyrene-AlCl$_3$ produces di(dicyclo-propylcarbinyl) ether in high yield (Scheme 15).[22]

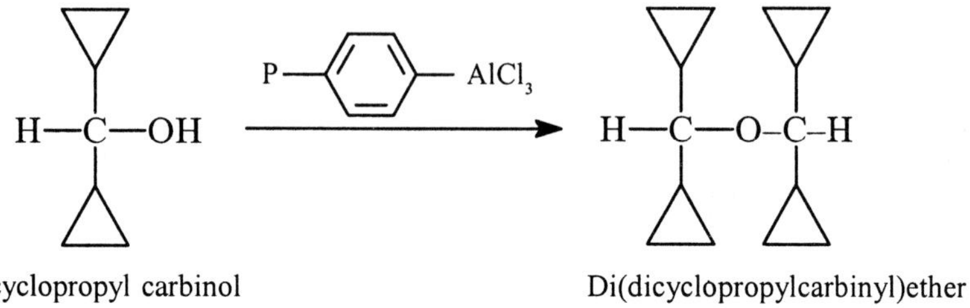

Dicyclopropyl carbinol   Di(dicyclopropylcarbinyl)ether

Scheme 15

Polystyrene-AlCl$_3$ is a useful catalyst for synthetic reactions, which require both a dehydrating agent and a Lewis acid. Thus, acetals are obtained in good yield by the reaction of aldehyde, alcohol and polymeric AlCl$_3$ in an organic inert solvent.

Polystyrene-AlCl$_3$ is also an effective catalyst for hydrolysis of acetals, e.g., heating the diethyl acetal of o-chlorobenzaldehyde with polymeric-AlCl$_3$ in benzene-methanol-water (2:6:1) for 17.5 hr gave o-chlorobenzaldehyde in 60% yield. Without the use of catalyst, the yield of aldehyde is only 4%.

7.4.2 Polymeric Super Acid Catalysts

A polymeric super acid catalyst is obtained[23] by binding aluminium chloride to sulfonated polystyrene. This polymeric super acid catalyst is used for the cracking and isomerisation of alkanes (e.g. n-hexane) at 357 °C at atmospheric pressure. Normally the above cracking and isomerisation is carried out in the presence of Lewis acid at high temperature and high pressure.

7.4.3 Polystyrene-metalloporphyrins

These catalysts[24] are useful for the oxidation of thiols to disulphide (a very facile reaction) in presence of base and is rapid even at room temperature (Scheme 16).

$$RSH + base \rightleftharpoons RS^- + H^+ + base$$
$$2RS^- + O_2 \longrightarrow 2RS^\cdot + O_2^-$$
$$2RS^\cdot \longrightarrow R_2S_2$$
$$O_2^- + H_2O \longrightarrow 2OH^- + \tfrac{1}{2}O_2$$

Scheme 16

7.4.4 Polymer Supported Photosensitizers

The photosensitizers supply molecular oxygen in a photochemical reaction. Examples of usual photosensitizers are Rose-Bengal, eosin-y, fluorescein. The photosensitizers are bound to Merifield type resin via ester bond (Scheme 17).[25]

A polymer-bound rose-bengal photosensitizer is commercially available. The efficiency of SENSITOX is about 65% of that of Rose-Bengal, but the high yields of the products and the ease of isolation more than compensate the slightly longer reaction periods (Scheme 17).

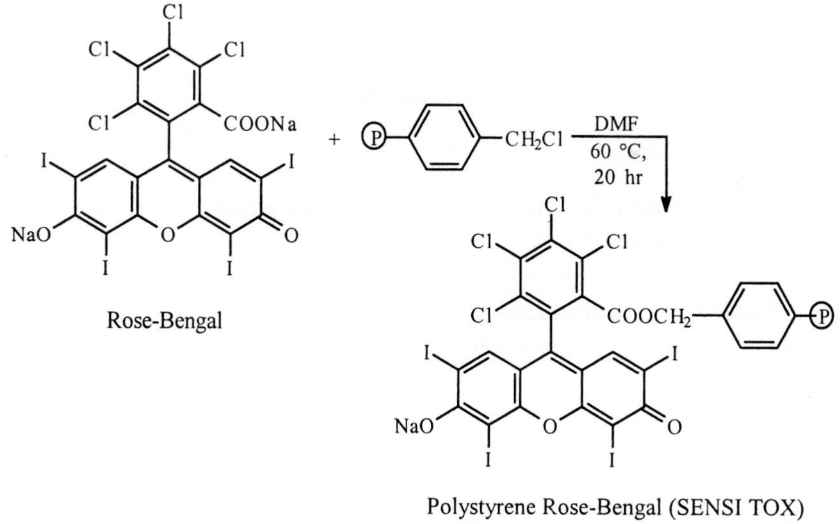

Rose-Bengal

Polystyrene Rose-Bengal (SENSI TOX)

Scheme 17

7.4.5 Polymer Supported Phase Transfer Catalysts

The use of phase transfer catalysts and crown ether will be discussed in Chapter 8. When the organophilic orium salts and crown ethers are immobilised on a polymer matrix, they retain most of their catalytic activity. However, the reaction rates are slower because of the difficulty in bringing the reacting species in contact with the catalytic site. This drawback is overcome by using a long chain to bind the catalyst to the polymer matrix. The advantage of using these polymer supported catalysts is that the product formed is separated from the reaction mixture by simple filtration and that the recycling of the catalyst is possible.

Using the above catalyst, 1-cyanooctane can be prepared by stirring a mixture of 1-bromooctane and potassium cyanide in the presence of resin P-$(CH_2)_6$-P$^+$ $(C_4H_9$-n$)_3$ Br$^-$ for 1.6 hr (Scheme 18).[26] Similarly, 1-iodooctane and n-octyl phenylsulphide are obtained as shown in (Scheme 18).

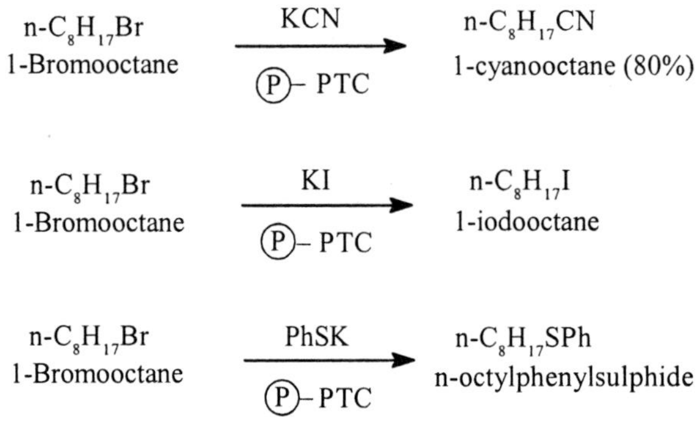

n-C_8H_{17}Br $\xrightarrow[\text{P-PTC}]{\text{KCN}}$ n-C_8H_{17}CN
1-Bromooctane   1-cyanooctane (80%)

n-C_8H_{17}Br $\xrightarrow[\text{P-PTC}]{\text{KI}}$ n-C_8H_{17}I
1-Bromooctane   1-iodooctane

n-C_8H_{17}Br $\xrightarrow[\text{P-PTC}]{\text{PhSK}}$ n-C_8H_{17}SPh
1-Bromooctane   n-octylphenylsulphide

Scheme 18

The preparation of various polymer supported phase transfer catalysts is described in literature.[27]

7.4.6 Miscellaneous Illustrations
Following are some typical illustrations using catalysts in green chemistry.

7.4.6.1 TiO$_2$ Photocatalyst in Green Chemistry
Titanium oxide-based photocatalytic systems have been developed[28] and are important for the purification of polluted water, the decomposition of offensive atmospheric odours as well as toxins, the fixation of CO_2 and the decomposition of chlorofluorocarbons on a huge global scale.

7.4.6.2 Solid Support Reagents
Using solid support reagents[29] the following synthesis have been carried out (Scheme 19).

(a) Cyanoethylation of alcohols and thioalcohols

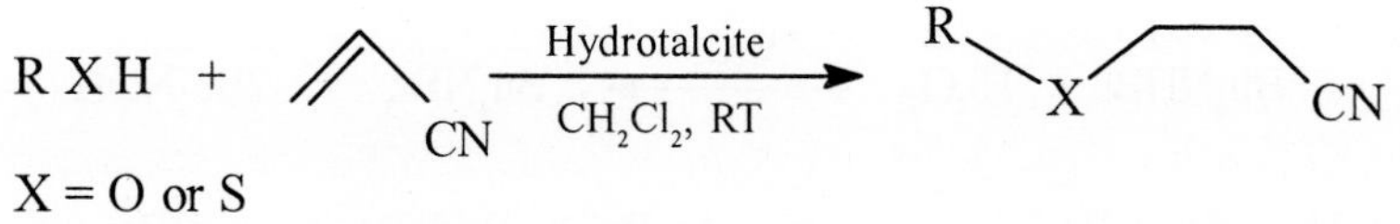

$$RXH \; + \quad \xrightarrow[\text{CH}_2\text{Cl}_2, \text{ RT}]{\text{Hydrotalcite}}$$

X = O or S

(b) Selective synthesis of β-nitroalkanols

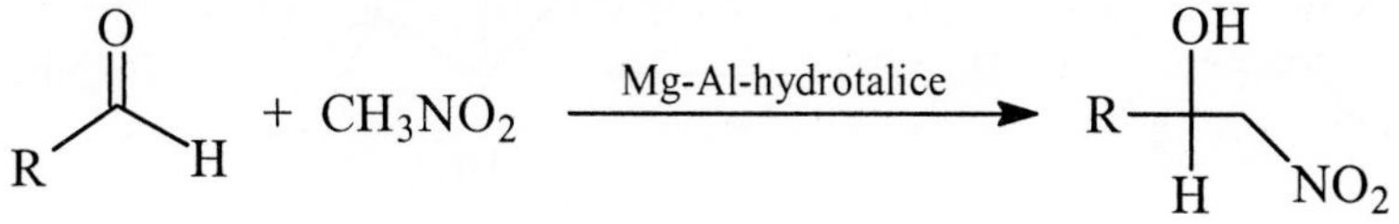

$$\text{(ketone)} + CH_3NO_2 \xrightarrow{\text{Mg-Al-hydrotalice}} $$

(c) Acylation of alcohols with carboxylic acids

$$R-OH + AcOH \xrightarrow{\text{Clay catalyst}} R-OAc + ROR + H_2O$$

(d) Formation of C–C bond

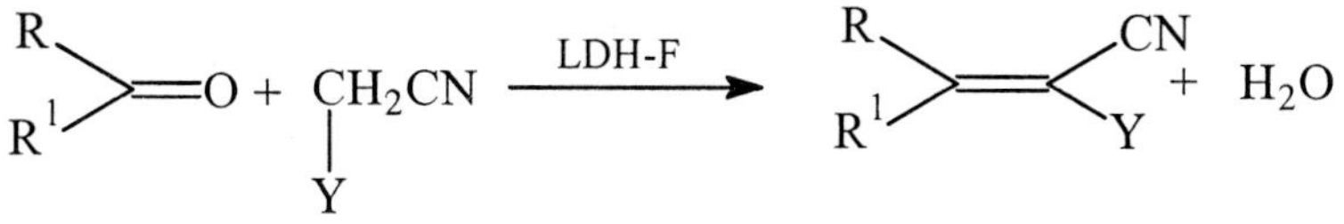

$$R_2C{=}O + CH_2CN{-}Y \xrightarrow{\text{LDH-F}} + H_2O$$

LDH-F is a layered double hydroxide fluoride-solid base catalyst

Scheme 19

7.4.6.3 Synthesis of Bromoorganics: Development of Newer and Ecofriendly Bromination Protocols and Brominating Agents

The bromoorganics are important precursors for the preparation of pharmaceutical, agrochemicals and other speciality chemicals.[30] The preparation of these compounds involve the use of toxic chemicals, especially Br_2, which has been a cause of great concern globally. It has been possible to develop newer and ecofriendly bromination protocols and brominating agents.[31] Some examples of bromination are given in (Scheme 20).

7.4.6.4 Synthesis of Pyridinium Fluorochromate (PFC)

PFC is one of the most useful oxidants for partial and selective oxidations of a variety of organic substrates.[32] PFC is obtained as shown in Scheme 21.[33]

$$3Bu_4NHBr_3 + H_2O_2 \xrightarrow{V_2O_5} Bu_4NBr_3 + 2Bu_4NOH$$

$$ArH + Bu_4NBr_3 \longrightarrow Ar–Br + Bu_4NBr + HBr$$

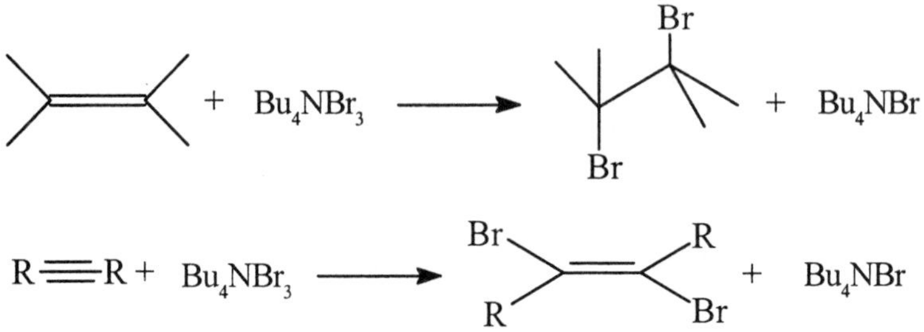

Scheme 20

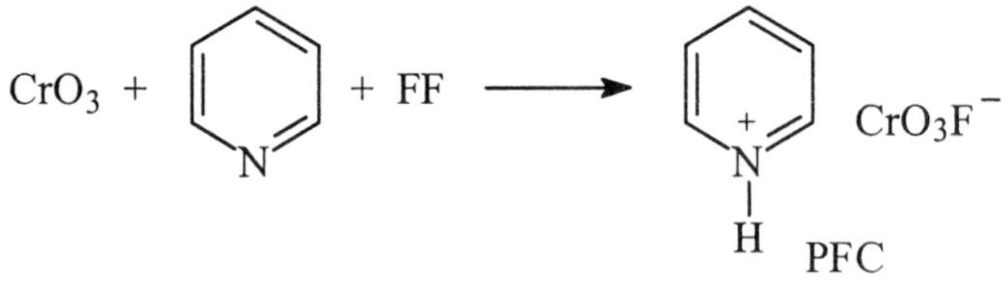

Scheme 21

7.4.6.5 Synthesis of Isooctane

Isooctane constitutes an integral part of gasoline. Its conventional method of synthesis requires large amount of highly toxic and corrosive acids. A convenient method[34] involves the use of suitable solid acids such as alumina, zeolites or nafion polymer (perfluoropolymer with –SO$_3$ side group) (Scheme 22).

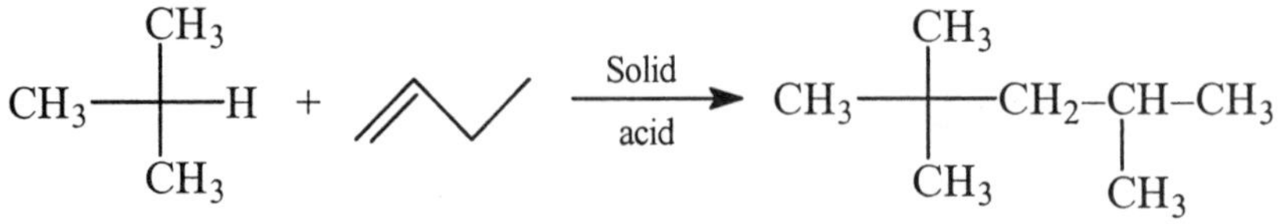

Isooctane

Scheme 22

References

1. P.T. Anastas and J.C. Warner, 'Green Chemistry, Theory and Practice', Oxford University Press (1998).
2. S. Kobayashi, I. Hachiya, H. Ishitani and M. Araki, *Synlett.*, 1993, 472.
3. A. Kawada, S. Mitamura and S. Kobayashi, *Synlett.*, 1994, 545.
4. S. Kobayashi, M. Araki and M. Yasuda, *Tetrahedron Lett.*, 1995, **36**, 5773.
5. S. Kobayashi, M. Araki, H. Ishitani, S. Nagayama and S. Hachiya, *Synlett.*, 1995, 233.
6. M. Misono, *Catal. Rev. Sci-Eng.*, 1987, **29**, 269; 1987, **30**, 339.
7. (a) T. Okuhara, N. Mizuno and M. Misono, *Adv. Catal.*, 1996, **41**, 113.
   (b) N. Mizuno and M. Misono, *Chem. Rev.*, 1998, **98**, 199.
8. M. Misono and N. Nojiri, *Appl. Catal.*, 1990, **64**, 1.
9. A. Corma, *Chem. Rev.*, 1995, **95**, 559.
10. A. Corma and H. Garcia, *Catal. Today*, 1997, **38**, 257.
11. P.D. Clark, A. Kirk and J.G.K. Yee, *J. Org. Chem.*, 1995, **60**, 1936.
12. H.P. Van Shaik, R.J. Vjin and F. Bickelhaupt, *Angew. Chem. Int. Ed. Engl.*, 1994, **33**, 1611.
13. Solid Supports and Catalysts in Organic Synthesis, ed. K. Smith, Ellis Horwood, Chichester, 1992.
14. J.H. Clark, 'Catalysis of Organic Reactions using Supported Inorganic Reagents', VCH, New York, 1994.
15. A.W. Ramaswamy, S. Sivbasanker and P. Ratnasamy, *Microporous Mater*, 1994, **2**, 451.
16. T. Selvam and A.P. Singh, *J. Chem. Soc. Chem. Commun.*, 1995, 883.
17. A.J. Butterworth, J.H. Clark, S.J. Barlow and T.W. Bastock, *U.K. Patent Appl.*, 1996.
18. J.H. Clark, A.P. Kybett and D.J. Macquarrie, 'Supported Reagents: Preparation, Analysis and Applications', VCH, New York, 1992.
19. T. Ando, S.J. Brown, J.H. Clark, D.G. Cork, T. Hanatusa, J. Lehira, J.M. Miller and M.J. Robertson, *J. Chem. Soc., Perkin Trans.*, 1986, **2**, 1133.
20. M. Held, A. Schmid, J.B. Van Beilen and B. Withold, *Pure Appl. Chem.*, 2000, **72(7)**, 1337.
21. M. Makosza, *Pure Appl. Chem.*, 2000, **72(7)**, 1399.
22. D.C. Neckers, D.A. Kooistra and G.W. Green, *J. Am. Chem. Soc.*, 1972, 9284.
23. V.L. Magnotta, B.C. Gates and G.C.A. Schuit, *J. Chem. Soc. Chem. Commun.*, 1976, 342; V.L. Magnotta and B.C. Gates, *J. Pol. Sci. Polym. Chem. Ed.*, 1977, **15**, 1341; V.L. Magnotta and B.C. Gates, *J. Catal.*, 1977, **46**, 266.
24. L.D. Rollmann, *J. Am. Chem. Soc.*, 1975, **97**, 2132.
25. E.C. Blossey, D.C. Neckers, A.L. Thayes and A.B. Schapp, *J. Am. Chem. Soc.*, 1973, **95**, 5820; A.P. Schapp, A.L. Thayer, E.C. Blossey and D.C. Neckers, *J. Am. Chem. Soc.*, 1975, **97**, 3741.
26. M. Cinouini, S. Colonna, H. Molinari and F. Montanari, *J. Chem. Soc. Chem. Commun.*, 1976, 396.
27. V.K. Ahluwalia and Renu Aggarwal, Organic Synthesis, Special Techniques, Narosa Publishing House, New Delhi, 2000, p. 150-190 and the references cited therein.
28. M. Anpo, *Pure Appl. Chem.*, 2000, **72**, 1265.
29. J.S. Yadav and H.M. Meshram, *Pure Appl. Chem.*, 2001, **73**, 199; B.M. Choudary, M. Lakshmi Kantam and B. Kavita, *Green Chem.*, 1999, **1**, 289; B.M. Choudary, V. Bhaskar, M. Lakshmi Kantam, K.K. Rao and K.V. Raghavan, *Green Chem.*, 2000, **2**, 67; B.M. Choudary, M. Lakshmi Kantam, V. Neeraja, K.K. Rao, F. Figneras and I. Delmotte, *Green Chem.*, 2001, **3**, 1.

30. G.W. Gribbe, *Chem. Soc. Rev.*, 1999, **28**, 335; A. Butler and J.V. Walker, *Chem. Rev.*, 1993, **93**, 1937.
31. M.K. Chaudhuri, A.T. Khan, B.K. Patel, D. Dey, W. Kharmawphlang, T.R. Lakshmiprabha and G.C. Mandel, *Tetrahedron Lett.*, 1998, **39**, 8163; U. Bora, G. Bose, M.K. Chaudhuri, S. Dhar, R. Gopinath, A.T. Khan and B.K. Patel, *Org. Lett.*, 2000, **2**, 247; B. Bora, M.K. Chaudhuri, D. Dey and S.S. Dhar, *Pure Appl. Chem.*, 2001, **73**, 93; G. Bose, P.M. Bujar Barua, M.K. Chaudhuri, D. Kalita and A.T. Khan, *Chem. Lett.*, 2001, 290.
32. J.H. Clark (ed.), Chemistry of Waste Minimization, Chapman and Hall, London, 1995.
33. U. Bora, M.K. Chaudhuri and S.K. Dehury, *Current Science*, 2002, **82**, 1427.
34. J.A. Cusumano, *Chemtech*, 1992, 485.

8. Phase-Transfer Catalysis in Green Synthesis

8.1 Introduction

Most of the pharmaceuticals or agricultural chemicals (insecticides, herbicides, plant growth regulators) are the result of organic synthesis. Most of the syntheses require a number of steps in which additional reagents, solvents and catalysts are used. In addition to the syntheses of the desired products, some waste material (by-products) is generated, the disposal of which causes problems and also environmental pollution. In view of this, attempts have been made to use procedures that minimise these problems. One of the most general and efficient methodologies that takes care of the above problems is to use a phase-transfer catalyst (PTC).[1-3]

Difficulties are often encountered in organic synthesis if the organic compound is soluble in organic solvent and the reagent in water. In such cases, the two reactants will react very slowly and the reaction proceeds only at the interface where these two solutions are in contact. The rate of the reaction can, of course, be slightly increased by stirring the reaction mixture and by using aprotic polar solvents, which solvate the cations so that the anions are free. Such solvents (like dimethylsulfoxide, dimethylformamide) are expensive and their removal is difficult. Also the use of strong bases (which are necessary for the reactions like Wittig etc.) create other problems and many side reactions take place. These problems can be overcome by using a catalyst, which is soluble in water as well as in the organic solvent. Such catalysts are known as phase-transfer catalysts (PTC).

The PTC reaction, in fact, is a methodology for accelerating the reaction between water insoluble organic compounds and water soluble reactants (reagent). The basic function of PTC is to transfer the anion (from the reagent) from the aqueous phase to the organic phase. As a typical example, the reaction of 1-chlorooctane with NaCN in water does not give 1-cyanooctane even if the reaction mixture is stirred for several days. However, if a small quantity of an appropriate PTC is added the product is formed in about 2 hr giving 95% yield (Scheme 1).

$$CH_3(CH_2)_6CH_2Cl \xrightarrow[\substack{PTC\ [CH_3(CH_2)_{15}P^+(n\text{-}Bu)_3] \\ 105\ °C,\ 2\ hr}]{NaCN,\ H_2O,\ decane} CH_3(CH_2)_6CH_2CN$$

1-Chlorooctane   1-Cyanooctane
                                                              95%

Scheme 1

The mechanism of the PTC reaction is well known and is described in relevant literature.[1-3]

Phase transfer catalysts used are the quaternary 'onium' salts, such as ammonium, phosphonium, antimonium and tertiary sulphonium salts. In practice, however, only a limited number of ammonium and phosphonium salts are widely used. Some of the PTCs normally used are:

(i)   Aliquat 336: methyl trioctyl ammonium chloride, $N^+CH_3 (C_8H_{17})_3Cl^-$

(ii) Benzyl trimethyl ammonium chloride or bromide (TMBA), $N^+(CH_3)_3CH_2C_6H_5X^-$ (X = Cl or Br)

(iii) Benzyl triethyl ammonium chloride or bromide (TEBA), $N^+ (C_2H_5)_3 CH_2C_6H_5X^-$ (X = Cl or Br)

(iv) Tetra-n-butyl ammonium chloride, bromide, chlorate or hydroxide, $N^+(n\text{-}Bu)_4X^-$ (X = Cl, Br, ClO_4, OH)

(v)   Cetyl trimethyl ammonium chloride or bromide (CTMAB for bromide), $N^+ (CH_3)_3 (CH_2)_{15}CH_3X^-$ (X = Cl or Br)

(vi) Benzyl tributyl ammonium chloride, $C_6H_5CH_2 (n\text{-}C_4H_9)_3 N^+ Cl^-$

(vii) Benzyl triphenyl phosphonium iodide, $C_6H_5CH_2(C_6H_5)_3P^+I^-$

Besides the above, another catalyst, crown ether is also widely used as PTC.

It is now well known that PTC reactions have considerable advantage over the conventional procedures. PTC reactions:

(i)   Are fast and do not require vigorous conditions.

(ii)   Do not require expensive aprotic solvents.

(iii) The reaction usually occurs at low temperature.

(iv)   The reaction is conducted in water and hence does not require anhydrous conditions.

(v)   With the help of PTC the anion is made available in the organic solvent and so the nucleophilicity increases.

(vi)   The work-up procedure is simple.

(vii) Use of strong bases (like alkoxide, sodamide, sodium hydride) in the reactions is avoided. The reaction proceeds even with OH$^-$ as it becomes strong nucleophile in presence of PTC.

(viii) Except the reactions which are sensitive to water, all other reactions can be carried out by PTC.

(ix)   The reactions which do not proceed in the normal way can be made to proceed in good yields.

A number of phase transfer catalysts are commercially available. However, these can be conveniently synthesised.[4] Some of the advantages of phase-transfer catalysts which are relevant to green synthesis are as follows:

(i) As the reaction is in two phases, a benign solvent may be used since PTC devoids the solubility for all the reactants like dipolar aprotic solvent and dimethylcarbonate. Moreover, in some cases organic solvents may not be required at all, the substrate forming the second phase.

(ii) The procedures of separation are simple resulting in less waste as the organic layer is mainly free from water soluble compounds and can easily be decanted off. It is important to vigil the concentration of anion in organic phase as it should not exceed the concentration of catalyst (unless it is soluble in absence of a catalyst).

(iii) PTC catalysed reactions are very rapid as the anions in the organic phase have very few water molecules associated with them making them highly reactive because of less activation energy which causes higher productivity.

(iv) These reactions can be run at a lower temperature owing to reduced activation energy that causes greater selectivity and lesser by-product formation.

8.2 Applications of PTC in Organic Synthesis

As already mentioned, PTC can be used in numerous types of organic reactions due to its advantages over conventional procedures. Some of the applications are as follows.

8.2.1 Nitriles from Alkyl or Acyl Halides

Reaction		Ref.
$CH_3(CH_2)_6CH_2Cl$ $\xrightarrow[\substack{PTC \\ C_{16}H_{33}P^+(n-C_4H_9)_3Br^- \\ 2\ hr,\ 105\ °C}]{NaCN/H_2O}$	$CH_3(CH_2)_6CH_2CN$ 94% (purity 97%)	5
$C_6H_5CH_2CH_2Cl$ $\xrightarrow[\substack{PTC \\ N^+(CH_3)_3CH_2C_6H_5Cl^- \\ 3\ hr,\ 90\text{-}95\ °C}]{NaCN/H_2O}$	$C_6H_5CH_2CH_2CN$ 91%	6
C_6H_5COCl $\xrightarrow[PTC\ Bu_4N^+X^-]{NaCN/H_2O}$	C_6H_5COCN 60-70%	7

8.2.2 Alkyl Fluorides from Alkyl Halides

The reaction of alkyl halides (chlorides or bromides) with KF in presence of a PTC give alkyl fluorides. Scheme 2 gives the preparation of 1-fluorooctane.[8]

$$CH_3(CH_2)_6CH_2Cl \; + \; KF \xrightarrow[C_{16}H_{33}P^+(C_4H_9)_3Br^-]{PTC} CH_3(CH_2)_6CH_2F$$

1-Chlorooctane   1-Fluorooctane
                                                                                                   77%

Scheme 2

The above procedure is far superior to the conventional methods[9] and can also be used for the preparation of labelled alkyl halides having ^{36}Cl by reacting the alkyl halide with labelled $Na^{36}Cl$. Scheme 3 illustrates[10] the synthesis of 1-chlorooctane ^{36}Cl.

$$CH_3(CH_2)_6CH_2Cl \; + \; Na^{36}Cl \xrightarrow[C_{16}H_{33}P^+(C_4H_9)_3Br^-]{PTC} CH_3(CH_2)_6CH_2{}^{36}Cl$$

1-Chlorooctane   1-Chlorooctane^{36}Cl

Scheme 3

8.2.3 Generation of Dihalocarbenes

Dihalocarbenes, synthetically useful intermediates, are normally generated by the action of a base on chloroform.

$$CHCl_3 \; + \; Base \longrightarrow CCl_3^- \xrightarrow{-Cl^-} \; :CCl_2$$
                                                                                   Dichlorocarbene

(a) The *in situ* generated carbene can add on to across the double bonds to give adducts. The use of PTC gives the adduct in 60-70% yield[11] (Scheme 4).

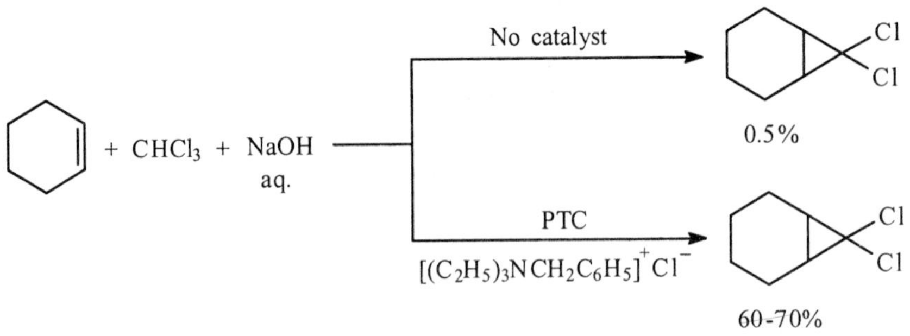

Scheme 4

(b) This method is used for making diazomethane (Scheme 5). The generated dichlorocarbene *in situ* is made to react with hydrazine to give diazomethane.[12]

$$NH_2NH_2 + CHCl_3 + NaOH \xrightarrow[\text{ether or } CH_2Cl_2]{\text{PTC } (C_4H_9)N^+OH^-} \begin{array}{c} CH_2N_2 \text{ in ether or } CH_2Cl_2 \\ \text{Diazomethane (35\%)} \end{array}$$

Scheme 5

However, use of crown ether gives better yield (48%).

(c) The dichlorocarbene generated by the PTC method reacts with primary amines to yield isonitriles (Scheme-6).[13]

$$RNH_2 + CHCl_3 + NaOH \xrightarrow[\text{aq.}]{C_6H_5CH_2N^+Et_3Cl^-} \begin{array}{c} R-N{\equiv}C \\ \text{Isonitrile} \\ \text{40-60\%} \end{array}$$

Scheme 6

This is a convenient method compared to the two step process.[14]

(d) The dichlorocarbene generated by PTC techniques reacts with amides, thioamides[15], aldoximes and amidines to give the corresponding nitriles (Scheme 7).

$$\left.\begin{array}{c} RCONH_2 \\ RCSNH_2 \\ RCH{=}NOH \\ RC{-}NH_2 \\ {\parallel} \\ NH \end{array}\right\} + CH_3Cl + NaOH \xrightarrow[\text{aq.}]{C_6H_5CH_2\overset{+}{N}Et_3Cl^-} R-CN$$

Scheme 7

(e) Alcohols on reaction with dichlorocarbene generated in a PTC catalysed system[16] gave good yield of chlorides (Scheme 8). In case of steroidal alcohols, the OH is replaced with Cl with retention of configuration.[17]

$$ROH + CHCl_3 + NaOH \xrightarrow[\text{aq.}]{C_6H_5CH_2\overset{+}{N}Et_3Cl^-} RCl + NaCl + H_2O$$

Scheme 8

(f) Reaction with aromatic aldehydes. Dichlorocarbene generated by PTC method reacts with aromatic aldehydes to give mandelic acids (Scheme 9).[18]

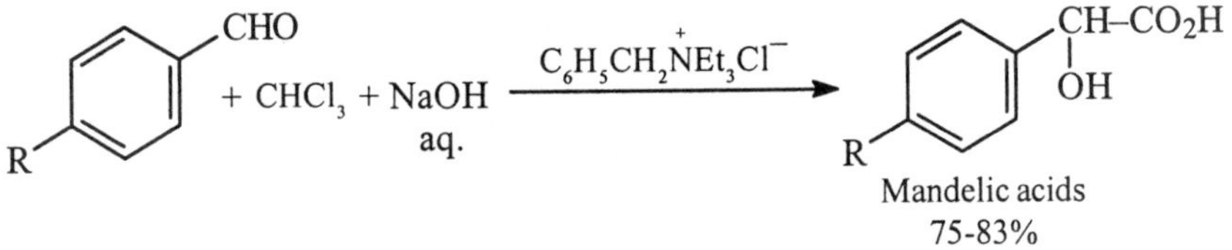

Scheme 9

8.2.4 Generation of Vinylidene Carbenes

These can be generated[19] from 3-chloro-3-methyl-1-butyne with base under vigorously anhydrous conditions. These add on to olefins *in situ* to give dimethyl vinylidinecyclopropanes (Scheme 10). Use of PTC technique with aqueous NaOH gives better yields and is more convenient.[20]

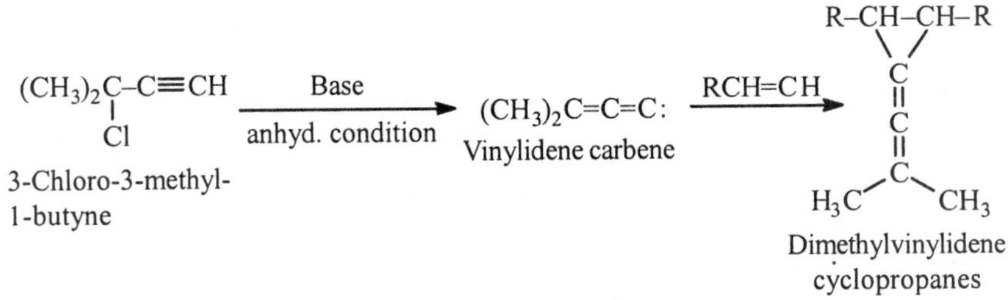

Scheme 10

8.2.5 Elimination Reactions

(a) Dehydrohalogenation can be achieved by the reaction of alkyl halides with aq. NaOH in presence of PTC.

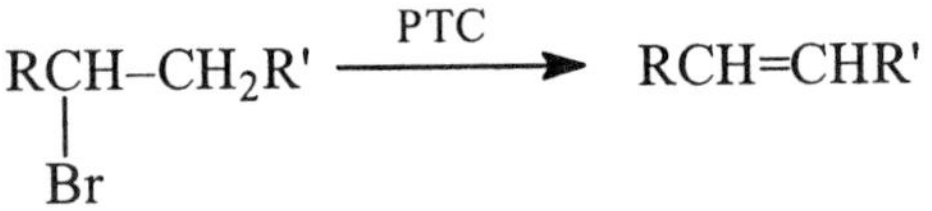

$$RCH-CH_2R' \xrightarrow{\text{PTC}} RCH=CHR'$$
$$\mid$$
$$Br$$

(b) The vic-dibromoalkanes can be neatly debrominated[21] by the PTC process using sodium thiosulphate with a catalytic amount of sodium iodide (Scheme 11).

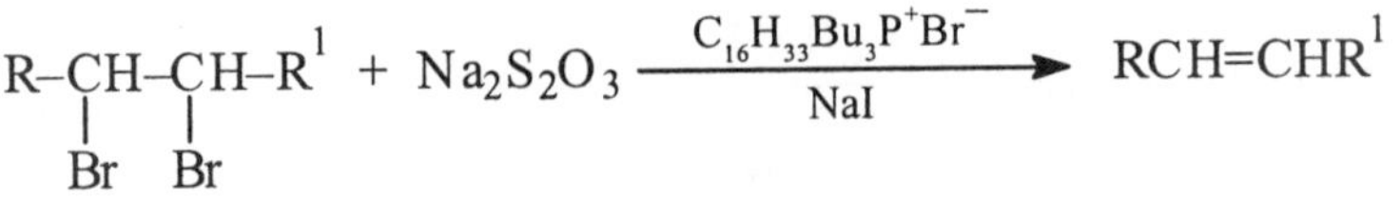

$$R-CH-CH-R^1 + Na_2S_2O_3 \xrightarrow[\text{NaI}]{C_{16}H_{33}Bu_3P^+Br^-} RCH=CHR^1$$
$$\mid \quad \mid$$
$$Br \quad Br$$

Scheme 11

Using this procedure the following compounds (Scheme 12) are obtained.[22]

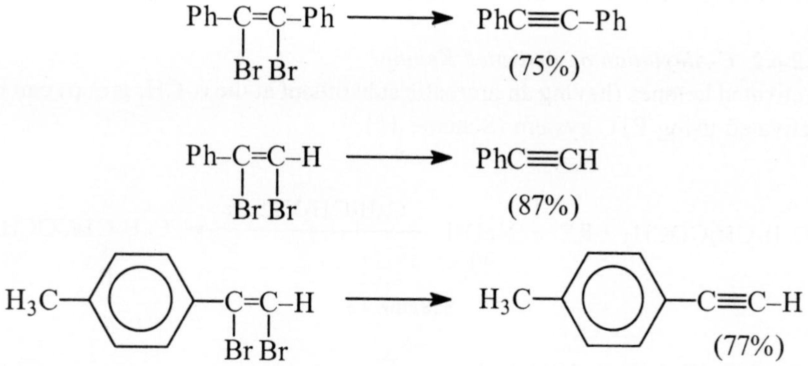
Ph–C=C–Ph
Br Br
PhC≡C–Ph
(75%)
Ph–C=C–H
Br Br
PhC≡CH
(87%)
H3C–
–C=C–H
Br Br
H3C–
–C≡C–H
(77%)

Scheme 12

8.2.6 C-Alkylations

8.2.6.1 C-Alkylations of Activated Nitriles

Normally C-alkylation involves the use of expensive condensing agents like sodamide, metal hydrides, potassium tertiary butoxide etc. and involves the use of anhydrous organic solvents. Due to the high selectivity of PTC, it is used for synthesis of monoalkyl derivatives of nitriles (Scheme 13).[23]

$$C_6H_5CH_2CN + C_2H_5Cl + \underset{aq.}{NaOH} \xrightarrow{C_6H_5CH_2\overset{+}{N}(C_2H_5)_3Cl^-} C_6H_5\text{–}\underset{\displaystyle C_2H_5}{\underset{|}{CH}}\text{–}CN + NaCl$$

Scheme 13

The C-alkyl derivatives of activated nitriles are useful intermediates for the manufacture of various pharmaceuticals and are commonly used in industry. Similarly, N-benzoyl-1,2-dihydroisoquinaldenitrile can be alkylated (Scheme 14)[24] in presence of NaOH (aq.) and TEBA.

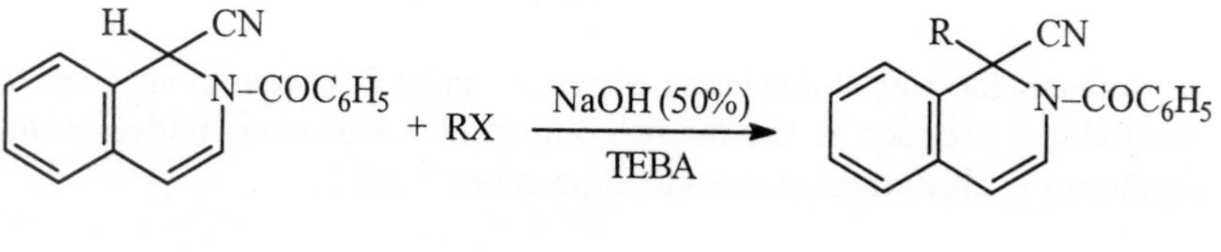
H
CN
N–COC6H5
+ RX
NaOH (50%)
TEBA
R
CN
N–COC6H5

Scheme 14

Alkaline hydrolysis of the alkylated product gives isoquinoline derivatives which are starting materials for the synthesis of alkaloids.

8.2.6.2 C-Alkylation of Activated Ketones

Activated ketones (having an aromatic substituent at the α-CH$_2$ group) can be activated using PTC system (Scheme 15).[25]

$$C_6H_5CH_2COCH_3 + RX + \underset{aq}{NaOH} \xrightarrow{C_6H_5CH_2\overset{+}{N}Et_3Cl^-} C_6H_5\underset{R}{\overset{|}{C}}HCOCH_3$$

Scheme 15

8.2.6.3 C-Alkylation of Aldehydes

Aldehydes containing only an α-hydrogen atom, e.g. isobutyraldehyde, can be alkylated with alkyl halides in presence of 50% aq. NaOH and catalytic amount of tetrabutyl ammonium ions (Scheme 16).[26]

$$(CH_3)_2CHCHO + RX + \underset{aq.}{NaOH} \xrightarrow{Bu_4\overset{+}{N}Y^-} (CH_3)_2-\overset{R}{\underset{}{\overset{|}{C}}}-CHO$$

Scheme 16

8.2.7 N-Alkylations

8.2.7.1 N-Alkylation of Aziridines

Aziridines cannot be easily N-alkylated using conventional conditions due to rapid ring opening. However alkylation of aziridine can be achieved using PTC conditions (Scheme 17).[27]

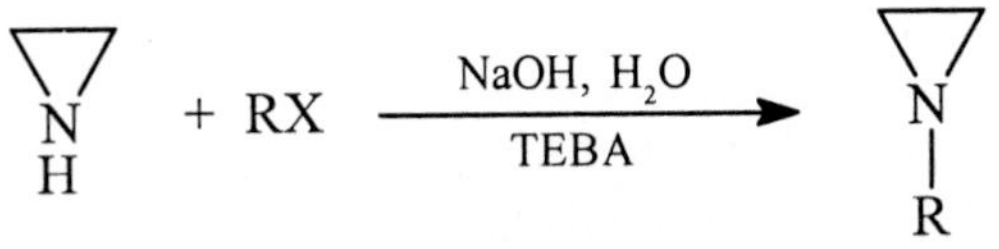

Scheme 17

In a similar way, N-alkylation of pyrrole under PTC conditions gives the N-alkylated product as the major compound.[28] However, under normal conditions C-alkylations also occur at positions 2 and 3.

8.2.7.2 N-Alkylation of β-Lactams

N-Alkylation of β-lactams has been carried out using PTC (Scheme 18).[29] This is an important step in the synthesis of norcardicin.

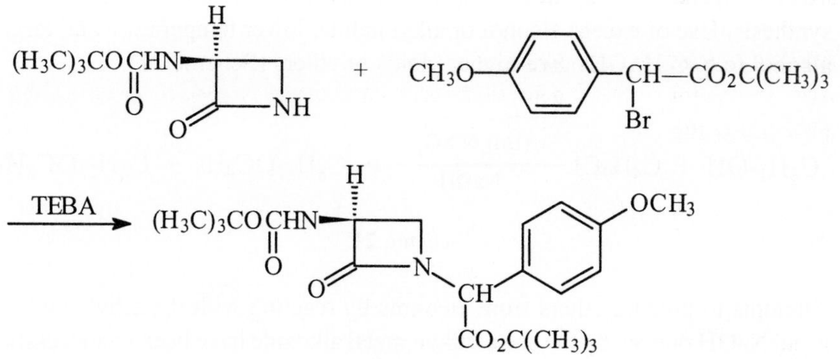

$$(H_3C)_3COCHN\text{---}... + CH_3O\text{---}\langle\rangle\text{---}CH\text{---}CO_2C(CH_3)_3$$

Scheme 18

8.2.8 S-Alkylation

The reaction of benzothiazole-2-thione with chlorobromomethane and PTC conditions gave the corresponding 2-chloromethylthio-products (Scheme 19).[30]

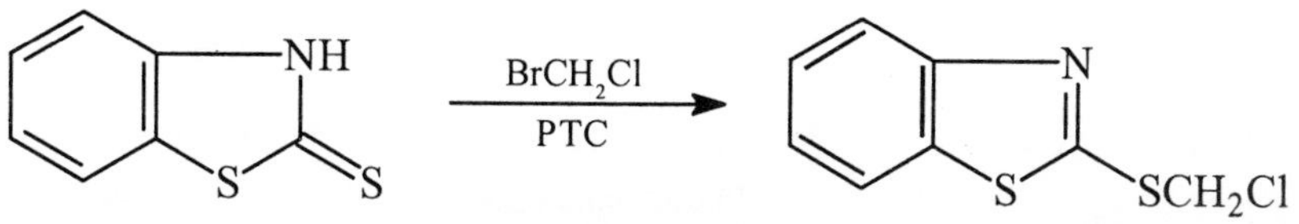

Scheme 19

S-Alkylation of 2-pyridinethiones, 2-quinolinethiones and pyrimidine derivatives has also been carried out.

8.2.9 Darzen's Reaction

The usual Darzen's reaction consists of the reaction of aldehydes or ketones with α-haloester or α-halonitriles in presence of a base like potassium tertiarybutoxide to give glycidic esters or nitriles, respectively (Scheme 20).

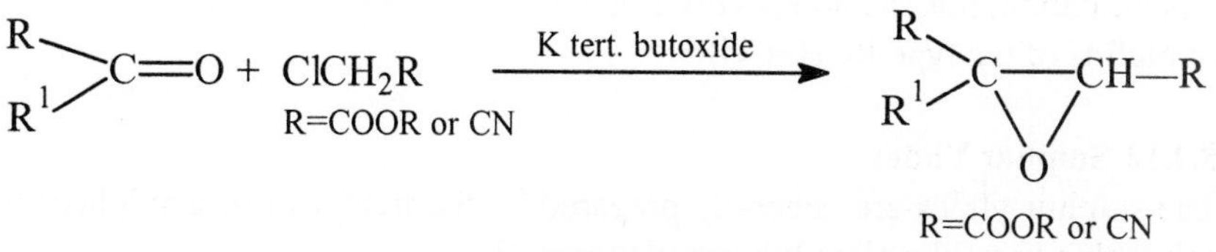

Scheme 20

This reaction has been found to occur in alkali solution in presence of PTC (benzyl triethylammonium chloride).[31]

8.2.10 Williamson's Ether Synthesis

The PTC technique provides a simple method for conducting Williamson ether synthesis. Use of excess alcohol or alkyl halide, lower temperature and larger alcohol (e.g. $C_8H_{17}OH$) give higher yields of ethers (Scheme 21).[32]

$$C_8H_{17}OH + C_4H_9Cl \xrightarrow[\text{NaOH}]{(Bu)_4N^+SO_4^-} C_8H_{17}OC_4H_9 + C_8H_{17}OC_8H_{17}$$

Byproduct

Scheme 21

Attempts to prepare ethers from alcohols by reacting with dimethyl sulphate in aq. NaOH or even by the use of alkali metal alkoxide have been unsuccessful. However, use of PTC (e.g. tetrabutyl ammonium salts) gives high yield of ethers.[33]

8.2.11 The Wittig Reaction

The usual Wittig reaction consists in the treatment of a phosphonium salt with a base (e.g. NaH) to give an ylide, which can react with aldehydes or ketones to give alkenes (Scheme 22).

$$Ph_3P + XCH_2R^1 \longrightarrow \underset{\text{Phosphonium salt}}{Ph_3\overset{+}{P}-CH_2R^1X^-} \xrightarrow{NaH} \underset{\text{Ylide}}{Ph_3P{=}CHR^1}$$

$$\xrightarrow{R^2C{=}O} \underset{\text{Alkene}}{R^1CH{=}CR^2} + Ph_3P{=}O$$

Scheme 22

It has been found that the formation of ylide from phosphonium salt (Scheme 22)[34] can be very conveniently effected by using a PTC in aq. NaOH. In PTC method the yield of the olefin increases. However, this PTC method is applicable only to aldehydes; no olefin is obtained from ketones. Even with this limitation, this method is very convenient for the preparation of a number of olefins of the type $RCH{=}CHR^1$.

8.2.12 Sulphur Ylides

The sulphur ylides are generally prepared by the treatment of a sulphonium salt with a base like alkyl lithium (Scheme 23).

$$(CH_3)_2S + CH_3I \longrightarrow (CH_3)_3\overset{+}{S}I^- \xrightarrow{\ RLi\ }$$

Dimethyl   Methyl   Sulphonium

sulphide   iodide   salt

$$(CH_3)_2\overset{+}{S}-\overset{-}{C}H_2 \longleftrightarrow (CH_3)_2S=CH_2$$

Sulphur ylide

Scheme 23

It has now been found that in place of strong base (like alkyl lithium), a PTC like $(C_6H_5)_4N^+I^-$ in presence of aq. NaOH can be used.[35]

8.2.13 Heterocyclic Compounds

8.2.13.1 3-Alkyl Coumarins

3-Alkyl Coumarins are increasingly being used as optical brightners[36] and were obtained in low yields using anhydrous conditions[37]. These have now been obtained (Scheme 24) in excellent yield and purity by the use of PTC in presence of aq. K_2CO_3 by the reaction of o-hydroxycarbonyl compounds with phenyl acetyl chlorides.[38]

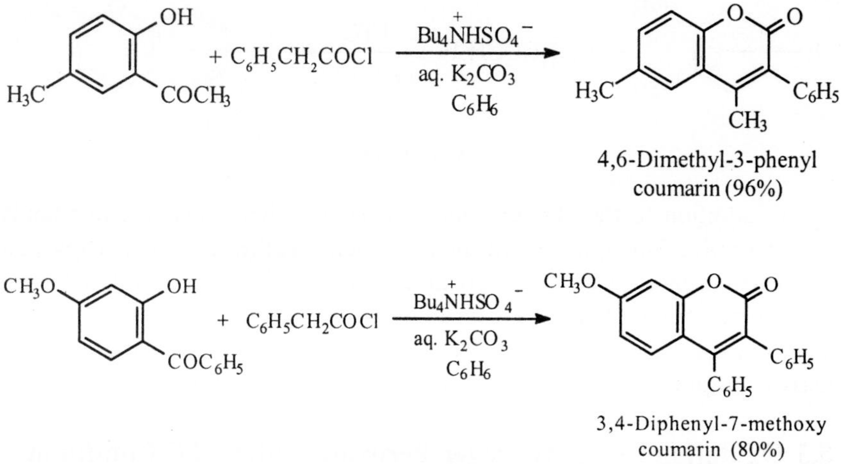

4,6-Dimethyl-3-phenyl coumarin (96%)

3,4-Diphenyl-7-methoxy coumarin (80%)

Scheme 24

8.2.13.2 Flavones

Flavones are an important class of natural products. These were synthesised in low yields by a number of methods.[39] These have now been obtained in excellent yields by the reaction of an appropriate o-hydroxyacetophenone with an appropriately substituted benzoyl chloride in benzene solution with a PTC in presence of NaOH or Na$_2$CO$_3$ followed by cyclisation of the formed o-hydroxydibenzoyl methane with p-toluene sulphonic acid (Scheme 25).[40]

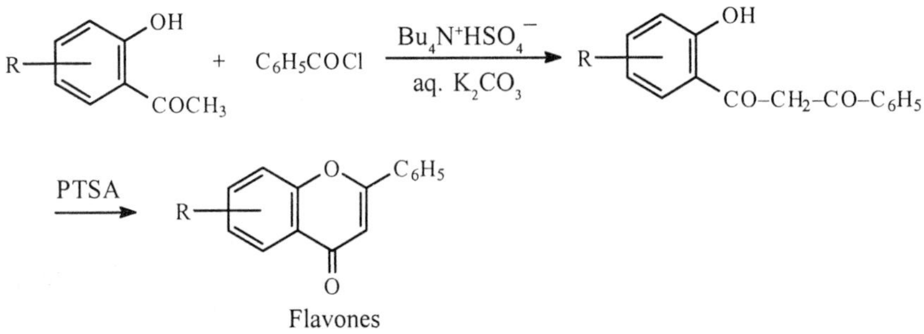

Flavones

Scheme 25

8.2.13.3 *3-Aryl-2H-1, 4-Benzoxazines*

The title compounds known for their anti-inflammatory activity were prepared earlier in low yields.[41] Polymer supported phase transfer catalysts have also been used (for details see sec 7.4.5) for various reactions. These have now been prepared[42] by the condensation of 2-aminophenols with phenacyl bromide in presence of a PTC in aq. K$_2$CO$_3$ (Scheme 26).

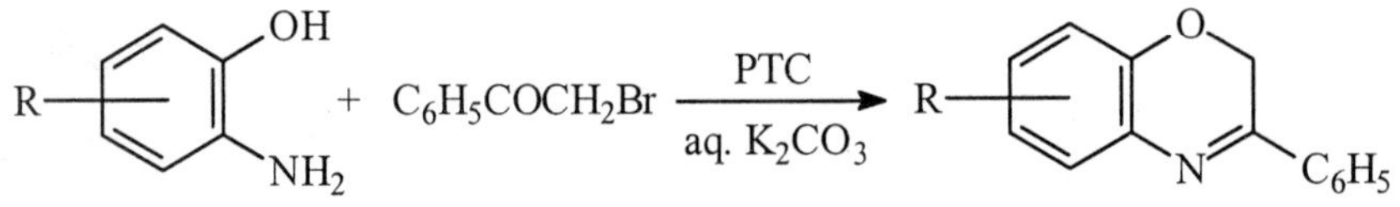

Scheme 26

In addition to the above, a number of other heterocyclic compounds, for example, 2-aroylbenzofurans, 1,3-benzoxathioles, dihydropyrans, 1,4-benzoxazines, hydantoin derivatives, piperazine-2,5-diones, thioethenes, 1-arylbenzimidazoles, benzofuran-1-oxides, piperazinones, thiazoles, 5-thiacyclohexanecarboxaldehyde, pyrroles, triazines, fused napthoquinone derivatives and β-lactams have been synthesised using PTC technique.[43]

8.3 Oxidation Using Hydrogen Peroxide Under PTC Condition

Hydrogen peroxide is an excellent environmentally benign oxidant that produces water as the only by-product. It is used in relatively dilute form (30 volume). Hydrogen peroxide could be used in place of higher waste generating peroxides such as peracetic acid or tertiary butyl hydrogen peroxide. There are two possible mechanisms for transferring peroxide to the organic phase. First transporting the HO_2^- ion. This anion is strongly hydrophilic and has a high hydration energy that does not exchange readily with other anions. It shows that the classical mechanism is followed. The second mechanistic approach seems to involve extraction via complexes of the type $R^+X^--H_2O_2$. Hydrophobic quaternary salts such as $(C_6H_{13})_4$ NBr are most widely used. In some cases hydrogen peroxide is not involved in direct oxidation, the main oxidant in these cases are metal complexes of W or Mo. In this case the role of hydrogen peroxide is to reoxidize the metal complexes *in situ* that make the process a catalytic one with respect to metal. The major environmental benefit for the usage of hydrogen peroxide is exemplified in alkene cleavage. Thus, cyclohexene on treating with 30% hydrogen peroxide in presence of catalytic amount of sodium tungstate and methyltricetylammonium hydrogen sulphate at 90 °C gives adipic acid in excellent yield.[44] The commercial viability of this route holds obviously high potential (Scheme 27).

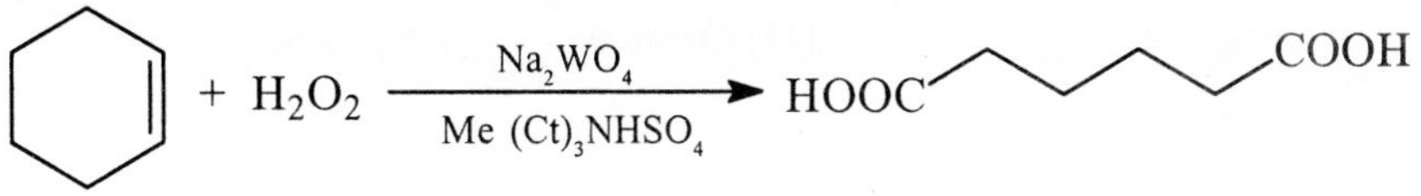

$$\text{cyclohexene} + H_2O_2 \xrightarrow[\text{Me (Ct)}_3\text{NHSO}_4]{\text{Na}_2\text{WO}_4} \text{HOOC} \diagdown\diagup\diagdown\diagup \text{COOH}$$

Scheme 27. Noyori synthesis of adipic acid

8.4 Crown Ethers

This is a group of cyclic polyethers which are used as phase transfer catalysts. These have been used for esterifications, saponifications, anhydride formation, oxidations, aromatic substitution reactions, elimination reactions, displacement reactions, generation of carbenes, alkylations etc. Some of the examples are as follows:

8.4.1 Esterification

Crown ethers are useful for esterification. Reaction of p-bromophenacyl bromide with potassium salt of a carboxylic acid using 18-crown-6 as the solubilizing agent have been used to prepare[45] p-bromophenacyl esters (Scheme 28).

8.4.2 Saponification

The main problem of using potassium hydroxide for saponification is its insolubility in organic solvents like toluene, but this can be solved by using hydrophobic and hydrocarbon soluble macrocyclic derivatives like dicyclohexyl 18-crown-6, it has been shown that potassium hydroxide is soluble in toluene. This special observation[46] has been used for the hydrolysis of sterically hindered esters using potassium hydroxide complex in toluene (Scheme 29).

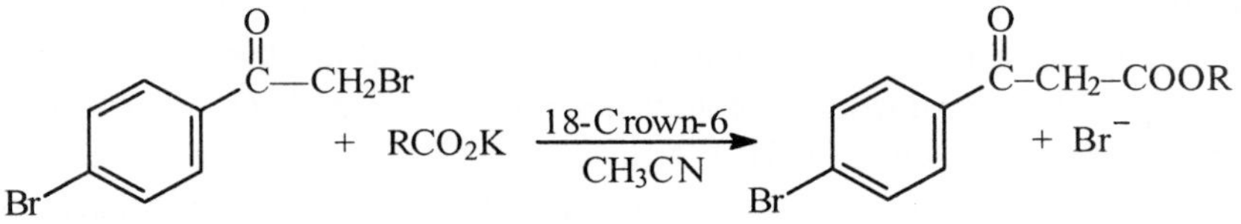

R = H, CH$_3$, CH$_3$CH$_2$, CH$_3$CH$_2$CH$_2$, C$_6$H$_5$, 2-CH$_3$C$_6$H$_4$,
2,4,6-trimethylbenzoyl, 4-t-butyl C$_6$H$_4$

Scheme 28

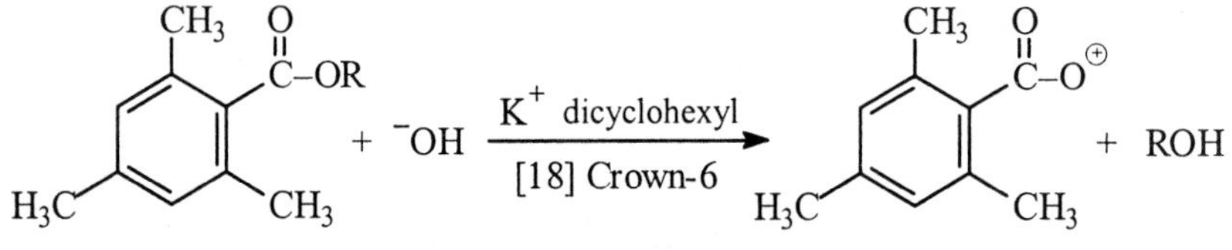

R = CH$_3$, t-Bu, neopentyl

Scheme 29

8.4.3 Anhydride Formation

A convenient synthesis of anhydrides has been described[47] by the reaction of potassium or sodium salts of carboxylic acids with activating halides (ethyl chloroformate, cyanuric chloride and benzyl chloroformate) in acetonitrile in the presence of 18-crown-6. Using this procedure, cinnamic acid, p-nitrobenzoic acid, benzoic acid, acetic acid and propionic acid are also converted into their anhydrides (Scheme 30).

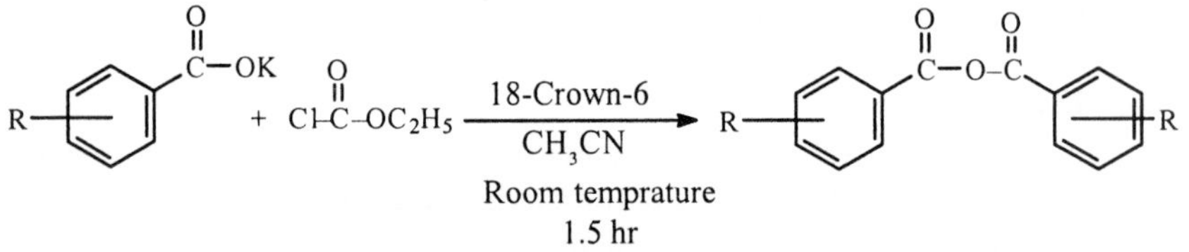

Scheme 30

8.4.4 Potassium Permanganate Oxidation

Potassium permanganate is the most widely used reagent for the oxidation of organic compounds. It is usually used in aqueous solution and this restricts its usefulness since many compounds are not sufficiently soluble in water and only a few organic solvents like acetic acid, t-butanol, dry acetone and pyridine are resistant to the oxidising action of the reagent. Alternatively, oxidation in presence of crown ether, dicyclohexano-18-crown-6 forms a permanganate complex. Under these conditions permanganate becomes soluble in benzene and the resulting solutions are excellent reagents for oxidation of a variety of organic substrate in organic solvents (Scheme 31).

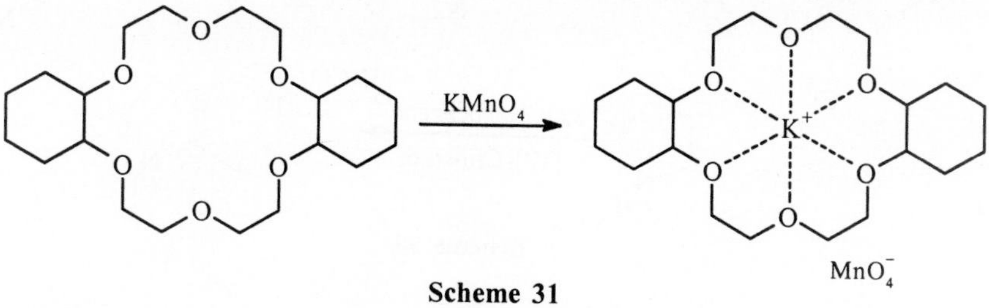

Scheme 31

Applying this technique, substituted catechols are converted into corresponding o-quinones in excellent yields[47] wherein only one equivalent of KMnO$_4$ is used (Scheme 32).

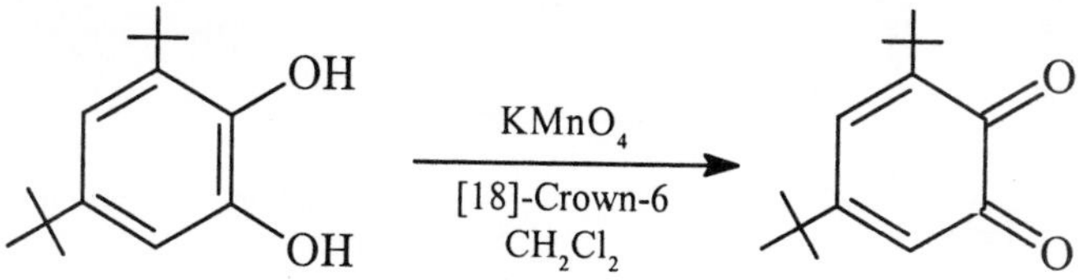

Scheme 32

Oxidation of α-pinene in the presence of crown ethers yield pinonic acid in 90% yield (Scheme 33).[48]

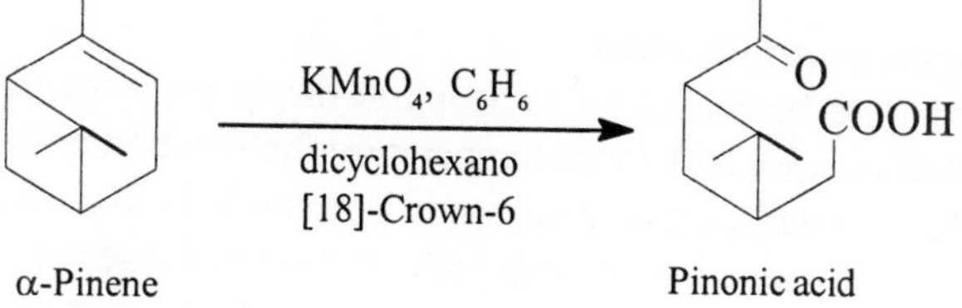

α-Pinene   Pinonic acid

Scheme 33

8.4.5 Aromatic Substitution Reaction

The substitution of 2-chloropyridine by 1,6-dihydroxyhexane in good yield was carried out using 18-crown-6 as a catalyst. Furthermore, the electron rich ring undergoes methoxide substitution in excellent yield using the same catalyst (Scheme 34).[49]

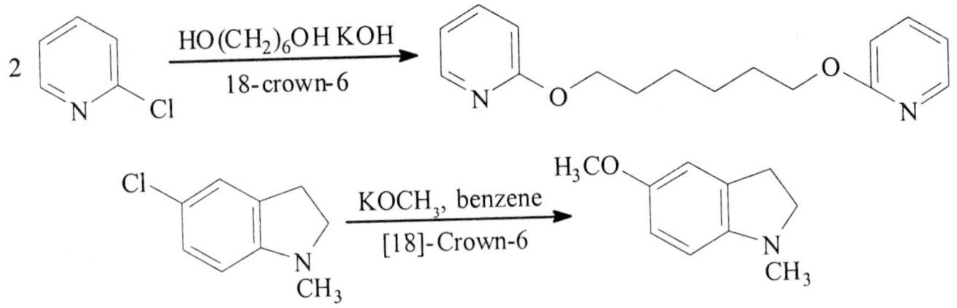

Scheme 34

8.4.6 Elimination Reaction

Crown ethers ability to enhance or alter the reaction is significantly important.[49] This is depicted in (Scheme 35).

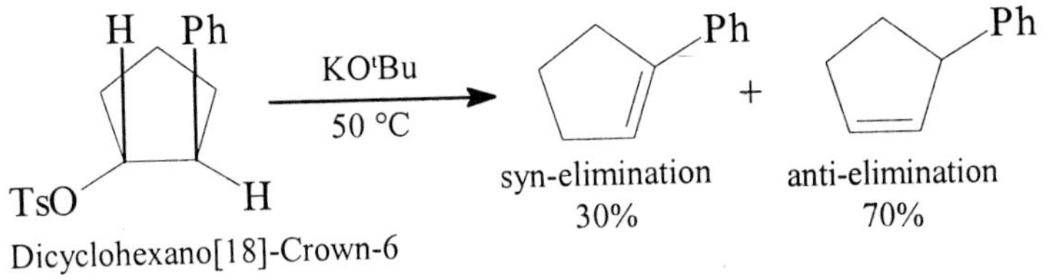

Scheme 35

In the absence of crown ether syn-elimination takes place in 91% yield along with 9% of anti-elimination.

8.4.7 Displacement Reaction

Crown ether has been used for nucleophilic displacement of chloride by cyanide at hindered position. At room temperature the reaction of 2-chloro-2-methyl cyclohexanone with potassium cyanide in acetonitrile in presence of 18-crown-6 affords the cyanide in excellent yield, but when the reaction is conducted at reflux temperature Favorskii rearrangement occurs to yield five membered compound in high yield (Scheme 36).

8.4.8 Generation of Carbene

Dicyclohexyl-18-crown-6 has been used[50] to convert cyclohexene and trans stilbene to the respective gem dihalocyclopropanes in 30-70% yields by the reaction of sodium hydroxide and chloroform at 40 °C. Dibenzo 18-crown-6 has also been used[51] as liquid-liquid phase transfer catalyst for carbene generation (Scheme 37).

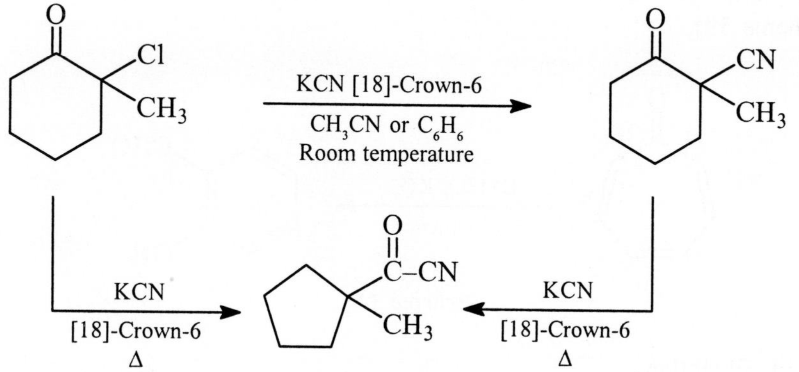

Scheme 36

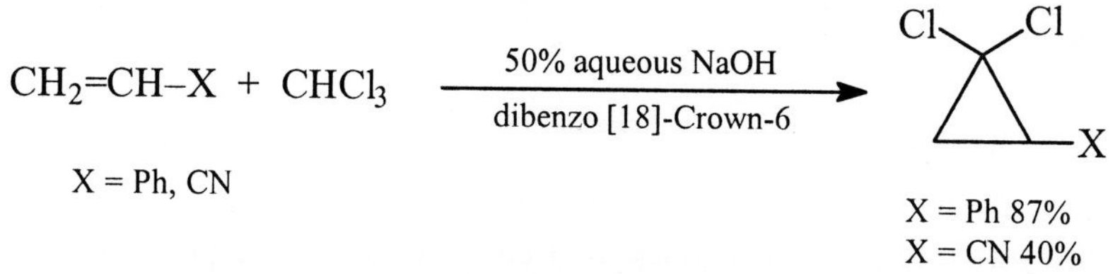

$$CH_2=CH-X + CHCl_3$$

$$X = Ph, CN$$

$$\xrightarrow[\text{dibenzo [18]-Crown-6}]{\text{50\% aqueous NaOH}}$$

X = Ph 87%
X = CN 40%

Scheme 37

8.4.9 Superoxide Anion Reaction

The most important application of crown ether is in superoxide chemistry. In fact the use of superoxide (K_2O and NaO_2) has been limited due to the solubility problem. Use of crown ether along with a superoxide for the oxidative dimerisation is a typical application. A cheap potassium superoxide (KO_2) is available from commercial source and is a source for the generation of superoxide radical anion readily available for the reaction (Scheme-38).[52]

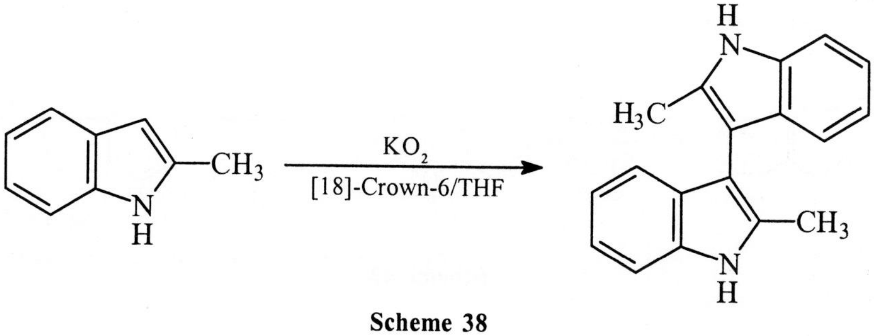

Scheme 38

When tropone is treated with KO_2 and 18-crown-6 in DMSO solution salicylaldehyde is obtained. This is an addition induced rearrangement (Scheme 39).

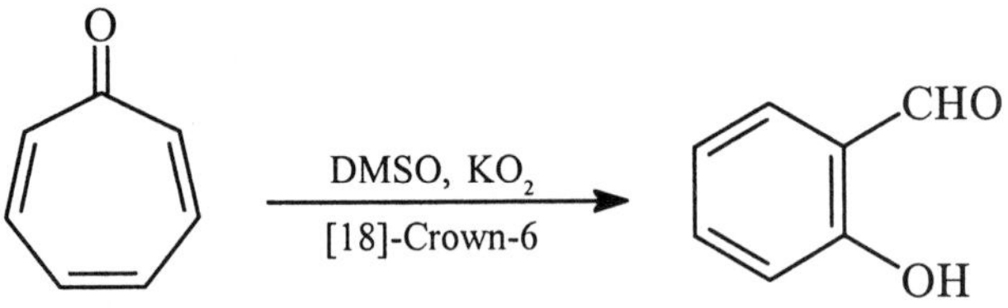

Scheme 39

8.4.10 Alkylation

Many aldehydes and ketones condense with acetonitrile in the presence of solid potassium hydroxide using 18-crown-6 as a catalyst (Scheme 40).[52]

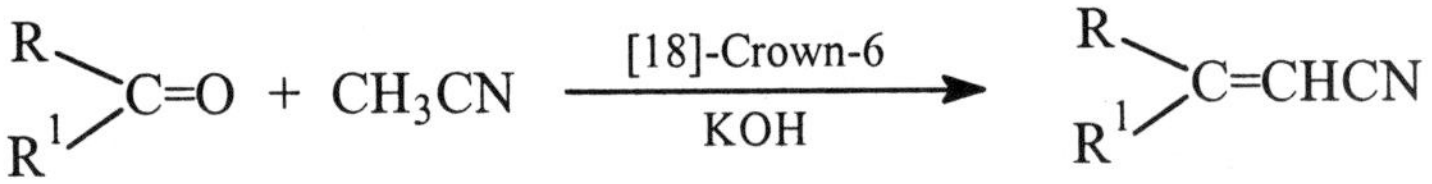

Scheme 40

Similarly, phenyl acetone can be alkylated[53] with n-butyl bromide in aqueous sodium hydroxide using dicyclohexyl-18-crown-6 as a catalyst (Scheme 41).

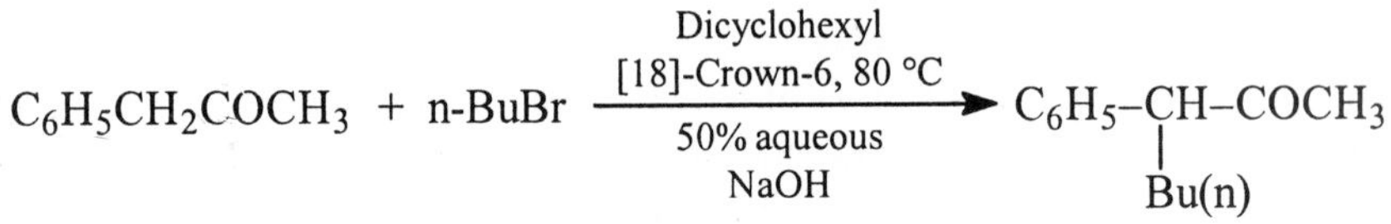

$$C_6H_5CH_2COCH_3 + n\text{-}BuBr \xrightarrow[\substack{\text{[18]-Crown-6, 80 °C} \\ \text{50% aqueous} \\ \text{NaOH}}]{\text{Dicyclohexyl}} C_6H_5\text{-}CH\text{-}COCH_3$$
$$\overset{|}{\underset{Bu(n)}{}}$$

Scheme 41

N-alkylation of pyrrole is carried out[54] using crown ether. The indolyl anion behaves as an ambident nucleophile and alkylation occurs[55] at nitrogen and at C-3 (Scheme 42).

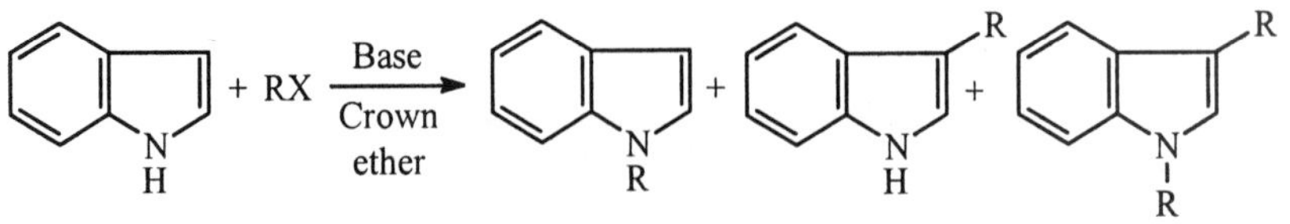

Scheme 42

References

1. E.V. Dehmlov and S.S. Dehmlov, Phase transfer catalyst, 3rd ed., Verlag Chemie, Weinheim, 1993.
2. C.M. Starks, C.L. Liotta and M. Halpen, Phase transfer catalysts: Fundamentals, Applications and Industrial Perspectives, Chapman & Hall, New York, 1994.
3. V.K. Ahluwalia and Renu Aggarwal, Organic Synthesis: Special Techniques, Narosa Publishing House, New Delhi, 2001.
4. V.K. Ahluwalia and Renu Aggarwal, Organic Synthesis: Special Techniques, Narosa Publishing House, New Delhi, 2001, pp. 1-58 and the references cited therein.
5. C.M. Starks, *J. Am. Chem. Soc.*, 1971, **93**, 195.
6. N. Sugemol, T. Fujita, N. Shigematsu and A. Ayadha, *Chem. Pharm. Bull.*, 1962, **10**, 427; *Japanese Patent* 1961/63.
7. K.E. Koening and W.P. Weber, *Tetrahedron Lett.*, 1974, 2275.
8. D. Landine, F. Mentanam and F. Rolla, *Synthesis*, 1974, 428.
9. H. Hudlicky, *Organic Fluorine Compounds*, Plenum, New York, 1971; C.M. Spaks, *J. Chem. Edu.*, 1968, **45**, 185.
10. D. Forster, *J. Chem. Soc. Chem. Commun.*, 1975, 918.
11. A.P. Kreshkov, E.N. Sugushkima and B.A. Krozdov, *J. Appl. Chem.*, USSR (English Trans.), 1965, **38**, 2357.
12. D.T. Sepp, K.V. Scherer and W.B. Weber, *Tetrahedron Lett.*, 1974, 2983; H. Stardinger and O. Kupfer, *Ber.*, 1912, **45**, 501.
13. W.B. Weber and G.W. Gukil, *Tetrahedron Lett.*, 1972, 1837.
14. I. Kugi, R. Miyr, M. Lipinski, F. Bodsheim and F. Rosendahl, *Org. Synth.*, 1961, **41**, 13; L. Fieser and M. Fieser, Reagents in Organic Synthesis, Wiley, N.Y., 1967, 405.
15. J. Graefe, *Z. Chem.*, 1975, **15**, 301.
16. I. Tabushe, Z. Yoshida and N. Takahashi, *J. Am. Chem. Soc.*, 1971, **93**, 1820.
17. R. Ikan, A. Makus and Z. Goldschmidt, *J. Chem. Soc.*, 1973, **11**, 591.
18. A. Merz., *Synthesis*, 1974, 724.
19. H.D. Martzler, *J. Am. Chem. Soc.*, 1961, **83**, 4990 and 4997; *J. Org. Chem.*, 1964, **72**, 3542.
20. G.F. Hennison, J.J. Sheehan and D.E. Maloneg, *J. Am. Chem. Soc.*, 1950, **72**, 3542; G.F. Hennison and A.P. Boisselle, *J. Org. Chem.*, 1961, **26**, 725.
21. D. Landini, S. Quici and F. Rolla, *Synthesis*, 1975, 397.
22. A. Gorgus and A. Lecog, *Tetrahedron Lett.*, 1876, 4723.
23. M. Makosza, *Tetrahedron Lett.*, 1968, **24**, 175.
24. M. Makosza, *Tetrahedron Lett.*, 1969, 677.
25. J. Jamouse and C.R. Hebd, *Scances Acad. Sci. Ser. C.*, 1951, **232**, 1424; J. Jonczyk, B. Serafin and M. Makosza, *Roz. Chem.*, 1971, **45**, 1027; J. Jinczyk, B. Serafin and E. Skulemowska, *Rocz. Chem.*, 1975, **45**, 2097.
26. H. Dietl and K.C. Brannock, *Tetrahedron Lett.*, 1973, 1273.
27. M. Maurette, A. Lopez, R. Martino and A. Lattes, *C.R. Acad. Sci. Ser. C.*, 1976, **282**, 599.
28. A. Jonezyk and M. Makosza, *Rocz. Chem.*, 1975, **49**, 1203.
29. P.G. Mattingly and M.J. Miller, *J. Org. Chem.*, 1981, **46**, 1557.
30. C.T. Gokalshi and G.A. Burk, *J. Org. Chem.*, 1977, **42**, 3094.
31. A. Jonczyk, M. Fedorynski and M. Makosza, *Tetrahedron Lett.*, 1972, 2395.
32. J. Jorrouse and C.R. Hebd, *Scances Acad. Sci. Ser. C.*, 1951, **232**, 1424; H.H. Freeman and R.A. Dubois, *Tetrahedron Lett.*, 1975, 3251.
33. A. Merz, *Angew. Chem. Int. Ed. Engl.*, 1973, **12**, 846.

34. G. Markl and A. Merz, *Synthesis*, 1975, 245; S. Hung and J. Stemmier, *Tetrahedron Lett.*, 1974, 315; W. Tagaki, I. Inouse, Y. Yano and T. Okanogi, *Tetrahedron Lett.*, 1974, 2587.
35. A. Merz and C. Markl, *Angew. Chem. Int. Ed. Engl.*, 1973, **12**, 846.
36. I. Dorlars, C.W. Schellhammer and J. Schroeder, *Angew. Chem. International Ed.*, 1975, **14**, 665.
37. P. Pulla Rao and G. Srimannaraya, Synthesis, 1981, **11**, 887; Olgialoro, *Gazz. Chem. Itial.*, 1879, **9**, 478; T.R. Seshadri and S. Varadarajan, *J. Sci. Ind. Res.*, 1952, **11B**, 48; *Proc. Indian Acad. Sci.*, 1952, **35A**, 75.
38. G. Sabitha and A.V. Subba Rao, *Syn. Commun.*, 1987, **17(3)**, 341; V.K. Ahluwalia, C.H. Khanduri, *Indian J.Chem.*, 1989, **28B**, 599.
39. J. Allen and R. Robinson, *J. Chem. Soc.*, 1924, **125**, 2192; W. Baker, *J. Chem. Soc.*, 1939, 1391; H.S. Mahal and K. Venkataraman, *Curr. Sci.*, 1933, **4**, 214; V.N. Gupta and T.R. Seshadri, *J. Sci. Ind. Res. (India)*, 1957, **16B**, 116.
40. V.K. Ahluwalia et al. unpublished results, P.K. Jain, J.K. Makrandi and S.K. Grover, *Synthesis*, 1982, 221.
41. D.R. Sridhar, C.V. Reddy Sastry, O.P. Bansal and R. Pulla Rao, *Indian J. Chem.*, 1983, **22B**, 297; 1981, **11**, 912; P. Battishoni, P. Brown and G. Fava, *Synthesis*, 1979, 220.
42. G. Sabita and A.V. Subba Rao, *Synth. Commun.*, 1987, **17(3)**, 341.
43. V.K. Ahluwalia and Renu Aggarwal, Organic Synthesis: Special Techniques, Narosa Publishing House, New Delhi, 2001, and the references cited therein.
44. R. Noyori and T. Ohkuma, 'Asymmetric Catalysis by Architectural and Functional Molecular Engineering. Practical Chemo- and Stereoselective Hydrogenation of Ketones', *Angew. Chem. Int. Ed. Engl.*, 2001, **40**, 40.
45. H.D. Durst, *Tetrahedron Lett.*, 1974, 2421.
46. C.J. Pedersen, *J. Am. Chem. Soc.*, 1967, **89**, 2485, 7017; 1970, **92**, 386, 391; C.J. Pedersen and H.K. Friensdorff, *Angew. Chem. Int. Ed. Engl.*, 1972, **11**, 16.
47. Charles M. Starks and Charles Liotta, Phase Transfer Catalysts, Academic Press, N.Y.
48. D. Sam and H.E. Simmons, *J. Am. Chem. Soc.*, 1972, **94**, 4024.
49. G. Gokel, Crown Ethers and Cryptands Monograph, Royal Society of Chemistry, Cambridge, 1991.
50. R.R. Kostikov and A.P. Molchanov, *Zh. Org. Khim.*, 1975, **11**, 1767.
51. M. Makosza and M. Ludwikow, *Angew. Chem. Int. Ed. End.*, 1974, 2983.
52. G.W. Gokel, S.A. Dibiase and B.A. Lipiska, *Tetrahedron Lett.*, 1976, 3495.
53. M. Cinquini, F. Montanori and P. Tundo, *Chem. Commun.*, 1974, 878.
54. E. Santaniello, C. Farachi and P. Pouti, *Synthesis*, 1970, 617.
55. J. Sundberg, The Chemistry of Indoles, Academic Press, New York, 1970.

9. Microwave Induced Green Synthesis

9.1 Introduction

Normally microwaves have wavelengths between 1 cm and 1 m (frequencies of 30 GHz to 300 Hz). These are similar to frequencies of radar and telecommunications. In order to avoid any interference with these systems, the frequency of radiation that can be emitted by household and industrial microwave oven is regulated, most of the appliances operate at a fixed frequency of 2.45 GHz.

The microwaves, as we know, are used for heating purposes. The mechanism of how energy is given to a substance which is subjected to microwave irradiation is complex. One view is that microwave reactions involve selective absorption of electromagnetic waves by polar molecules, non-polar molecules being inert to microwaves. When molecules with a permanent dipole are submitted to an electric field, they become aligned and as the field oscillates their orientation changes, this rapid reorientation produces intense internal heating. The main difference between classical heating and microwave heating, lies in core and homogenous heating associated with microwaves, whereas classical heating is all about heat transfer by preheated molecules.

The preferred reaction-vessel for microwave induced organic reaction, is a tall beaker (particularly for small scale reactions in the laboratory), loosely covered and the capacity of the beaker should be much greater than the volume of the reaction mixture. Alternatively, teflon and polystyrene containers can be used.[1,2] These materials are transparent to microwaves. Metallic containers should not be used as reaction vessels.

In microwave induced organic reactions, the reactions can be carried out in a solvent medium or on a solid support in which no solvent is used. For reactions in a solvent medium, the choice of the solvent is very important.[1,2] The solvent to be used must have a dipole moment so as to absorb microwaves and a boiling point at least 20-30 °C higher than the desired reaction temperature. An excellent solvent in a domestic microwave oven is N,N-dimethylformamide (DMF) (b.p. 160 °C, $\varepsilon = 36.7$). The solvent can retain water formed in a reaction, thus, obviating the need for water separation. Some other solvents of choice are given as follows:

Solvent	b.p. (°C)	Dielectric constant (ε)
Formamide	216	11.1
Methanol	65	32.7
Ethanol	78	24.6
Chlorobenzene	214	5.6
1,2-Dichlorobenzene	180	1.53
1,2,4-Trichlorobenzene	214	1.57
1,2-Dichloroethane	83	10.19
Ethylene glycol	196	37.7
Dioxane	101	2.20
Diglyme	162	7.0
Triglyme	216	1.42

Hydrocarbon solvents, for example, hexane ($\varepsilon = 1.9$), benzene ($\varepsilon = 2.3$), toluene ($\varepsilon = 2.4$) and xylene are unsuitable because of less dipole moment and also because these solvents absorb microwave radiations poorly. However, addition of small amounts of alcohol or water to these solvents can lead to dramatic coupling effects. Liquids which do not have a dipole moment cannot be heated by microwaves. By adding a small amount of a dipolar liquid to a miscible non-dipolar liquid, the mixture will rapidly achieve a uniform temperature under irradiation.

Microwaves may be considered as a more efficient source of heating than conventional steam (or oil heated vessels), since the energy is directly imparted to the reaction medium rather than through the walls of a reaction vessel. In fact, the rapid heating capability of the microwave leads to considerable saving in dissolution or the reaction time. The smaller volume of solvent required contributes to saving in cost and diminishes the waste disposal problem.[3,4]

Microwave procedures are limited[5] by the presence of solvents which reach their boiling points within a very short time (~ 1 min) of exposure to microwave. Consequently, high pressures are developed, leading to damage to the vessels material or the microwave oven itself and may occasionally lead to explosion.

Well-designed industrial microwave ovens are available now. Consideration of safety aspects coupled with the limitations of the solvents imposed by microwave heating, has led to many reactions being carried out in water or more commonly under solvent free conditions. This is a major green advantage of microwave reactions. It is believed that due to high polarity and non-volatility, ionic liquids (see also Chapter 14) might be ideal for carrying out high temperature reactions efficiently, since temperatures of over 200 °C can be readily attainable.

9.2 Applications

It is possible to carry out a number of microwave organic synthesis. These syntheses are grouped in the following three categories:

(i) Microwave-assisted reactions in water.

(ii) Microwave-assisted reactions in organic solvents.

(iii) Microwave solvent-free reactions (solid state reactions).

Some important microwave assisted organic synthesis are:

9.2.1 Microwave Assisted Reactions in Water

9.2.1.1 Hofmann Elimination

In this method, normally quaternary ammonium salts are heated at high temperature and the yield of the product is low. Use of microwave irradiation has led to high-yielding synthesis of a thermally unstable Hofmann elimination product (Scheme 1). In this water-chloroform system is used.

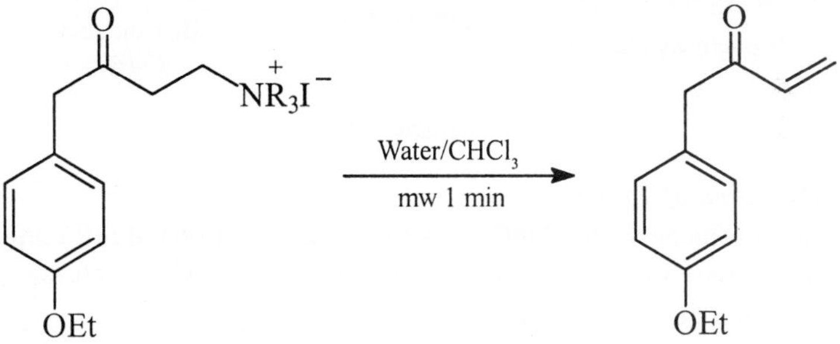

Scheme 1

9.2.1.2 Hydrolysis

Microwave reactions have been extensively used for hydrolysis.

9.2.1.2.1 Hydrolysis of Benzyl Chloride

Hydrolysis of benzyl chloride with water in microwave oven gives 97% yield[6] of benzyl alcohol in 3 min (Scheme 2). The usual hydrolysis in normal way takes about 35 min.

$$C_6H_5CH_2Cl + H_2O \xrightarrow[\text{3 min}]{\text{mw}} C_6H_5CH_2OH$$

Benzyl chloride   Benzyl alcohol
                                                          (97%)

Scheme 2

9.2.1.2.2 Hydrolysis of Benzamide

The usual hydrolysis of benzamide takes 1 hr. However, under microwave conditions, the hydrolysis is completed in 7 min giving[5] 99% yield of benzoic acid (Scheme 3).

$$C_6H_5CONH_2 \xrightarrow[\text{mw 7 min}]{20\% \text{ H}_2\text{SO}_4} C_6H_5COOH$$

Benzamide   Benzoic acid
                                                (99%)

Scheme 3

9.2.1.2.3 Hydrolysis of N-phenyl Benzamide

The acid hydrolysis of N-phenylbenzamide usually takes 18-20 hr. However, under microwave conditions the reaction is completed in 12 min giving[6] 74% of benzoic acid (Scheme 4).

$$C_6H_5CONHC_6H_5 \xrightarrow[\text{mw 12 min}]{20\% \text{ H}_2\text{SO}_4} C_6H_5COOH$$

N-phenylbenzamide   Benzoic acid
                                                (74%)

Scheme 4

9.2.1.3 Oxidation of Toluene

Oxidation of toluene with $KMnO_4$ under normal conditions of refluxing takes 10-12 hr compared to reaction in microwave conditions[5], which takes only 5 min and the yield is 40% (Scheme 5).

$$C_6H_5CH_3 \xrightarrow[\substack{\text{aq.KMnO}_4 + \text{aq. KOH} \\ \text{mw 5 min}}]{[O]} C_6H_5COOH$$

Toluene   Benzoic acid
                                                (40%)

Scheme 5

9.2.1.4 Oxidation of Alcohols

A number of primary alcohols can be oxidised to the corresponding carboxylic acid (Scheme 6) using sodium tungstate as catalyst in 30% aqueous hydrogen peroxide.

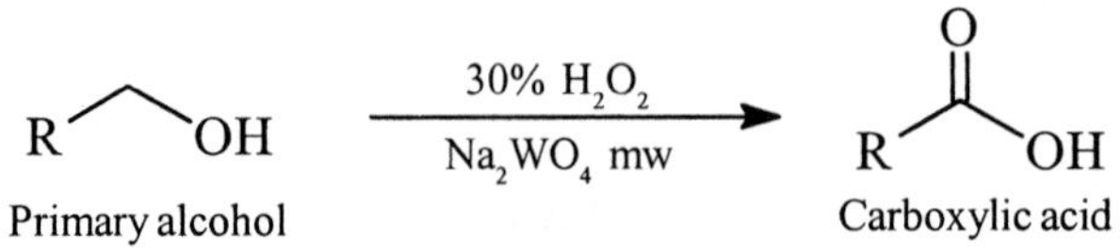

Primary alcohol   Carboxylic acid

Scheme 6

Similarly, secondary alcohols have been oxidised[5a] under microwave irradiation by using doped supports like clayfen (montmorillonite K10 + iron (III) nitrate), silica manganese dioxide, claycop (montmorillonite K10 + copper(II) nitrate)-H_2O_2, CrO_3-wet alumina, iodobenzenediacetate-alumina, $CuSO_4$-alumina, oxone-wet alumina (Scheme 6a).

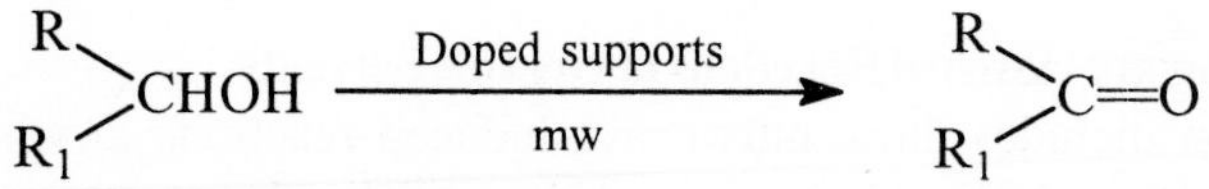

$$\underset{R_1}{\overset{R}{>}}CHOH \xrightarrow[\text{mw}]{\text{Doped supports}} \underset{R_1}{\overset{R}{>}}C=O$$

R, R_1 = various aromatic, aliphatic and heterocyclic groups

Scheme 6a

Also oxidation of linear and cyclic secondary alcohols and benzylic alcohols to the corresponding carbonyl compounds under microwave irradiation conditions can be achieved.

Arenes on oxidation with $KMnO_4$ impregnated on alumina under microwave irradiation in dry media (instead of several days under classical conditions) gave[5b] ketones (CH_2 group is oxidised to keto) (Scheme 6b).

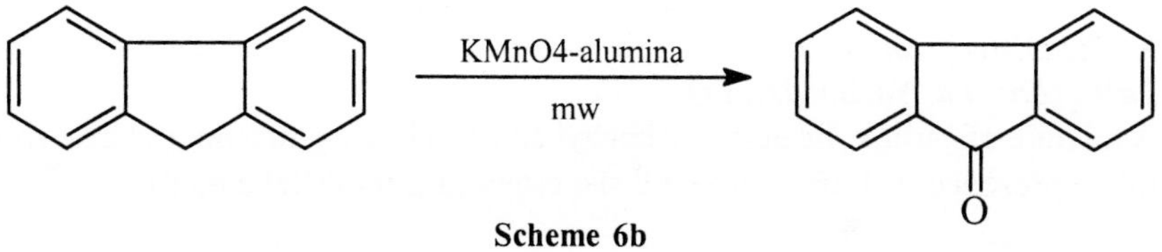

Scheme 6b

Thiols have been oxidised[5b] to disulphides on mineral supports like silica, celite, florisel, alumina (Scheme 6c).

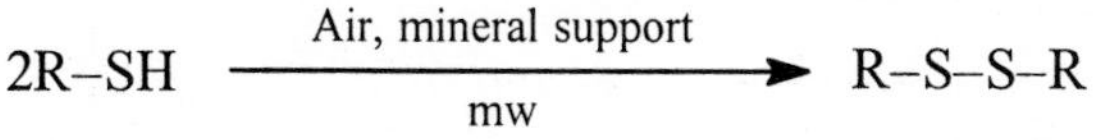

$$2R-SH \xrightarrow[\text{mw}]{\text{Air, mineral support}} R-S-S-R$$

R–SH = p-nitrophenylmercaptan, 2-mercaptopyridine, thiosalicylic acid

Scheme 6c

9.2.1.5 *Hydrolysis of Methylbenzoate to Benzoic Acid (Saponification)*
Saponification of methylbenzoate in aqueous sodium hydroxide under microwave conditions (2.5 min) gives[5] 84% yield of the benzoic acid (Scheme 7).

$$C_6H_5COOCH_3 \xrightarrow[\text{mw 2.5 min}]{\text{aq. NaOH}} C_6H_5COOH$$

Methylbenzoate   Benzoic acid
                                                          (84%)

Scheme 7

9.2.2 Microwave-Assisted Reactions in Organic Solvents

This section includes those microwave induced reactions in which one or both the reactants (if liquid) act as a solvent and also those reactions in which organic solvent is used to assist the reaction.

9.2.2.1 Esterification: Reaction of Carboxylic Acid and Alcohol

A mixture of benzoic acid and n-propanol on heating in a microwave oven for 6 min in presence of catalytic amount of conc. sulphuric acid gives[5,6,7] propylbenzoate (Scheme 8).

$$C_6H_5COOH + nC_3H_7OH \xrightarrow[\text{mw 6 min}]{\text{conc. H}_2\text{SO}_4} C_6H_5COOC_3H_7$$

Benzoic acid   Propylbenzoate
                                                             70%

Scheme 8

9.2.2.2 Esterification: Reaction of Carboxylic Acids and Benzyl Ethers Using LnBr₃ (Ln = La, Nd, Sm, Dy, Er)

A mixture of carboxylic acid and benzyl ether on heating in a microwave oven in the presence of LnBr₃ afforded[8] the esters in 2 min (Scheme 9).

$$ArCH_2OR + R^1COOH \xrightarrow[\substack{\text{Ln = Ld, Nd Sm, Dy, Er} \\ \text{mw 2 min}}]{\text{Ln Br}_3} ArCH_2OCOR^1 + ROH$$

Scheme 9

9.2.2.3 Fries Rearrangement

Fries rearrangement is a useful method for the preparation of phenolic ketones and is usually carried out by heating a mixture of substrate and aluminium chloride.

There is considerable rate enhancement of Fries rearrangement by commercial microwave ovens over conventional methods. Thus, a mixture of p-cresyl acetate and anhydrous aluminium chloride are heated in dry chlorobenzene in a sealed tube in a microwave oven for 2 min to give[9] 85% yield of the product (Scheme 10).

9.2.2.4 Orthoester Claisen Rearrangement

In the usual conventional procedure, a mixture of allyl alcohol, triethyl orthoacetate and propanoic acid is heated in a sealed tube for 48 hr. However, under microwave conditions[10] a mixture of allyl alcohol, triethyl orthoacetate and propanoic acid in dry dimethylformamide is heated in microwave oven for 10 min. The product (Scheme 11) is obtained in 83% yield.

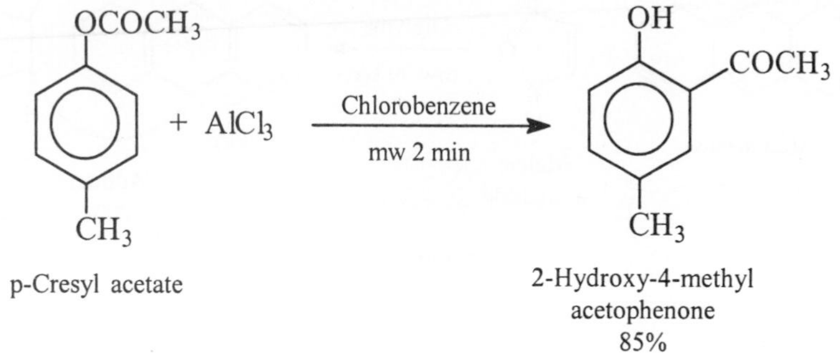

p-Cresyl acetate

2-Hydroxy-4-methyl
acetophenone
85%

Scheme 10

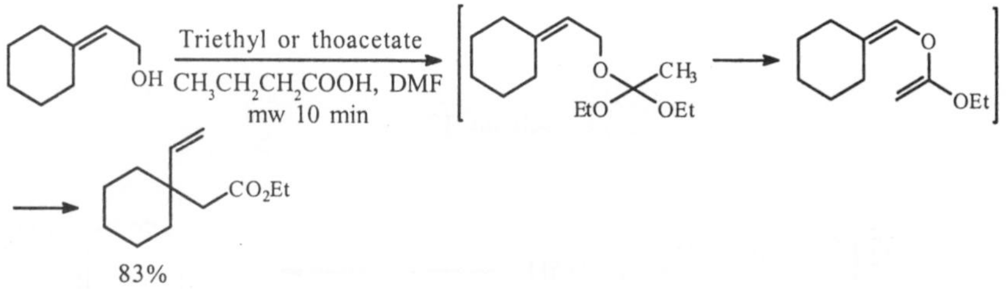

83%

Scheme 11

9.2.2.5 Diels Alder Reaction

The reaction involves 1,4-addition of an alkene (e.g., maleic anhydride) to a conjugated diene (e.g. anthracene) to form an adduct of six membered ring. Under usual conditions[11] the reaction requires a reflux period of 90 min. However, under microwave conditions[3,12] diglyme is used as a solvent and 80% yield of the adduct is obtained in 90 sec (Scheme 12).

9.2.2.6 Synthesis of Chalcones

Microwaves have been used for the synthesis[13] of chalcones and related enones. Considerable rate enhancement is observed, bringing down the reaction time from hours to minutes in improved yield (Scheme 13).

9.2.2.7 ***Decarboxylation***

Conventional decarboxylation of carboxylic acids involve refluxing in quinoline in presence of copper chromite and the yields are low. However, in the presence of microwaves[14], decarboxylation takes place in much shorter time as illustrated in Scheme 14.

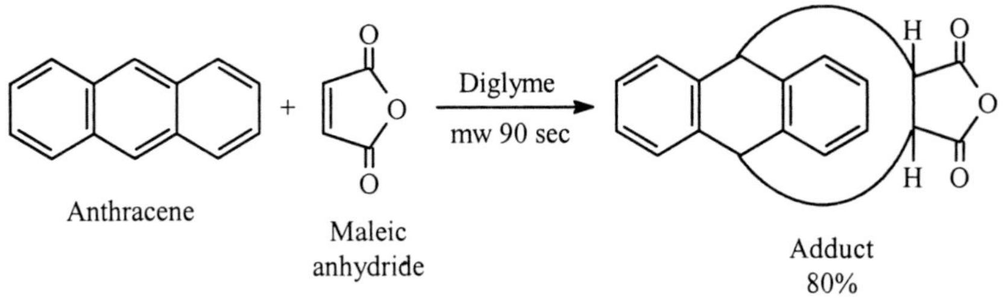

O
O
O
Diglyme
mw 90 sec
H O
O
H O
Anthracene
Maleic anhydride
Adduct
80%

Scheme 12

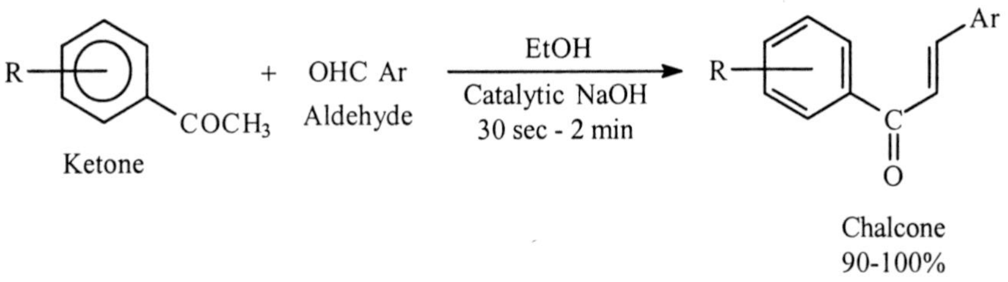

R
COCH3
OHC Ar
Aldehyde
EtOH
Catalytic NaOH
30 sec - 2 min
R
Ar
C
O
Ketone
Chalcone
90-100%

Scheme 13

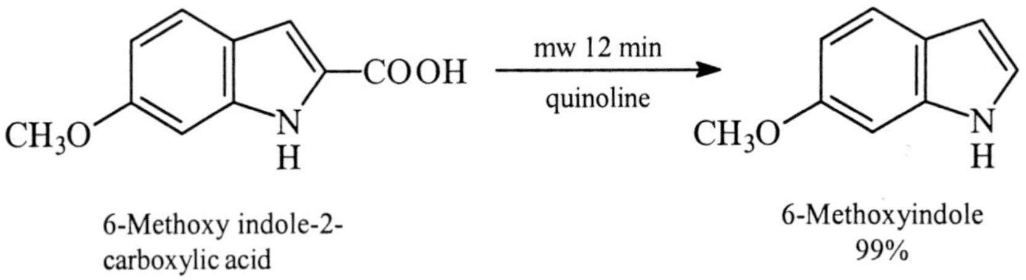

CH3O
N
H
COOH
mw 12 min
quinoline
CH3O
N
H
6-Methoxy indole-2-carboxylic acid
6-Methoxyindole
99%

Scheme 14

9.2.3 Microwave Solvent Free Reactions (Solid State Reactions)

Application of microwave irradiation in organic reactions has added a new dimension to solid phase synthesis. By the use of this technique, it is now possible to carry out reactions without the use of toxic or other solvents, which is one of the main problems associated with green synthesis. In these, the reactants are dissolved in a suitable solvent like water, alcohol, methylene chloride etc. and the solution stirred with a suitable adsorbent or solid support like silica gel, alumina or phyllosilicate (M^{n+}-montomorillonite). After stirring, the solvent is removed *in vacuo* and the dried solid support on which the reactants have been adsorbed are used for carrying out the reaction under microwave irradiation. Following are some of the important applications of solid support synthesis.

9.2.3.1 Deacetylation

Aldehydes[15], phenols[16] and alcohols[16] are protected by acetylation. After the reaction, the deacetylation of the product is carried out usually under acidic or basic conditions; the process takes long time and the yields are low. Use of microwave irradiation reduces the time of deacetylation and the yields are good. Some examples are (Scheme 15) as follows:

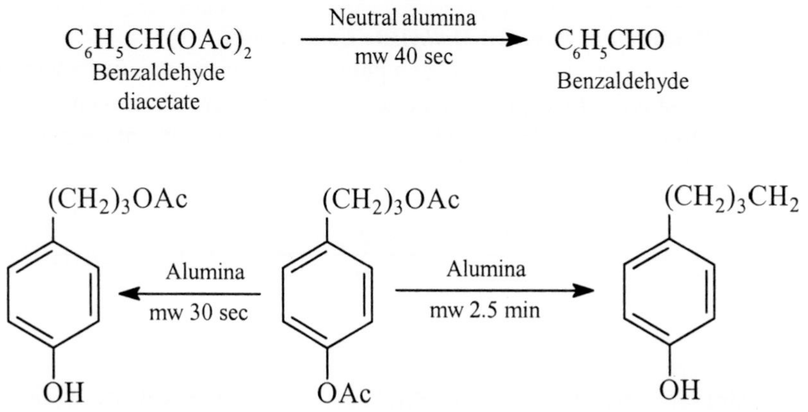

Scheme 15

9.2.3.2 Deprotection

The carboxylic function is generally protected by the benzyl protecting group. After the reaction sequence, the deprotection of benzyl ester is carried out by using potassium carbonate[17], aluminium chloride[18], Na-NH$_3$[19] etc. Most of the deprotection procedures give moderate yields and a longer reaction time is required. The microwave irradiation procedure[20] is completed in 3-10 min and yields are high (89-92%) (Scheme 16).

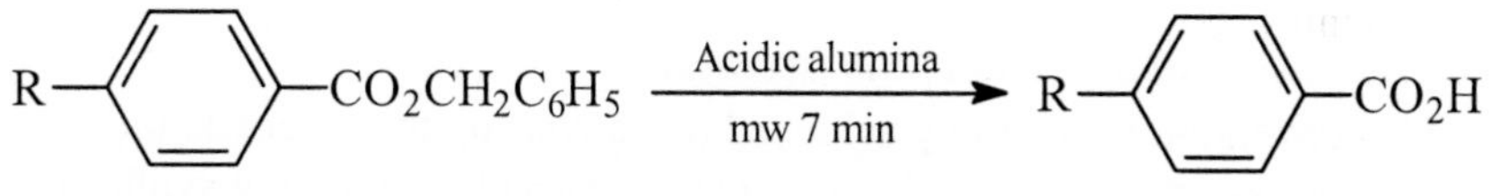

R= H or CH$_3$

Scheme 16

In a similar way, t-butyl dimethylsilyl group can be removed.[21]

9.2.3.3 Saponification of Esters

Hindered esters which take 5 hr under classical[22,23] heating with alkali can be easily saponified under microwave irradiation[24] using KOH-Aliquat (Scheme 17).

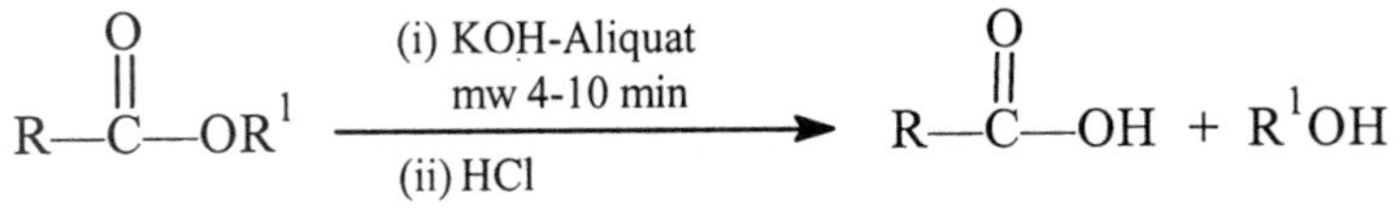

$$R-\overset{\overset{\displaystyle O}{\|}}{C}-OR^1 \quad \xrightarrow[\text{(ii) HCl}]{\begin{array}{c}\text{(i) KOH-Aliquat}\\ \text{mw 4-10 min}\end{array}} \quad R-\overset{\overset{\displaystyle O}{\|}}{C}-OH + R^1OH$$

Scheme 17

9.2.3.4 Alkylation of Reactive Methylene Compounds

Efficient and rapid alkylation of compounds containing reactive methylene group (e.g., ethylacetoacetate) can be achieved in a microwave oven[24,25], using tetrabutyl ammonium chloride (TBAC) as PTC without solvent (Scheme 18).

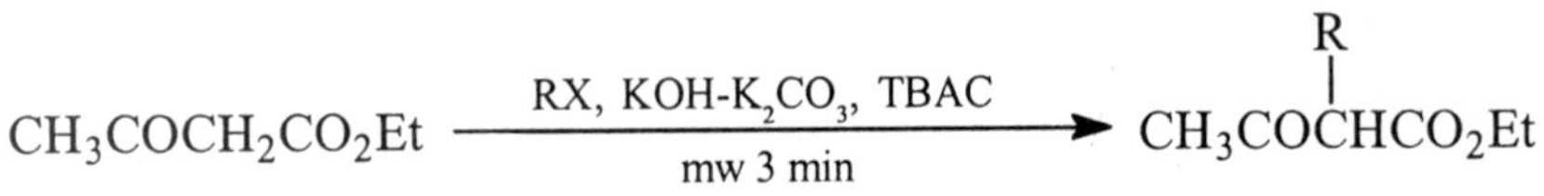

$$CH_3COCH_2CO_2Et \quad \xrightarrow[\text{mw 3 min}]{RX,\ KOH\text{-}K_2CO_3,\ TBAC} \quad CH_3CO\overset{\overset{\displaystyle R}{|}}{C}HCO_2Et$$

Scheme 18

In a similar way, alkylation of ethyl mercaptoacetate can be achieved[26] (Scheme 19).

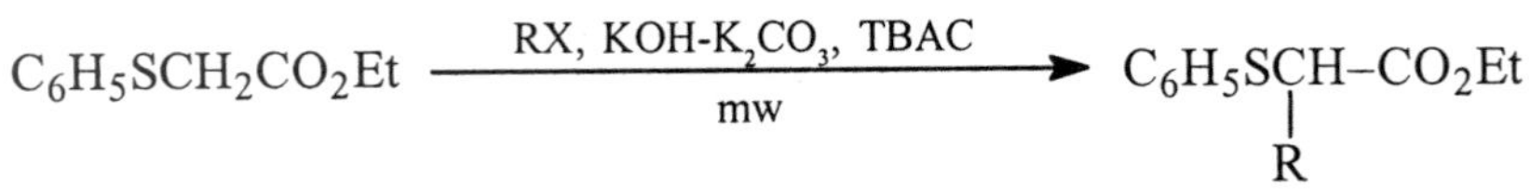

$$C_6H_5SCH_2CO_2Et \quad \xrightarrow[\text{mw}]{RX,\ KOH\text{-}K_2CO_3,\ TBAC} \quad C_6H_5S\overset{}{C}H-CO_2Et \atop R$$

Scheme 19

9.2.3.5 Condensation of Active Methylene Compounds with Aldehydes

Condensation of compounds containing active methylene group can be made to react with the aromatic aldehydes to give corresponding arylidene derivatives[27,28,29] (Scheme 20).

9.2.3.6 Synthesis of Nitriles from Aldehydes

This conversion is generally achieved by converting the aldehyde into oxime followed by its dehydration by a variety of reagents.[30] One pot synthesis has also been developed[31,32], but these methods suffer from a number of drawbacks. In the microwave assisted reaction, the aldehyde is converted into oxime by reaction with hydroxylamine hydrochloride and potassium fluoride on alumina

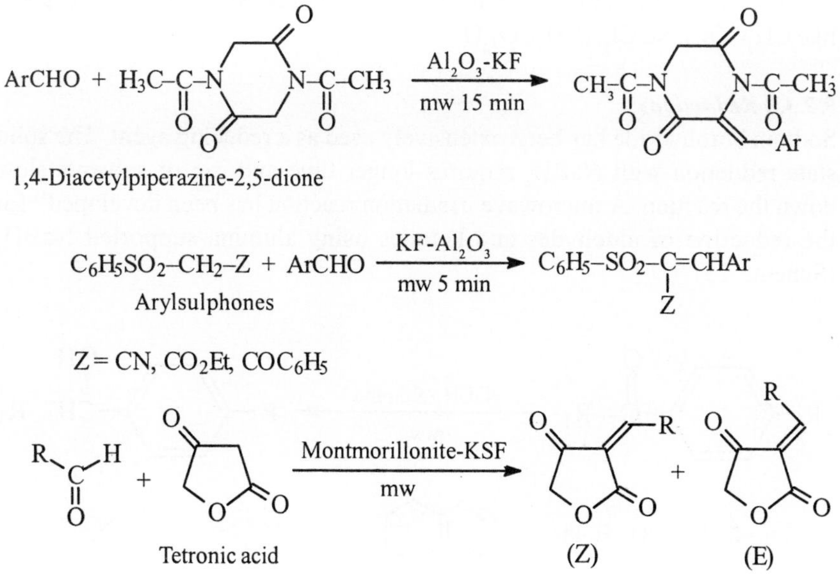

ArCHO + H₃C–C–N N–C–CH₃ $\xrightarrow[\text{mw 15 min}]{\text{Al}_2\text{O}_3\text{-KF}}$ CH₃–C–N N–C–CH₃

1,4-Diacetylpiperazine-2,5-dione

C₆H₅SO₂–CH₂–Z + ArCHO $\xrightarrow[\text{mw 5 min}]{\text{KF-Al}_2\text{O}_3}$ C₆H₅–SO₂–C=CHAr
                                                                                                Z

Arylsulphones

Z = CN, CO₂Et, COC₆H₅

Tetronic acid   Montmorillonite-KSF / mw   (Z)   (E)

Scheme 20

without a solvent. The absorbed oxime is converted into nitrile in the second step by treatment with carbon disulphide (Scheme 21). This method is quick, simple and convenient[33] and gives nitriles in good yield.

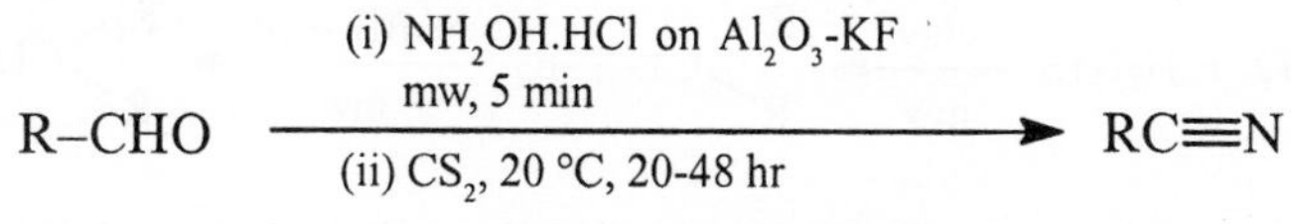

R–CHO $\xrightarrow[\text{(ii) CS}_2\text{, 20 °C, 20-48 hr}]{\substack{\text{(i) NH}_2\text{OH.HCl on Al}_2\text{O}_3\text{-KF} \\ \text{mw, 5 min}}}$ RC≡N

Scheme 21

9.2.3.7 Synthesis of Anhydrides from Dicarboxylic Acid

Dicarboxylic acids can be converted into anhydrides (Scheme 22) in the presence of isopropenyl acid (which acts as a water scavenger) under microwave irradiation[34] using montmorillonite-KSF. The driving force is the formation of acetone.

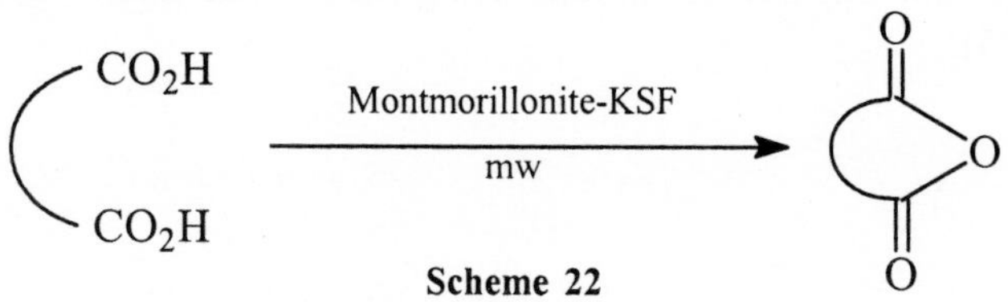

Scheme 22

This method is rapid and convenient and avoids using corrosive reagents like CH_3COCl, $SOCl_2$, $(CH_3CO)_2O$.

9.2.3.8 Reductions

Sodium borohydride has been extensively used as a reducing agent. The solid state reduction with $NaBH_4$ requires longer time and use of solvents slow down the reaction. A microwave irradiation reaction has been developed[5a] for the reduction of aldehydes and ketones using alumina supported $NaBH_4$ (Scheme 23).

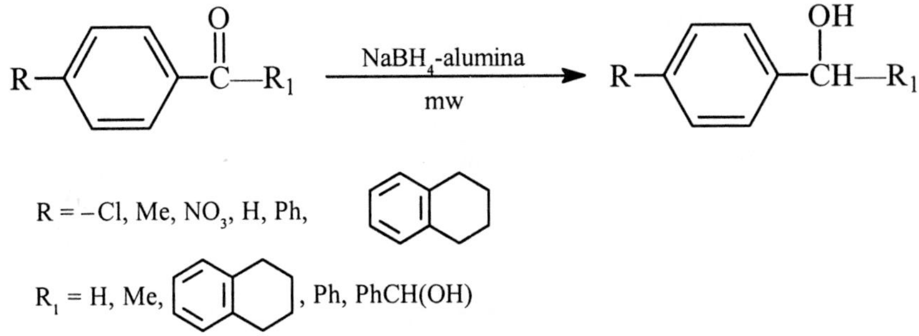

$R = -Cl$, Me, NO_3, H, Ph,

$R_1 = $ H, Me, [image], Ph, PhCH(OH)

Scheme 23

Sodium borohydride in combination with wet montmorillonite K10 clay has been used[5a] for reductive amination of carbonyl compounds (Scheme 24).

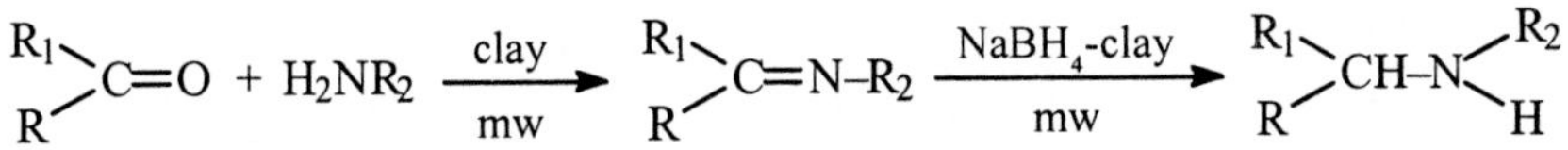

$R = $ i-Pr, Ph, o-OHC$_6$H$_4$-, p-MeOC$_6$H$_4$-, p-NO$_2$C$_6$H$_4$-, p-ClC$_6$H$_4$-, Et, n-C$_5$H$_{11}$
$R_1 = $ H, -(CH$_2$)$_5$-, -(CH$_2$)$_6$-, n-Pr
$R_2 = $ Ph, n-Pr, morpholine, piperidine, n-C$_{10}$H$_{21}$

Scheme 24

Leukart reductive amination of carbonyl compounds was considerably enhanced[5a] (in comparison to conventional heating) by a specific microwave effect under solvent-free conditions using monomode microwave reactor (Scheme 25).

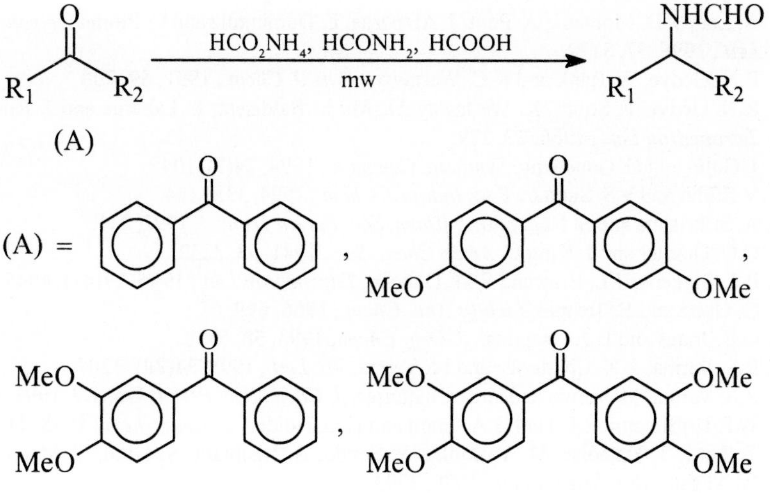

Scheme 25

9.2.3.9 Synthesis of Heterocyclic Compounds

Heterocyclic compounds are immensely important due to their wide spectrum of biological activity. A number of diverse heterocyclic systems have been synthesised. For details see Sec. 13.3.

9.3 Conclusion

The use of microwaves has led to substantial savings in time for many syntheses in the laboratory as well as in industry. Microwave induced reactions can be carried out in water or organic solvents. The organic solvents if used, are required in very small quantities. The most important feature of microwaves induced reactions is that these can also be carried out in the solid state, i.e., without the use of any solvent.

References

1. D. Michael, P. Mingos and D.R. Baghrust, *Chem. Soc. Rev.*, 1991, **20**, 1.
2. A.K. Bose, M.S. Manhas, B.K. Banik and E.W. Robb, *Res. Chem. Intermed.*, 1994, **20(1)**, 1.
3. S.S. Bari, A.K. Bose, A.G. Chaudhary, M.S. Manhas, V.S. Raju and E.W. Robb, *J. Chem. Ed.*, 1992, **69(11)**, 938.
4. Shui-Teinchem, Shyh-Horngchiou and Kung-Tsungwang, *J. Chem. Soc. Chem. Commun.*, 1990, 807.

5.   M.N. Gedye, F.E. Smith and K.C. Westaway, *Can. J. Chem.*, 1988, **66**, 17.
5a. R.S. Verma and R.K. Saini, *Tetrahedron Lett.*, 1997, **38**, 2623; R.S. Verma and D. Kumar, *Synth. Commun.*, 1999, **29**, 1333.
5b. A. Loupy, D. Monteux, A. Petit, J. Aizpurua, E. Domingulz and C. Plomu, *Tetrahedron Lett.*, 1996, **37**, 8177.
6.   R.N. Gedye, W. Rank and K.C. Westaway, *Can. J. Chem.*, 1991, **69**, 700.
7.   R.N. Gedye, F. Smith, K. Westaway, H. Ali, L. Baldisera, L. Laberge and J. Rousell, *Tetrahedron Lett.*, 1986, **27**, 279.
8.   J. Gulin and G. Guncheng, *Synthetic Commun.*, 1994, **24(7)**, 1045.
9.   V. Sridar and V.S. Sundara Rao, *Indian J. Chem.*, 1994, **33B**, 184.
10. A. Srikrishna and S. Nagaraju, *J. Chem. Soc. Perkin Trans. I*, 1992, 311.
11. O.C. Dermer and J. King, *J. Amer. Chem. Soc.*, 1941, **63**, 3232.
12. R.J. Giguene, T.L. Bray and S.M. Duncan, *Tetrahedron Lett.*, 1986, **27(41)**, 4945.
13. G. Opitz and E. Tempel, *Liebigs Ann. Chem.*, 1966, **699**, 68.
14. G.B. Jones and B.J. Chapman, *J. Org. Chem.*, 1993, **58**, 5558.
15. R.S. Varma, A.K. Chatterjee and M. Verma, *Tet. Lett.*, 1993, **34(20)**, 3207.
16. R.S. Varma, M. Varma and A.K. Chatterjee, *J. Chem. Soc. Perkin Trans. I*, 1993, 999.
17. W.F. Huffimann, R.F. Hall, J.A. Grant and H.G. Holden, *J. Med. Chem.*, 1978, **21**, 413.
18. T. Tsuji, T. Kataoka, M. Yoskioka, Y. Sendo, S. Nishitant, S. Hirai, T. Maeda and W. Nagata, *Tetrahedron Lett.*, 1979, 2793.
19. C.W. Roberts, *J. Am. Chem. Soc.*, 1954, **76**, 6203.
20. R.S. Varma, A.K. Chaterjee and M. Varma, *Tetrahedron Lett.*, 1993, **34**, 4603.
21. R.S. Varma, J.B. Lamture and M. Verma, *Tet. Lett.*, 1993, **34(19)**, 3029.
22. A. Loupy, M. Pedoussaut and J. Sanoulet, *J. Org. Chem.*, 1986, **51**, 740.
23. A. Dietrich and J.M. Lehn, *Tetrahedron Lett.*, 1973, 1225.
24. Loupsy, P. Pigeon, M. Ramdani and P. Jaequault, *Synthetic Commun.*, 1994, **24(2)**, 159.
25. D. Ruhua, W. Guliang and J. Yao Zhing, *Synthetic Commun.*, 1994, **24(1)**, 111.
26. D. Ruhua, W. Yuliang and Y. Yaozhong, *Synthetic Commun.*, 1994, **24(13)**, 1917.
27. D. Villemin and A.B. Alloum, *Synthetic Commun.*, 1990, **20(21)**, 3325.
28. D. Villemin and A.B. Alloum, *Synthetic Commun.*, 1991, **21(1)**, 63.
29. S. Gelin and P. Pollut, *Synthetic Commun.*, 1980, **10**, 805; D.G. Schmidt, P.D. Seemuth and H. Zimmer, *J. Org. Chem.*, 1983, **48**, 1914.
30. H.G. Forey and D.R. Dalton, *J. Chem. Soc. Chem. Commun.*, 1973, 628; J.C. Krause and S. Shaikh, *Synthesis*, 1974, 563; H. Shinogaki, M. Imaizumi and M. Tajima, *Chem. Lett.*, 1983, 929.
31. S.N. Karmarkar, S.L. Kelkar and Wadia, *Synthesis*, 1985, 510; D. Dauzonne, P. Demerseman and R. Royer, Ibid., 1981, 739; M. Miller and G. Loudon, *J. Org. Chem.*, 1975, **40**, 126.
32. C. Fizet and J. Streith, *Tetrahedron Lett.*, 1974, 3187.
33. D. Villemin, M. Lalaoui and A.B. Alloum, *Chem. Ind.*, 1991, 176.
34. D. Villemin, B. Labiad and A. Loupy, *Synthetic Commun.*, 1993, **23(4)**, 419.

10. Ultrasound Assisted Green Synthesis

10.1 Introduction

The word 'ultrasound' has become common knowledge due to the widespread use of ultrasound scanning equipments in medical applications. Ultrasound refers to sound waves having frequencies higher than those to which the human ear can respond (μ, > 16 KHz) (Hz = Hertz = cycles per second). High frequency ultrasound waves are used in medical equipments. The ultrasound frequencies of interest for chemical reactions (about 20-100 KHz) are much lower than those used for medical applications, but the power used is higher.

The ultrasound is generated with the help of an instrument having an ultrasonic transducer, a device by which electrical or mechanical energy can be converted into sound energy. The most commonly used are the electromechanical transducers which convert energy into sound — they are mostly made of quartz and are commonly based on the piezoelectric effect. When equal and opposite electrical charges are applied to opposite faces of a crystal of quartz, expansion or contraction occurs. Application of rapidly reversing charges sets up a vibration that emits ultrasonic waves called the *piezoelectric effect*. In modern ultrasonic equipments, the piezoelectric transducers are made from ceramic impregnated barium titanate. Such devices convert over 95% of the electrical energy into ultrasound. In practice, the devices only have an optimum operating frequency.

When a sound wave, propagated by a series of compression and refraction cycles, pass through a liquid medium, it causes the molecules to oscillate around their mean position. During the compression cycle, the average distance between the molecules is reduced and during refraction, the average distance between the molecules is increased. In the refraction cycle, under appropriate conditions, the attractive forces of the molecules of the liquid may be overcome, causing formation of bubbles. In case the internal forces are great enough to ensure collapse of these bubbles, very high local temperature (around 5000°C) and pressure (over 1000 bar) may be created. It is this very high temperature and pressure that initiate chemical reactions.

The term 'sonochemistry' is used to describe the effect of ultrasound waves on chemical reactivity. A number of reviews on the chemical applications of ultrasound have been published.[1-5]

10.2 Applications of Ultrasound

Following are some of the important applications of ultrasound in chemical synthesis. Most of the reactions/syntheses reported are carried out at room temperature unless otherwise specified. The symbol ⟯⟯⟯ is used for reactions carried out on exposure to ultrasound.

10.2.1 Esterification

This is generally carried out in presence of a catalyst like sulphuric acid, p-toluenesulphonic acid, tosylchloride, polyphosphoric acid, dicyclo-hexylcarbodiimide etc. The reaction takes longer time and yields are low. A simple procedure for the esterification of a variety of carboxylic acids with different alcohols at ambient temperature using ultrasound has been reported[6] (Scheme 1).

$$RCOOH \ + \ R^1OH \ \xrightarrow[\text{⟯⟯⟯}]{H_2SO_4, \ R.T.} \ RCOOR^1$$

Scheme 1

10.2.2 Saponification

Saponification can be carried out under milder conditions using sonification.[7,8] Thus, methyl 2,4-dimethylbenzoate on saponification (20 KHz) gives the corresponding acid in 94% yield (Scheme 2), compared to 15% yield by the usual process of heating with aqueous alkali (90 min).

Scheme 2

10.2.3 Hydrolysis

Nitriles can be hydrolysed[9] to carboxylic acids (Scheme 3) under basic condition on sonication.

$$ArCN \xrightarrow[\text{)))}]{OH^- / H_2O} ArCOOH$$

Scheme 3

10.2.4 Substitution Reactions

Halides can be converted into cyanides. Thus, the reaction of benzyl bromide in toluene with potassium cyanide, catalysed by alumina, on sonication gives[10] the substitution product, viz. benzyl cyanide in 76% yield. In the absence of ultrasound alkylation is the preferred pathway (Scheme 4). The difference is because ultrasound forces cyanide into the surface of alumina, enhancing cyanide nucleophilicity and reducing the lewis acid character.

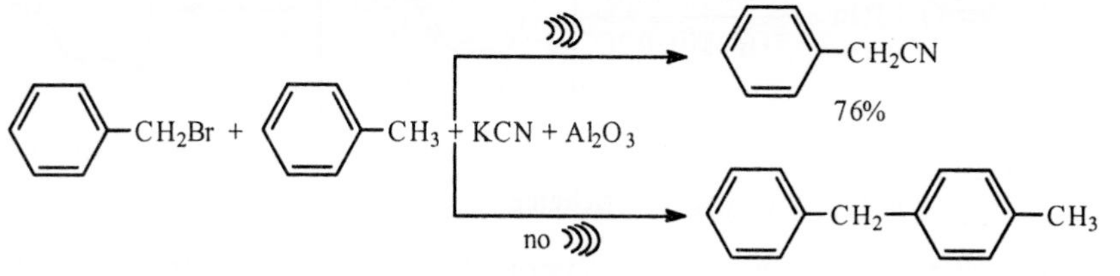

Scheme 4

Sonication of acyl chloride and potassium cyanide in acetonitrile gave[11] the corresponding acyl cyanides (Scheme 5).

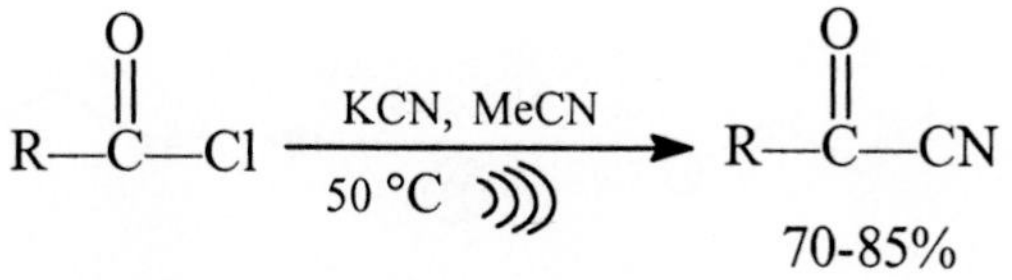

$$R-\overset{\overset{\displaystyle O}{\|}}{C}-Cl \xrightarrow[\text{50 °C)))}]{KCN, MeCN} R-\overset{\overset{\displaystyle O}{\|}}{C}-CN$$

70-85%

R-H, 2-CH$_3$C$_6$H$_4$-, 3-CH$_3$C$_6$H$_4$-, 4-CH$_3$C$_6$H$_4$-, 4-OCH$_3$C$_6$H$_4$

Scheme 5

Similarly, alkyl bromides on sonication with KSCN in presence of a PTC catalyst gave[12] the corresponding sulphocyanide in 62% yield (Scheme 6).

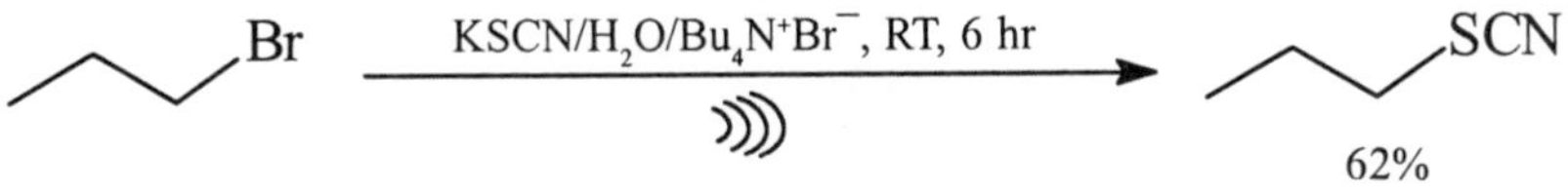

Scheme 6

10.2.5 Addition Reactions

1,4-addition to α,β-unsaturated carbonyl compounds is generally[12] carried out by organocopper reagents. However, improvement in yields, rates and ease of experimental techniques is observed[13] when organocopper compounds are generated *in situ* by sonication of copper (I) compounds, organic halides and lithium in diethyl-ether-THF at 0°C (Scheme 7).

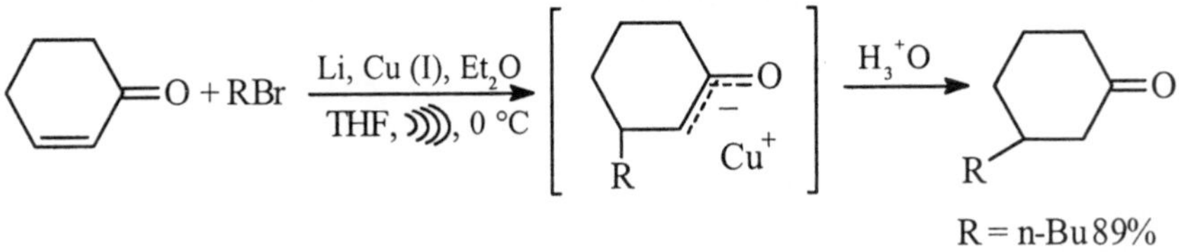

$$R = n\text{-}Bu\,89\%$$

Scheme 7

In a similar way, aryl zinc compounds are generated ultrasonically[14] (Scheme 8).

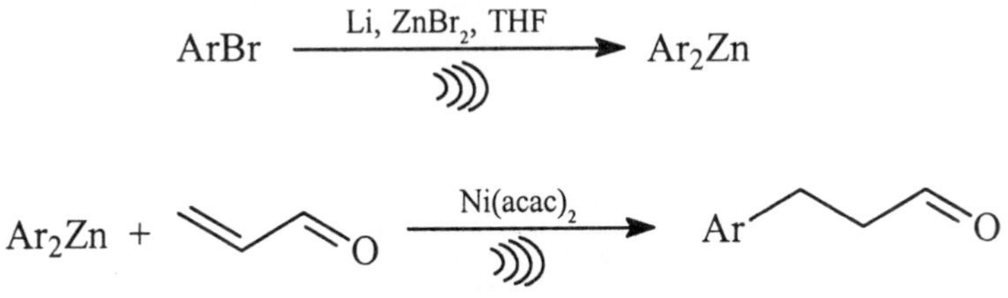

Scheme 8

1,3-Dipolar cycloaddition of nitrones to olefins gave excellent yield[15] of the product in much shorter time (Scheme-9) compared to the usual reaction conditions.[16]

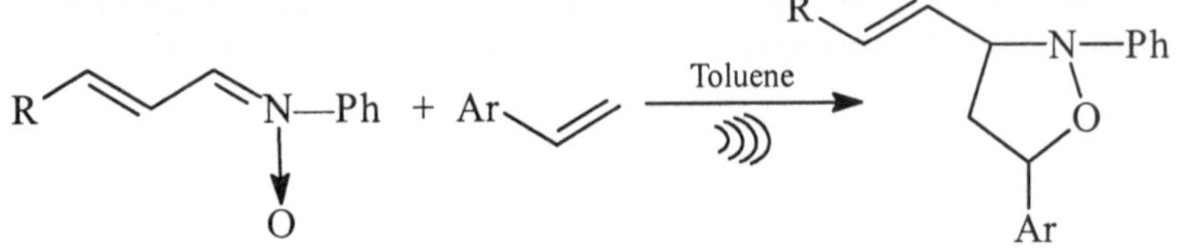

Scheme 9

10.2.6 Alkylations

N-Alkylation of secondary amine takes place under sonication in the presence of a PTC reagent, polyethylene glycol monomethyl ether. Similarly N-alkylation of diphenyl amine is accomplished[17] under sonication (Scheme 10).

This reaction does not take place in the absence of ultrasound.

C-Alkylation of isoquinoline derivatives can be effected[18] using sonication under PTC conditions (Scheme 11).

The O-Alkylation of primary alcohols with benzyl bromide in the presence of silver oxide gives 72% yield of the O-benzylated product[19] (Scheme 12).

Under normal conditions without sonication, the yield is very low.

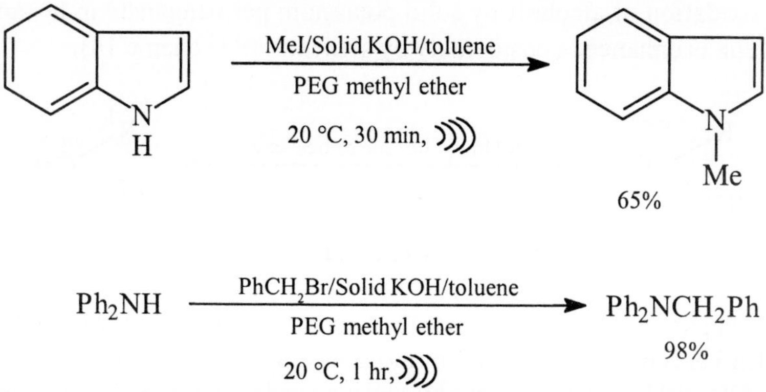

Scheme 10

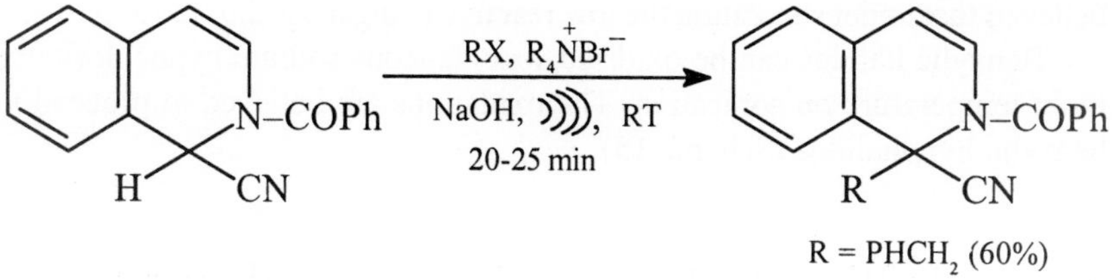

$R = PHCH_2$ (60%)

Scheme 11

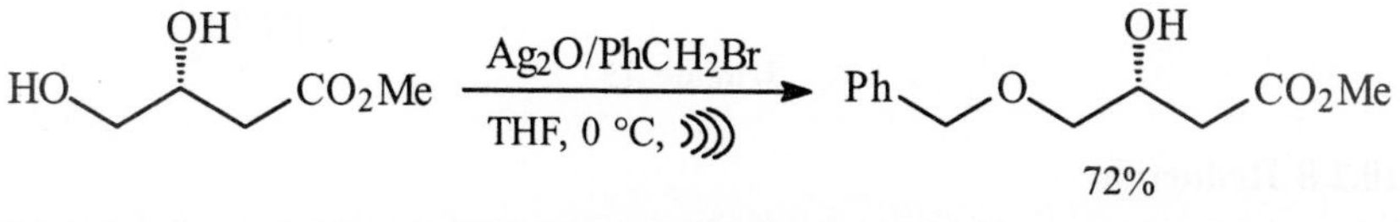

Scheme 12

S-alkylation is accelerated[20] under sonication (Scheme 13).

$$RSH + R^1X \xrightarrow[\text{)))}]{K_2CO_3/DMF} RSR^1$$

Scheme 13

In the above S-alkylation, K_2CO_3 is broken into small particles in the DMF solvent which liberates a high energy by cavitation.

10.2.7 Oxidation

The oxidation of alcohols by solid potassium permanganate in hexane or benzene is enhanced considerably by sonication[21] (Scheme 14).

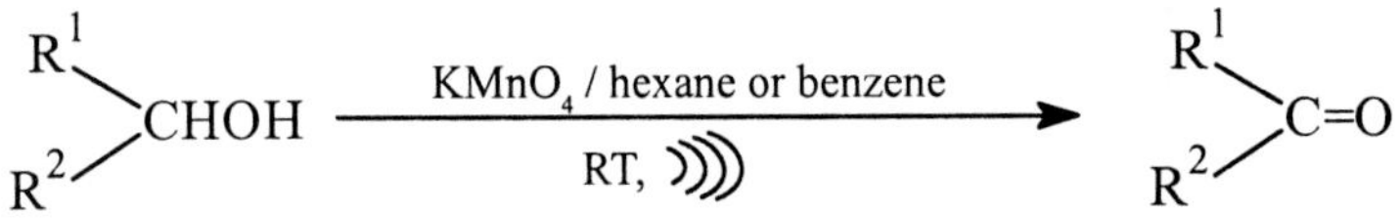

$$\begin{matrix} R^1 \\ \\ R^2 \end{matrix}\!\!CHOH \xrightarrow[\text{RT,)))}]{KMnO_4 / \text{hexane or benzene}} \begin{matrix} R^1 \\ \\ R^2 \end{matrix}\!\!C{=}O$$

Scheme 14

Using the above method, octan-2-ol gives corresponding ketone in 92.8% yield in 5 hr compared to 2% yield by mechanical stirring. Similarly, cyclohexanol gave 53% yield of cyclohexanone by oxidation under sonication (5 hr) compared to the 4% yield under usual conditions.

Oxidation of cinnamyl alcohol with manganese dioxide in a suitable solvent (like hexane or octane) gives[22] the corresponding aldehyde on sonication. It is believed that under sonication the low reactive manganese dioxide is activated.

Benzylic halides can be oxidised with aqueous sodium hypochlorite[23] at room temperature on sonication. The oxidations are believed to proceed via benzylic hypohalides (Scheme 15).

$$ArRCHX + NaOCl \xrightarrow[\text{)))}]{CH_3CN, \, RT} \left[ArRCHOCl \right]$$

$$\downarrow$$

$$ArRCO$$

Scheme 15

10.2.8 Reduction

Sonication increases considerably[24] the reactivity of platinum, palladium and rhodium black in formic acid medium making easier the hydrogenation[25] of a wide range of alkenes at room temperature by sonication. Also, hydrazine-palladium on copper couple is useful for the hydrogenation of alkenes in ethanol at room temperature using an ultrasonic bath[26]. A commercially used example of a sonochemically enhanced catalytic reaction is the ultrasonic hydrogenation of soyabean oil.[27]

Sonication also increases the activation of nickel powder[28] which is used for the reduction of alkenes.

Sonochemical reduction of nickel salts such as chloride with zinc powder gives catalytically active nickel.[29] Under these conditions the excess of metallic zinc also gets activated and reduces the water present in the medium producing hydrogen gas. In this way, not only the catalyst but also the reagent is produced *in situ* with maximum efficiency and safety. This process has been used for the reduction of carbon-carbon double bonds in α,β-unsaturated carbonyl compounds; the C–C double bond is reduced much faster than the carbonyl group. The variation in the conditions, especially pH permits[30] the selective reduction of C = C in preference to C = O (Scheme 16).

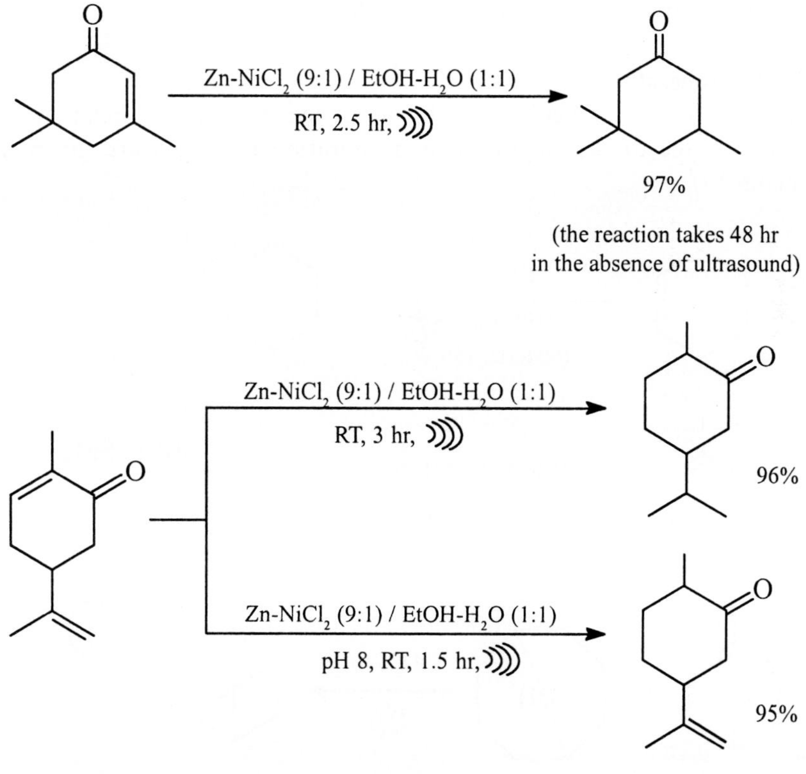

Zn-NiCl$_2$ (9:1) / EtOH-H$_2$O (1:1)

RT, 2.5 hr,)))

97%

(the reaction takes 48 hr in the absence of ultrasound)

Zn-NiCl$_2$ (9:1) / EtOH-H$_2$O (1:1)

RT, 3 hr,)))

96%

Zn-NiCl$_2$ (9:1) / EtOH-H$_2$O (1:1)

pH 8, RT, 1.5 hr,)))

95%

Scheme 16

Carbonyl groups can be reduced by sonication using metal and THF. Thus camphor on sonication[31] in THF gives a mixture of endo and exo borneol in the same ratio as by using metal in ammonia solution (Scheme 17).

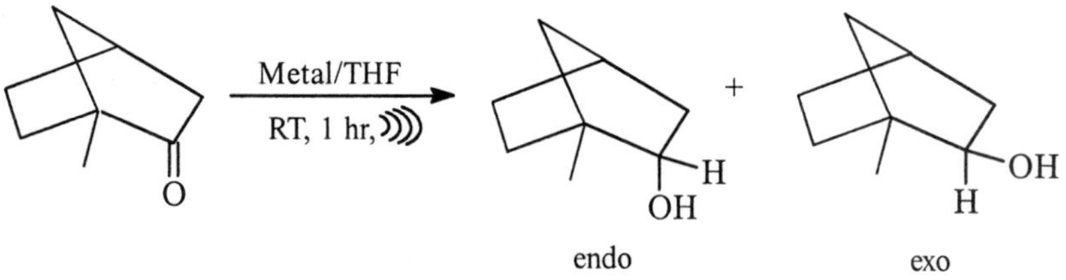

Scheme 17

The endo product is obtained in 73, 68 or 42% yield by the use of Li, Na or K, respectively, as the metal in the above reduction.

Carbonyl groups as in quinones or α-diketones can be reduced on sonication with zinc in presence of trimethylchlorosilane.[32] The Clemmensen reduction can also be carried out by sonication in better yields.[33]

10.2.9 Hydroboration

Hydroboration could be enhanced[34] by ultrasound, especially in heterocyclic systems. Some of the important hydroboration reactions are given in (Scheme 18).

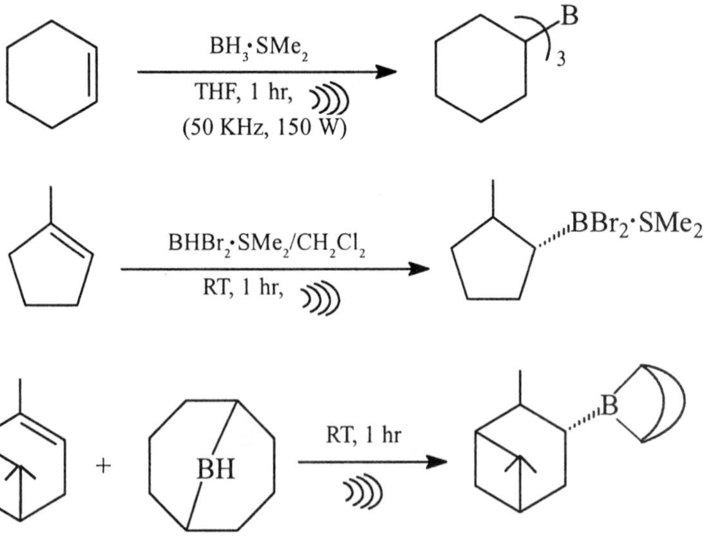

Scheme 18

10.2.10 Coupling Reactions

Homocoupling of organometallic generated *in situ* by the reaction of alkyl, aryl of vinyl halides with lithium in THF takes place on sonication[35,36] (Scheme 19). No reaction takes place in absence of ultrasound.

$$C_6H_5Br \xrightarrow[\text{)))}]{\text{Li, THF}} C_6H_5{-}C_6H_5$$

Scheme 19

In a similar way, coupling of benzyl halides in presence of copper or nickel powder generated by the lithium reduction of the corresponding halides in the presence of ultrasound give high yields[37] of dibenzyl (Scheme 20).

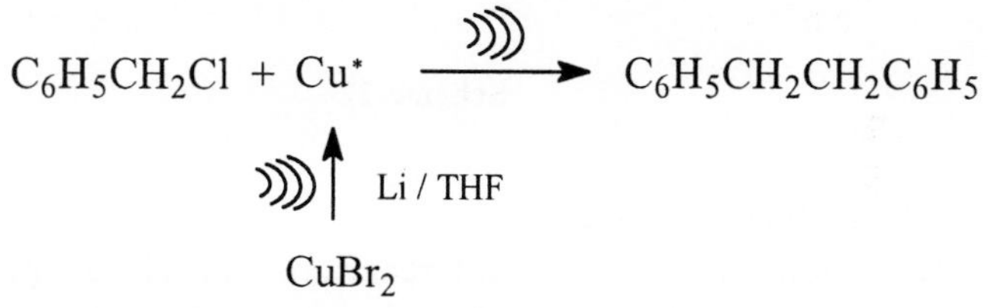

$$C_6H_5CH_2Cl + Cu^* \xrightarrow{\text{)))}} C_6H_5CH_2CH_2C_6H_5$$

$$\text{)))} \uparrow \text{ Li / THF}$$

$$CuBr_2$$

Scheme 20

The classical Ullmann's coupling takes place at high temperature giving low yields. However, in sonication, the size of the metal powder is considerably reduced.[38] Breaking of the particles brings them in contact with the reactive solutions a fresh surface, the reactivity of which is not hindered by the usual oxide layer. The coupling of o-iodonitrobenzene with copper powder is given in (Scheme 21).

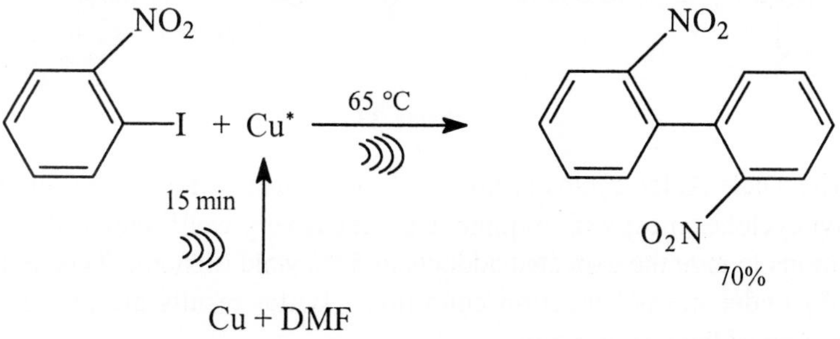

Scheme 21

10.2.11 Friedel-Crafts Reaction

The Friedel-Crafts acylation of aromatic compounds is facilitated[39] by ultrasound (Scheme 22).

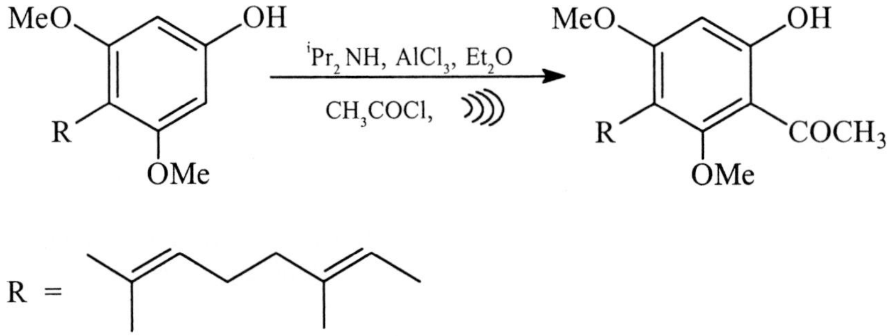

Scheme 22

10.2.12 Diels-Alder Reaction

Sonication facilitates Diels-Alder reaction. Therefore, the addition of dimethylacetylene dicarboxylate to furan in water at 22-45 °C gives quantitative yield of the adduct (Scheme 23).

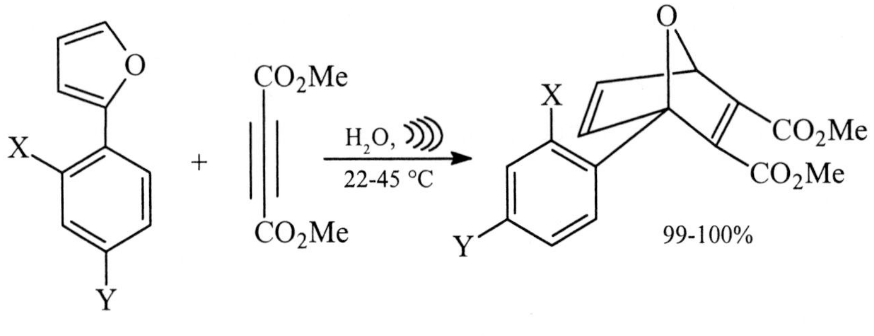

Scheme 23

The Diels-Alder cycloaddition of various dienes (mostly belonging to 1-vinyl cyclohexenes) with o-quinone proceeds very well[40] under ultrasound conditions to give the expected adducts in 59% yield (Scheme-24) compared to 30% under normal reaction conditions. Better results are obtained by soniciation of the neat mixture.

10.2.13 Simmons-Smith Reaction

In this reaction, sonochemically activated zinc and methylene iodide are used. The generated carbene adds on to an olefinic bond to give 91% yield of the cyclopropane derivative compared (Scheme 25) to 51% yield by the normal route.

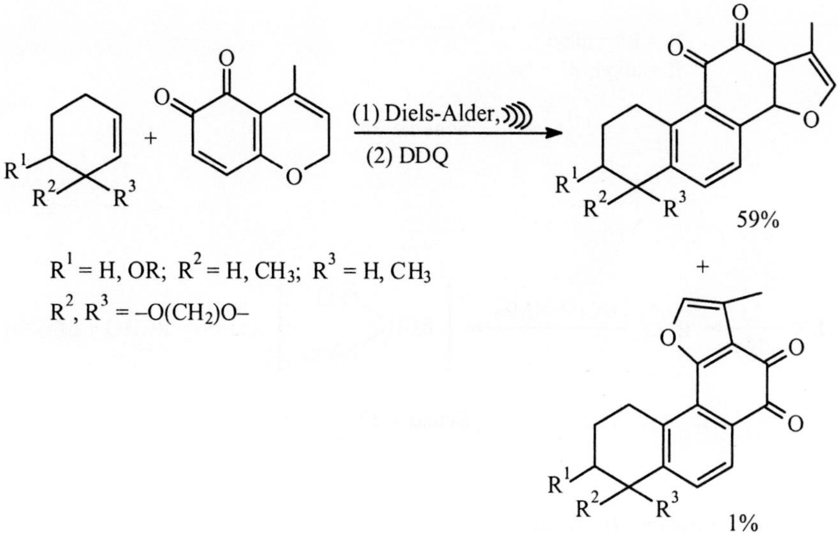

R^1 = H, OR; R^2 = H, CH$_3$; R^3 = H, CH$_3$

R^2, R^3 = –O(CH$_2$)O–

Scheme 24

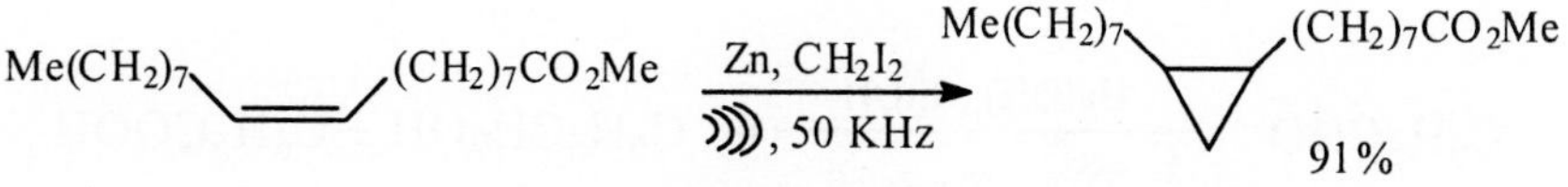

Scheme 25

The above method can be scaled up[41] and has several advantages. The reagent used, Zn/CH$_2$I$_2$ is known as Simmons-Smith reagent.

Ketones on reaction with Simmons-Smith reagent results in methylenation[42] of carbonyl group (Scheme 26). Normally such methylenation of carbonyl group requires complex reagents. It can now be accomplished by sonication.

10.2.14 Bouveault Reaction

Sonication of aryl halides with lithium with low intensity ultrasonic gives organolithium reagents. These are used in Bouveault reaction giving higher yields of aldehydes (Scheme 27) than the traditional methods.[42]

The non-ultrasonic Bouveault reaction suffers from numerous side reactions.

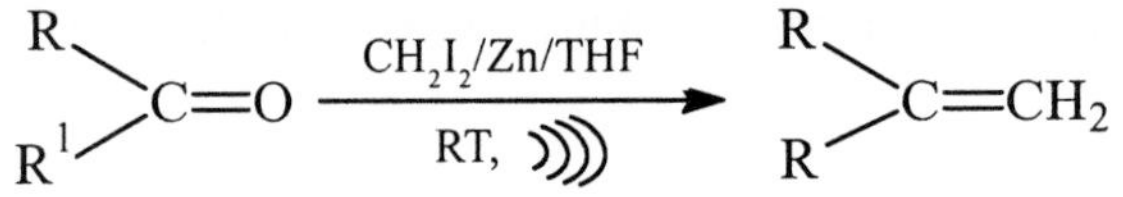

$$\underset{R^1}{\overset{R}{>}}C{=}O \quad \xrightarrow[\text{RT, }))]{\text{CH}_2\text{I}_2/\text{Zn/THF}} \quad \underset{R}{\overset{R}{>}}C{=}CH_2$$

R = R^1 = alkyl
R = alkyl, R^1 = H

Scheme 26

$$RX \xrightarrow[))]{\text{Li}} \overset{-}{R}\overset{+}{Li} \xrightarrow{\text{HC(O)NMe}_2} \left[RCH\overset{\overset{-}{O}\overset{+}{Li}}{\underset{NMe_2}{<}} \right] \xrightarrow{H_3O^+} RCHO + Me_2NH$$

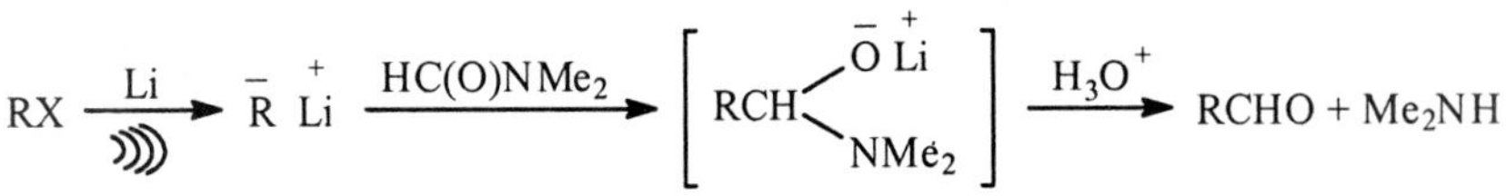

Scheme 27

10.2.15 Cannizaro Reaction

The Cannizarro reaction under heterogenous conditions catalysed by barium hydroxide is considerably accelerated (Scheme 28) by sonication. The yields are 100% after 10 min, whereas no reaction is observed during this period with the use of ultrasound.[43]

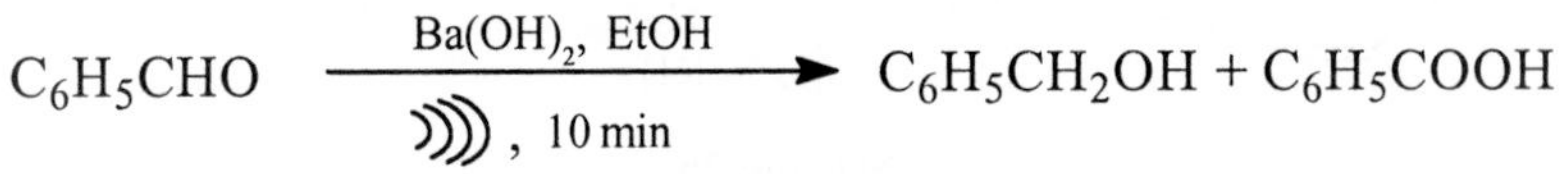

$$C_6H_5CHO \xrightarrow[)), \ 10\,\text{min}]{\text{Ba(OH)}_2, \ \text{EtOH}} C_6H_5CH_2OH + C_6H_5COOH$$

Scheme 28

10.2.16 Strecker Synthesis

The Strecker synthesis of aminonitriles is possible using sonication[44] (Scheme 29).

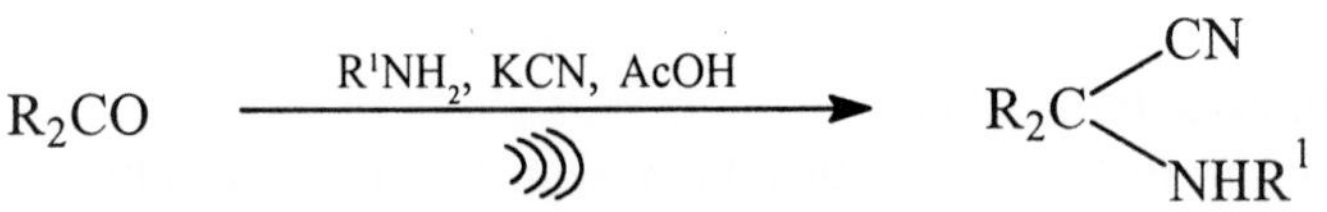

$$R_2CO \xrightarrow[))]{\text{R}^1\text{NH}_2, \ \text{KCN, AcOH}} R_2C\overset{\diagup CN}{\underset{\diagdown NHR^1}{}}$$

Scheme 29

10.2.17 Reformatsky Reaction

The Reformatsky reaction using sonication gave excellent yields compared to the traditional methods using activated zinc or trimethyl borate as a cosolvent.[45] In the sonication procedure the metal zinc is activated by iodine and the reaction is done in dioxane (Scheme 30).

Reformatsky reaction with nitriles leads to the formation of imine, which readily hydrolyse to the ketones. Using appropriate nitrile, keto-γ-butyrolactone is obtained in good yield[46] (Scheme 31).

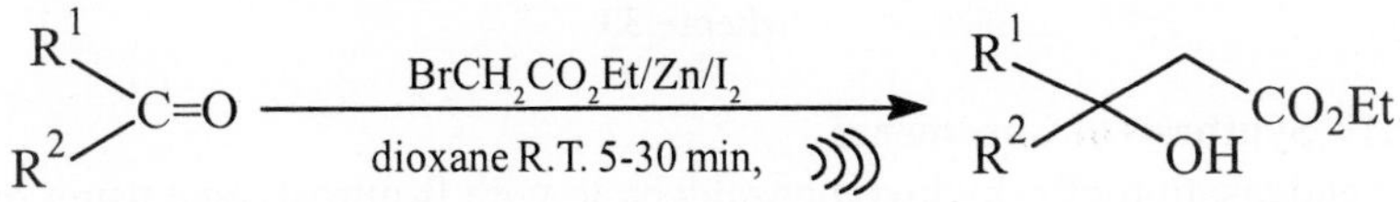

$R^1 = H$; $R^2 = Ph$ or $(CH_3)_2CH$
R^1–$R^2 = -(CH_2)_4-$

Scheme 30

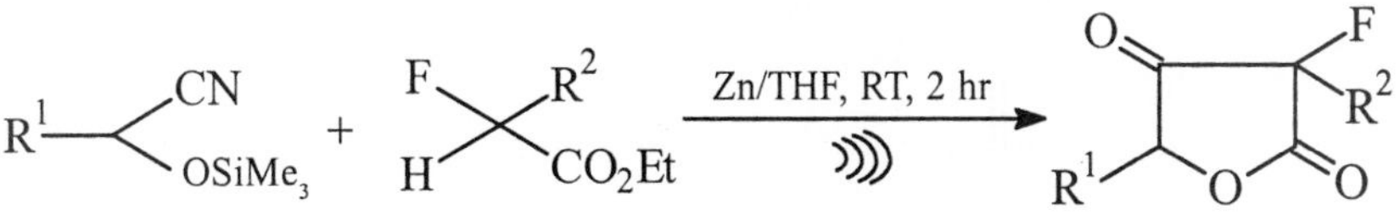

$R^1 = CH_3$, Ph; $R^2 = H$, CH_3

Scheme 31

10.2.18 Conversion of Ketones into Tertiary Alcohols

Ketones are generally converted into tertiary alcohols by treatment with separately prepared Grinard reagent. However, under sonication conditions, ketones on treatment with alkyl halide and lithium in THF give good yields of the tertiary alcohols[47] (Scheme 32).

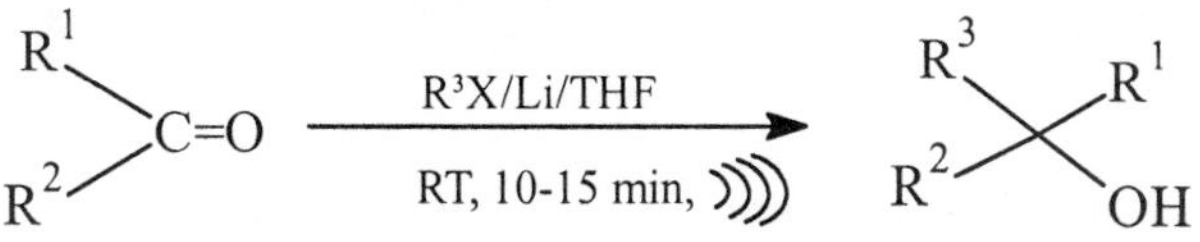

(70-100%)

Scheme 32

In the above reaction, known as Barbier reaction, even non-reactive aldehydes can be used. Even reactive allyl or benzylhalides which normally undergo Wurtz coupling can also be used (Scheme 33).

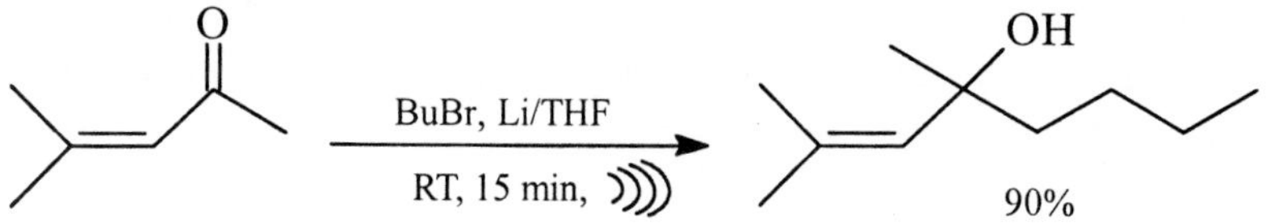

Scheme 33

10.2.19 Synthesis of Chromenes

The condensation of o-hydroxybenzaldehyde with β-nitrostyrene using basic alumina catalyst gives good yield of 3-nitro-2H-chromene on sonication[48] (Scheme 34).

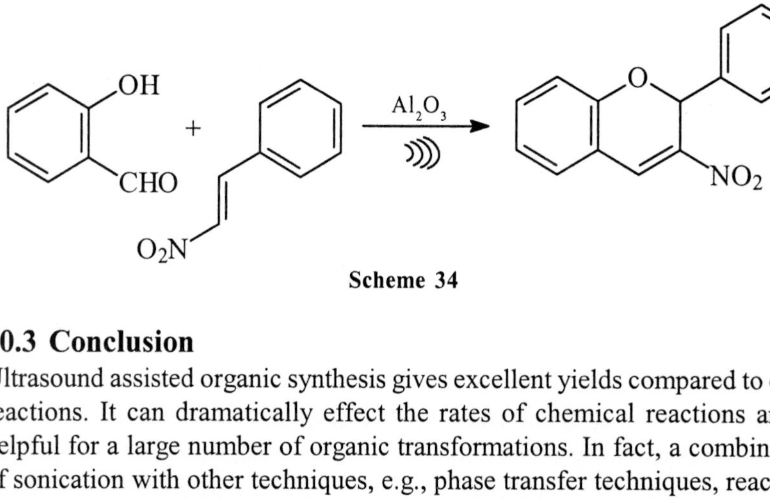

Scheme 34

10.3 Conclusion

Ultrasound assisted organic synthesis gives excellent yields compared to other reactions. It can dramatically effect the rates of chemical reactions and is helpful for a large number of organic transformations. In fact, a combination of sonication with other techniques, e.g., phase transfer techniques, reactions in aqueous media etc. give best results. Sonication has also been shown to stimulate microbiological reactions.

References

1. J.P. Lorimer and T.J. Mason, *Chem. Soc. Rev.*, 1987, **16**, 239-274.
2. J.L. Lindly and T.J. Mason, *Chem. Soc. Rev.*, 1987, **16**, 275-311.
3. C. Einhorn, J. Einhorn and J.L. Luche, *Synthesis*, 1987, 787-813.
4. J.M. Khurana, Chemistry Education, 1990, 24-29.
5. V.K. Ahluwalia and Renu Aggarwal, Organic Synthesis: Special Techniques, Narosa Publishing House, New Delhi, 2001, 116-149.
6. J.M. Khurana, P.K. Sahoo and G.C. Markop, *Synth. Commun.*, 1990, 2267.
7. S. Moon, L. Duclin and J.V. Craney, *Tetrahedron Lett.*, 1979, 3917.
8. D.S. Krislol, H. Klotz and R.C. Parker, *Tetrahedron Lett.*, 1981, **22**, 407.

9. J. Elguero, P. Goya, J. Lissavestzky and A.M. Valdeomillos, *C.R. Acad. Sci. Paris*, 1984, **298**, 877.
10. T. Ando, S. Smith, T. Kaweta, J. Jehihara and T. Haatusa, *J. Chem. Soc. Chem. Commun.*, 1984, 439.
11. T. Ando, J. Kawate and T. Hanatusa, *Synthesis*, 1983, 637.
12. G.H. Posner, *Org. React. (NY)*, 1972, **19**, 1.
13. J.L. Lunche, C. Petrier, A.L. Gemal and N. Zirk, *J. Org. Chem.*, 1982, **47**, 3805.
14. J.C.S. Barboza, C. Petrier and J.L. Luche, *Tetrahedron Lett.*, 1985, **26**, 829.
15. P.A. Griew and P. Garner, *Tetrahedron Lett.*, 1983, **24**, 1897.
16. D.R. Borthakur and J.S. Sandhu, *J. Chem. Soc. Chem. Commun.*, 1988, 1444.
17. R.S. Davidson, A.M. Patil, A. Safdar and D. Thornthwalite, *Tetrahedron Lett.*, 1983, **24**, 5907.
18. J. Ezquema and J. Alvarez-Bullis, *J. Chem. Soc. Chem. Commun.*, 1984, 54.
19. R.D. Walkup and R.T. Cunningham, *Tetrahedron Lett.*, 1987, **28**, 4019.
20. J.M. Khurana and P.K. Sahoo, *Syn. Commun.*, 1992, 1691.
21. J. Yamakawi, S. Sumi, T. Ando and J. Hanatusa, *Chemistry Lett.*, 1983, 379.
22. T. Kimura, M. Fujila and T. Ando, *Chemistry Lett.*, 1988, 137.
23. J.M. Khurana, P.K. Sahoo, S.S. Titus and O.L. Mailap, *Synth. Commun.*, 1900, 1357.
24. A.W. Mattsev, *Russ. J. Phys. Chem.*, 1976, **50**, 993.
25. J. Jurezak and R. Ostaszewki, *Tetrahedron Lett.*, 1988, **29**, 959.
26. D.H. Shin and B.H. Han, *Bull. Korean Chem. Soc.*, 1985, **6**, 247.
27. K.J. Moulton, S. Koritala and E.N. Frankel, *J. Am. Oil Chem. Soc.*, 1983, **60**, 1257.
28. K.S. Suslick and D.J. Casadonte, *J. Am. Chem. Soc.*, 1987, **109**, 3459.
29. C. Petrier and J.L. Luchi, *Tetrahedron Lett.*, 1989, **28**, 2347.
30. C. Petrier and J.L. Luchi, *Tetraderon Lett.*, 1987, **28**, 2351.
31. K.S. Suslick and D.J. Casadonli, *J. Am. Chem. Soc.*, 1987, **109**, 3459.
32. P. Boudjouk and J. Su, *Synthetic Commun.*, 1986, **16**, 775.
33. W.P. Reeves, J.A. Murry, D.W. Willoughby and W.J. Friedrid, *Synthetic Commun.*, 1988, **18**, 1961.
34. H.C. Brown and U.S. Racherla, *Tetrahedron Lett.*, 1985, **26**, 2187.
35. B.H. Han and P. Boudjouk, *Tetrahedron Lett.*, 1981, **22**, 2757.
36. T. Kitazumi and N. Ishikawa, *J. Am. Chem. Soc.*, 1985, **107**, 5186.
37. P. Boudjouk, D.P. Thompson, W.H. Ohrborm and B.H. Hans, *Organometallics*, 1986, **5**, 1257.
38. J. Lindly, T.J. Mason and J.P. Lorimer, *Ultrasonics*, 1987, **25**, 45; L. Lindley, J.P. Lorimer and T.J. Mason, *Ultrasonic*, 1986, **24**, 292.
39. D.M. Trose and B.P. Coppola, *J. Am. Chem. Soc.*, 1982, **104**, 6879.
40. J. Lee and J.K. Sayder, *J. Am. Chem. Soc.*, 1989, **111**, 1522.
41. H. Tso, T. Chou and H. Hung, *J. Chem. Soc. Chem. Commun.*, 1887, 1552.
42. C. Petrier, A.L. Gemal and J.L. Luck, *Tetrahedron Lett.*, 1982, **23**, 3361.
43. A. Fuentes and V.S. Sinisterra, *Tetrahedron Lett.*, 1986, **27**, 2967.
44. J. Menedez, G.G. Trigo and M.M. Solhuber, *Tetrahedron Lett.*, 1986, **27**, 3285.
45. B. Han and P.J. Boudjouk, *J. Org. Chem.*, 1982, **47**, 5030.
46. T. Kitazume, *Synthesis*, 1986, 853.
47. J.L. I uche and J.C. Damianu, *J. Am. Chem. Soc.*, 102, 7926.

11. Biocatalysts in Organic Synthesis

11.1 Introduction

The most important conversions in the context of green chemistry is with the help of enzymes. Enzymes are also referred to as biocatalysts and the transformations are referred to as biocatalytic conversions. Enzymes are now easily available and are an important tool in organic synthesis. The earliest biocatalytic conversion known to mankind is the manufacture of ethyl alcohol from molasses, the mother liquor left after the crystallisation of cane sugar from concentrated cane juice. This transformation is brought about by the enzyme 'invertase' which converts sucrose into glucose and fructose and finally by the enzyme zymase which converts glucose and fructose into ethyl alcohol. It is well known that most of the antibiotics have been prepared using enzymes (enzymatic fermentation).

The biocatalytic conversions have many advantages in relevance to green chemistry. Some of these are given below:

- Most of the reactions are performed in aqueous medium at ambient temperature and pressure.
- The biocatalytic conversions normally involve only one step.
- Protection and deprotection of functional groups is not necessary.
- The reactions are fast reactions.
- The conversions are stereospecific.

One of the most common examples is the biocatalytic conversion of Penicillin into 6-APA by the enzyme 'Penacylase' (one step process). However, the chemical conversion requires a number of steps (Scheme 1).

A special advantage of the biochemical reactions is that they are chemoselective, regioselective and stereoselective. Also, some of the biochemical conversions are generally not possible by conventional chemical means. Two such examples in heterocyclic compounds are given in (Scheme 2).[1]

A number of diverse reactions are possible by biocatalytic processes, which are catalysed by enzymes. The major six classes of enzymes and the type of reactions they catalyse are discussed as follows:

1. **Oxidoreductases:** These enzymes catalyse oxidation-reduction reactions. This class includes oxidases (direct oxidation with molecular oxygen) and dehydrogenases (which catalyse the removal of hydrogen from one substrate and pass it on to a second substrate).

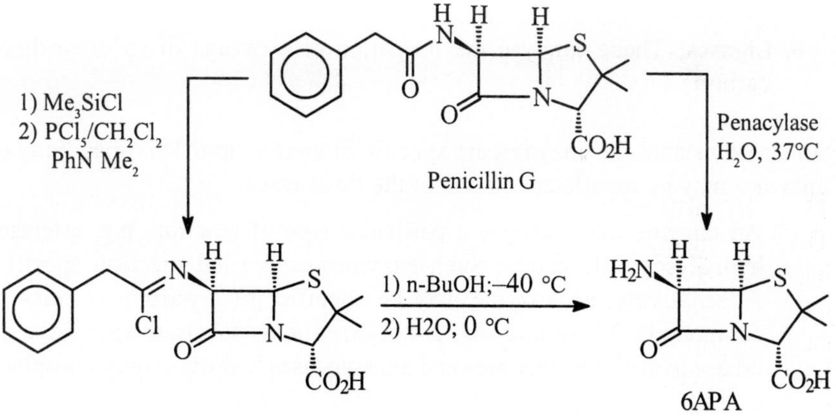

1) Me₃SiCl
2) PCl₅/CH₂Cl₂
PhN Me₂

Penicillin G

Penacylase
H₂O, 37°C

1) n-BuOH;–40 °C
2) H2O; 0 °C

6AP A

Scheme 1

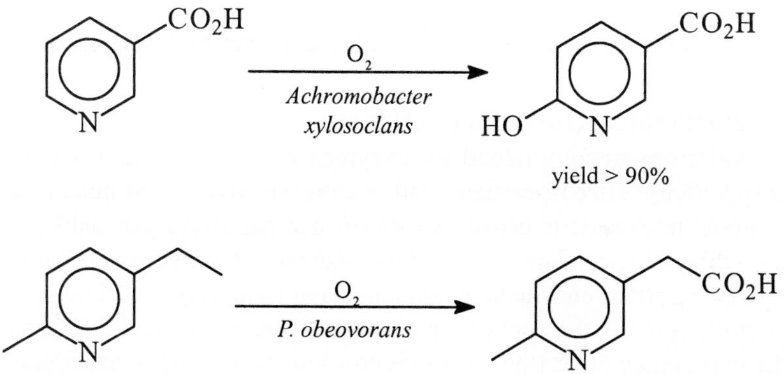

O₂

*Achromobacter
xylosoclans*

yield > 90%

O₂

P. obeovorans

Scheme 2

2. Transferases: These enzymes catalyse the transfer of various functional groups, e.g. transaminase.

3. Hydrolases: This group of enzymes catalyse hydrolytic reactions, e.g. penteases (proteins), esterases (esters) etc.

4. Lyases: These are of two types, one which catalyses addition to double bond and the other which catalyses removal of groups and leaves double bond. Both addition and eliminations of small molecules are on sp³-hybridized carbon.

5. Isomerases: These catalyse various types of isomerisation, e.g. racemases, epimerases etc.

6. Ligases: These catalyse the formation or cleavage of sp^3-hybridized carbon.

As already stated the enzymes are specific in their action. This specificity of enzymes may be manifested in one of the three ways:

(i) An enzyme may catalyse a particular type of reaction, e.g. esterases hydrolyses only esters. Such enzymes are called reaction specific. Alternatively, an enzyme may be specific for a particular class of compounds. These enzymes are referred to as substrate specific, e.g., urease hydrolyses only urea and phosphatases hydrolyse only phosphate esters.

(ii) An enzyme may exhibit kinetic specificity. For example, esterases hydrolyse all esters but at different rates.

(iii) An enzyme may be stereospecific. For example, maltase hydrolyses α-glycosides but not β-glycosides. On the other hand emulsin hydrolyses the β-glycosides but not the α-glycosides.

It should be noted, that a given enzyme could exhibit more than one specificities.

11.2 Biochemical (Microbial) Oxidations

The oxidations accomplished by enzymes or microorganisms excel in regiospecificity, stereospecificity and enantioselectivity. The optical purity (enantiomeric excess) is usually very high nearing 100%. An unbelievably large number of enzymatic (or microbial) oxidations have been accomplished.

Two important enzymatic oxidations have been very well known since early times. One is the conversion of alcohol into acetic acid by bacterium acetic in presence of air (the process is now known as quick-vinegar process) (Scheme-3) and the second one is the conversion of sucrose into ethyl alcohol by yeast (Scheme-4) (this process is used for the manufacture of ethyl alcohol).

$$CH_3CH_2OH + O_2 \xrightarrow{\text{Bacterium acetic}} CH_3COOH + H_2O$$

Ethyl alcohol   Acetic acid

Scheme 3

In a similar way, lactose can be converted into lactic acid (Scheme 5). The above enzymatic oxidations are referred to as fermentation.

Microbial oxidations occur under very mild conditions, usually around 70°C and in dilute solution. They are slow and often take days.

$$C_{12}H_{22}O_{11} + H_2O \xrightarrow[\text{yeast}]{\text{Invertase}} 2C_6H_{12}O_6$$

Sucrose   Glucose & fructose

$$C_6H_{12}O_6 \xrightarrow[\text{yeast}]{\text{Invertase}} 2C_2H_5OH + 2CO_2$$

                                        Ethyl alcohol

Scheme 4

$$C_{12}H_{22}O_{11} + H_2O \xrightarrow[\text{Lactic}]{\text{Bacillus acidic}} 4CH_3CH(OH)CO_2H$$

Lactose   Lactic acid

Scheme 5

Considerable amount of work has been reported in the hydroxylation of aromatic rings. Thus, benzene on oxidation with Pseudomonas putida in presence of oxygen gives the cis-diol (Scheme 6).[2] The cis-diol obtained could be converted by four steps into 1,2,3,4-tetrahydroxy compound, conduritol-F[3] and by five steps into the hexahydroxy compound, pinitol, an antidiabetic agent (Scheme 6).[4]

However, Micrococcus spheroids like organism converts benzene into trans, trans-muconic acid (Scheme 7).[5]

In a similar way, toluene, halogensubstituted benzenes, halogensubstituted toluene gave the corresponding cis diols (Scheme 8).[6]

The cis-diol obtained from chlorobenzene is converted into 2,3-isopropyliden-L-ribose-γ-lactone in four steps (Scheme 9).[7]

Enzymatic conversion of ketones to esters is commonly encountered in microbial degradation.[8] A typical transformation in the enzymatic Baeyer-Villiger oxidation, which converts cyclohexanone into the lactone (Scheme 10) using a purified cyclohexanone oxygenase enzyme.[9, 10] Some more examples of Baeyer-Villiger oxidation are given (see pages 95 and 96).

This enzyme also converts phenylacetaldehyde into phenylacetic acid in 65% yield.

Similarly, 4-methylcyclohexanone can be converted into the corresponding lactone (Scheme 11) in 80% yield[11] with > 98% ee with cyclohexanone oxygenase, obtained from *Acineto bacter*.

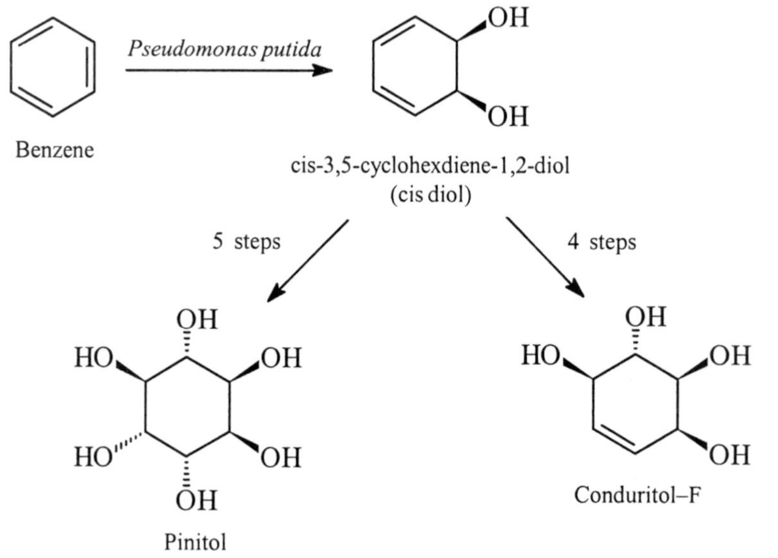
Pseudomonas putida
OH
OH
Benzene
cis-3,5-cyclohexdiene-1,2-diol
(cis diol)
5 steps
4 steps
OH
HO
OH
HO
OH
OH
Pinitol
OH
HO
OH
OH
Conduritol–F

Scheme 6

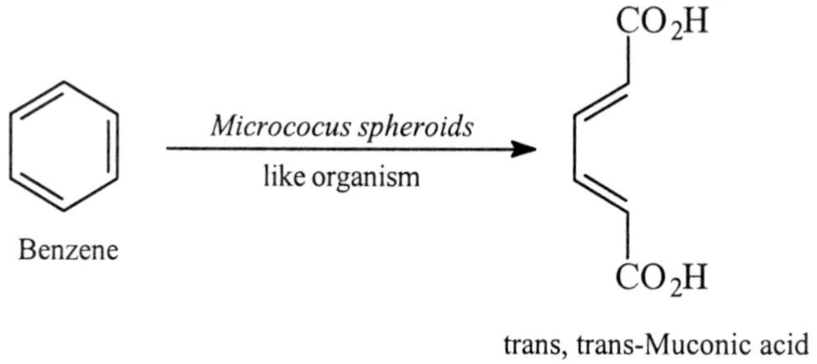
Micrococus spheroids
like organism
Benzene
CO_2H
CO_2H
trans, trans-Muconic acid

Scheme 7

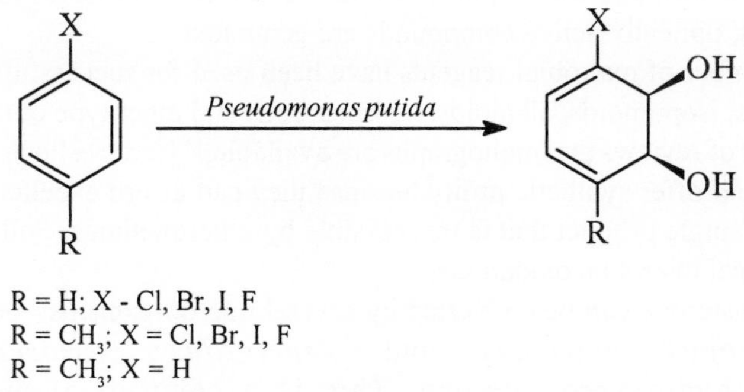
X
Pseudomonas putida
OH
OH
R
R = H; X - Cl, Br, I, F
R = CH3; X = Cl, Br, I, F
R = CH3; X = H

Scheme 8

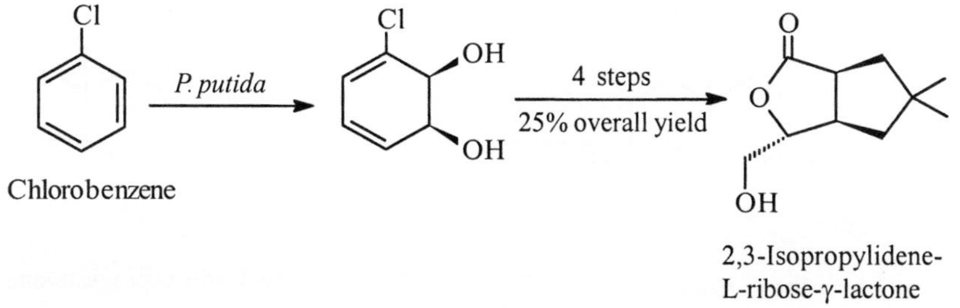
Cl
P. putida
OH
OH
4 steps
25% overall yield
O
OH
Chlorobenzene
2,3-Isopropylidene-
L-ribose-γ-lactone

Scheme 9

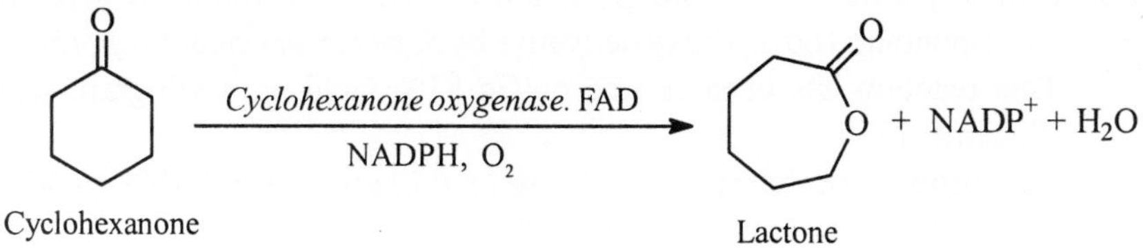
O
Cyclohexanone oxygenase. FAD
NADPH, O2
+ NADP+ + H2O
Cyclohexanone
Lactone

Scheme 10

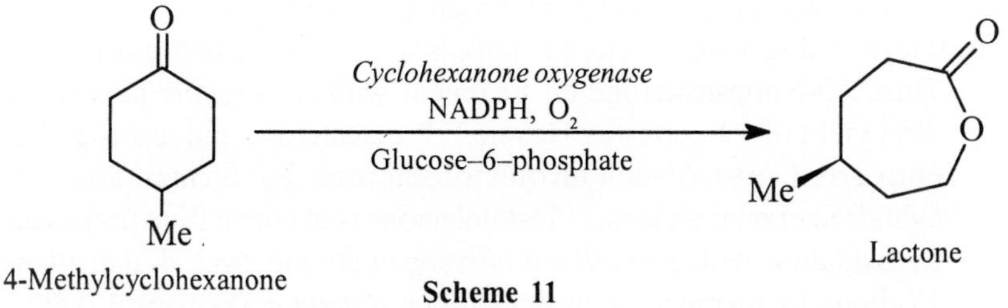
O
Cyclohexanone oxygenase
NADPH, O2
Glucose–6–phosphate
Me
4-Methylcyclohexanone
Lactone

Scheme 11

In case of steroids, many different positions can be hydroxylated by different microorganisms, and usually, only one diastereomer is formed. From achiral molecules, optically active compounds are generated.

A number of microbial reagents have been used for successful oxidation of steroids, isoprenoids, alkaloids, hydrocarbons and other type of molecules. A number of reviews and monographs are available.[12] Here we have given few cases which offer synthetic utility because they can afford excellent yield or they give single product that is inaccessible by other methods. Following are some typical microbial oxidations:

1. Progesterone can be converted by several microorganisms[13] particularly *Rhizopus nigrioans* and *Aspergillus ochraceus* into 11α-hydroxyprogesterone. This is a commercial method of manufacturing of 11α-hydroxyprogesterone as raw material for medicinally important steroids (Scheme 12).

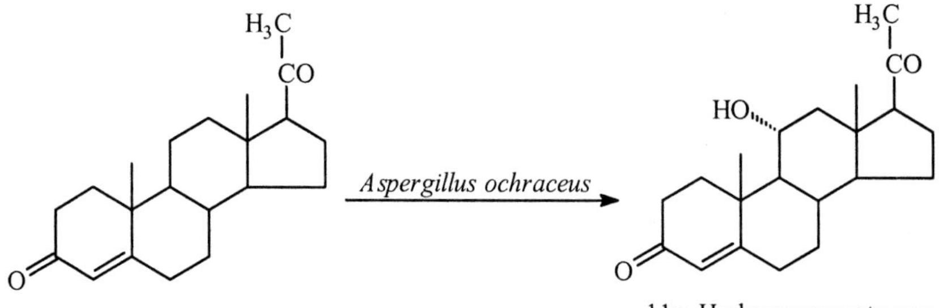

Progesterone   11α-Hydroxyprogesterone

Scheme 12

2. Hydroxylation of 9β-10α-pregna-4,6-diene-3,20-dione to give the corresponding 16α-hydroxy derivative by *Sepedonium ampullosporium*.[14] This reaction has been carried out in 81% yield on a kilogram scale (Scheme 13).
3. Oxidation of oesterone by *Gibberella fujikuroi* gives[15] 75% yield of 15α-hydroxyoesterone (Scheme 14).
4. Hydroxylation of the cholesterol by *Mycobacterium* sp. gives[16] cholest-4-en-3-one (Scheme 15).
5. Allylic oxidation of 17-methyltestosterone by *Gibberella saubinetti* gives[17] the corresponding 6β-hydroxy product (Scheme 16).
6. Baeyer-Villiger oxidations of steroids are accomplished biochemically. Thus, 19-Nortestosterone on treatment with *Aspergillus tamarii* gives 70% yield of 19-nortestololactone.[18] Progresterone and testosterene are converted into Δ'-dehydrotestololactone by fermentation with cylindrocarpon radicicola.[19] Testololactone is obtained from progesterone by oxidation with *Penicillium chrysogenum* and from 4-androstene-3, 17-dione by treatment with *Penicillium lilacinum* (Scheme 17).[20]

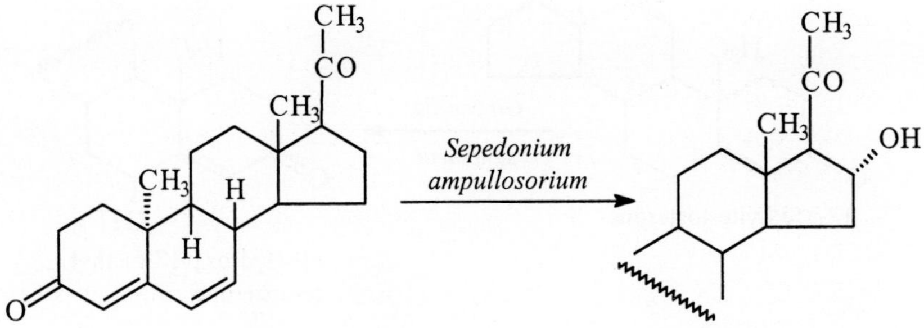

9β, 10α-Pregna-4, 6-diene-3, 20-dione   16α-Hydroxy product
81%

Scheme 13

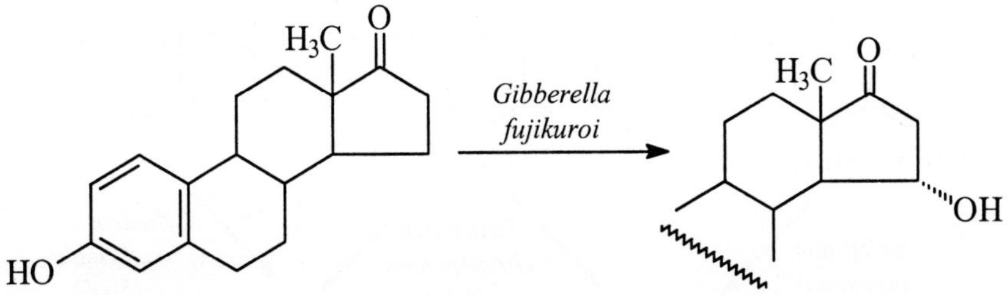

Oestrone   15 α-Hydroxy oestrone

Scheme 14

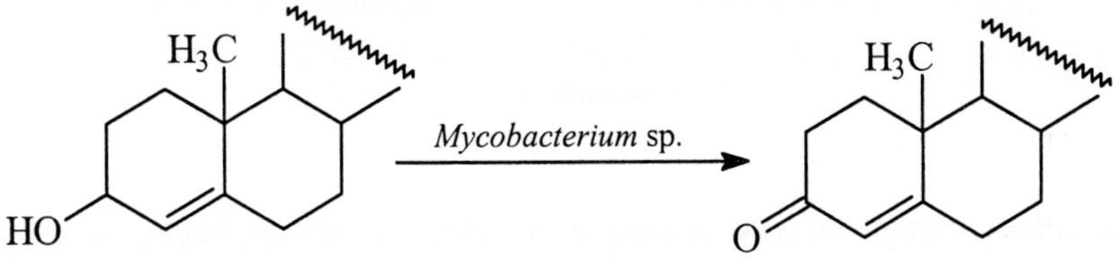

Cholesterol   Cholest-4-en-3-one

Scheme 15

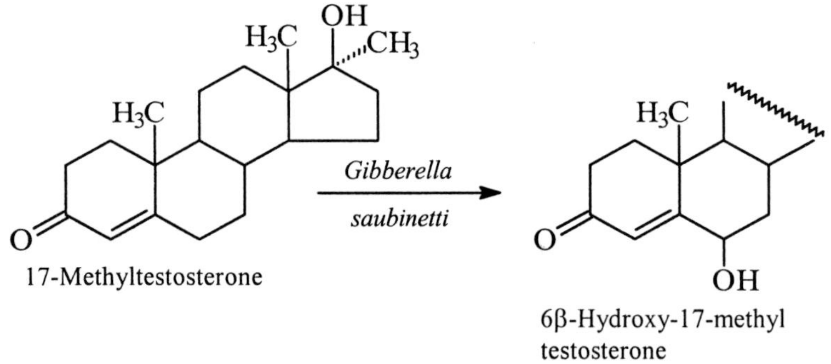

17-Methyltestosterone

Gibberella saubinetti

6β-Hydroxy-17-methyl testosterone

Scheme 16

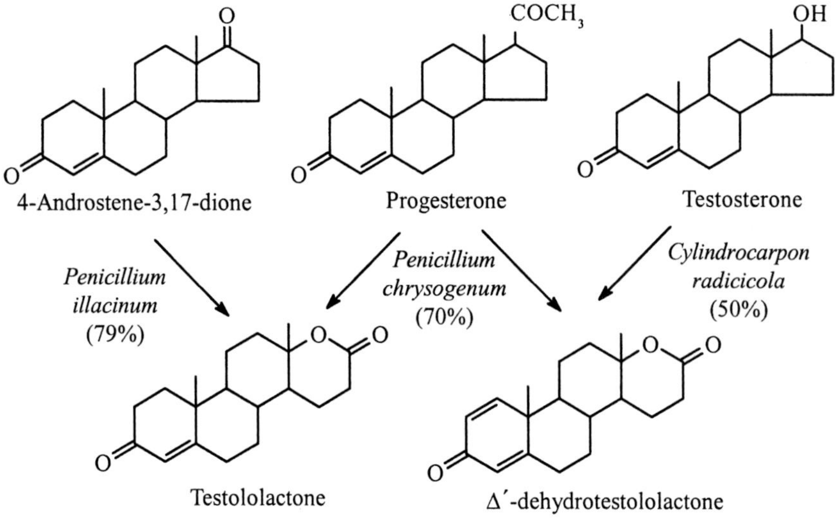

4-Androstene-3,17-dione

Progesterone

Testosterone

Penicillium illacinum (79%)

Penicillium chrysogenum (70%)

Cylindrocarpon radicicola (50%)

Testololactone

Δ′-dehydrotestololactone

Scheme 17

Sometimes, a single enzyme is capable of many oxidations. Some examples are:

(i) Cyclohexanone oxygenase from Acinotobactor strain NCIB 9871 in presence of NADH (reduced nicotinamide adenine dinucleotide) converts aldehydes into esters (Baeyer-Villiger reaction) ; phenylboronic acids into phenols; sulphides into optically active sulfoxides; and selenides into selenoxides (Scheme 18).[21]

(ii) Horse liver dehydrogenase oxidises primary alcohols to acids (esters)[22] and secondary alcohols to ketones[23] (Scheme 19).

(iii) Horseradish peroxidase catalyses dehydrogenative coupling[24] and oxidation of phenol to quinones[25] (Scheme 20).

(iv) Mushroom polyphenol oxidase hydroxylates phenols and oxidizes them to quinones[26] (Scheme 21).

Besides the above, there are a number of other examples involving the use of enzymes in oxidations.[27]

(v) Some important transformations (oxidations) of diols with horse liver alcohol - dehydrogenase (LHADH) using NAD are given (Table 1):

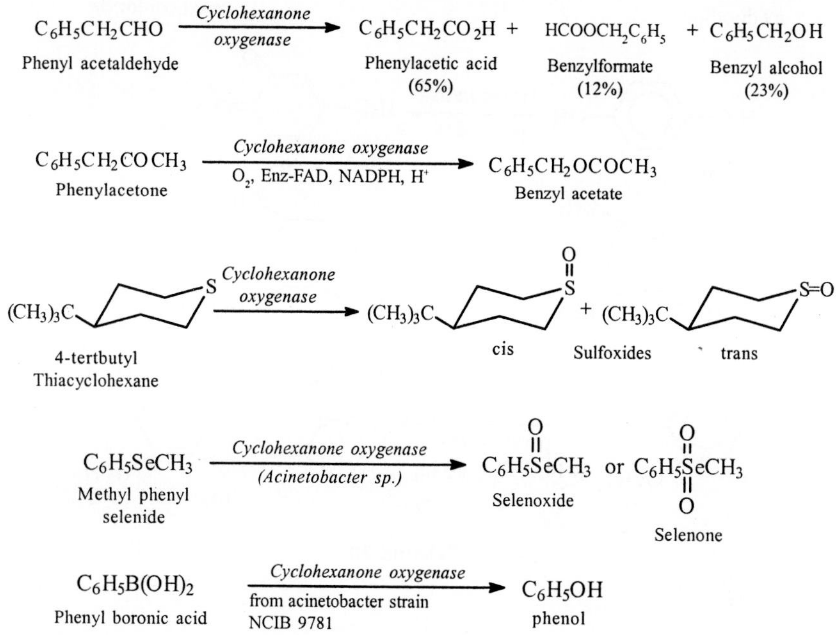

$$C_6H_5CH_2CHO \xrightarrow[oxygenase]{Cyclohexanone} C_6H_5CH_2CO_2H + HCOOCH_2C_6H_5 + C_6H_5CH_2OH$$

Phenyl acetaldehyde   Phenylacetic acid   Benzylformate   Benzyl alcohol
                         (65%)   (12%)   (23%)

$$C_6H_5CH_2COCH_3 \xrightarrow[O_2, Enz\text{-}FAD, NADPH, H^+]{Cyclohexanone\ oxygenase} C_6H_5CH_2OCOCH_3$$

Phenylacetone   Benzyl acetate

4-tertbutyl Thiacyclohexane   cis   Sulfoxides   trans

$$C_6H_5SeCH_3 \xrightarrow[(Acinetobacter\ sp.)]{Cyclohexanone\ oxygenase} C_6H_5SeCH_3\ or\ C_6H_5SeCH_3$$

Methyl phenyl selenide   Selenoxide   Selenone

$$C_6H_5B(OH)_2 \xrightarrow[\substack{from\ acinetobacter\ strain \\ NCIB\ 9781}]{Cyclohexanone\ oxygenase} C_6H_5OH$$

Phenyl boronic acid   phenol

Scheme 18

Part. oxidation product   72-77%

(±)-trans-3-methycyclohexanol $\xrightarrow[\text{alcohol dehydrogenate}]{\text{Horse liver}}$ (−)-(S)-3-methylcyclohexanone
                                                 50% yield ee 100%

(±)-cis-3-methycyclopentanol $\xrightarrow[\text{alcohol dehydrogenate}]{\text{Horse liver}}$ (+)-(S)-2-methylcyclopentanone
                                                  55% yield ee 96%

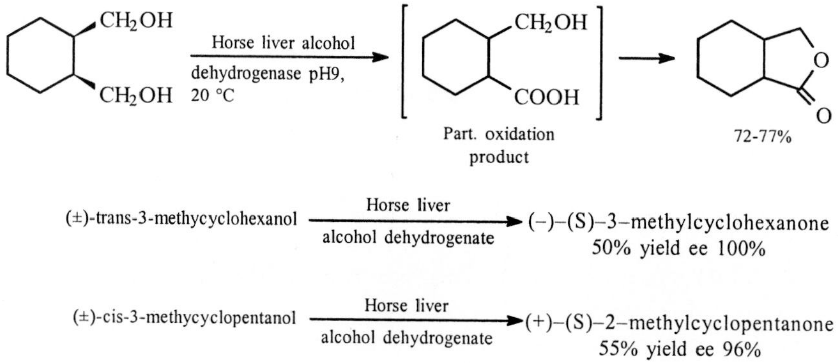

Scheme 19

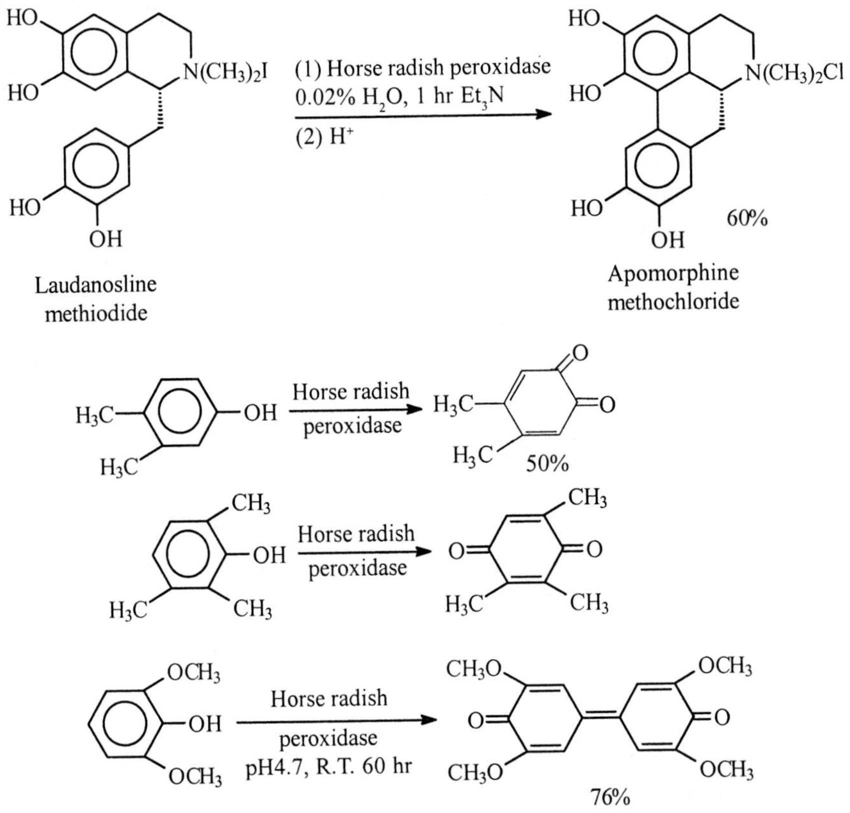
HO
HO
N(CH3)2I
HO
OH
Laudanosline methiodide
(1) Horse radish peroxidase
0.02% H2O, 1 hr Et3N
(2) H+
HO
HO
N(CH3)2Cl
HO
OH
60%
Apomorphine methochloride
H3C
H3C
OH
Horse radish peroxidase
O
O
H3C
H3C
50%
CH3
OH
H3C
CH3
Horse radish peroxidase
CH3
O
O
H3C
CH3
OCH3
OH
OCH3
Horse radish peroxidase
pH4.7, R.T. 60 hr
CH3O
O
CH3O
OCH3
O
OCH3
76%

Scheme 20

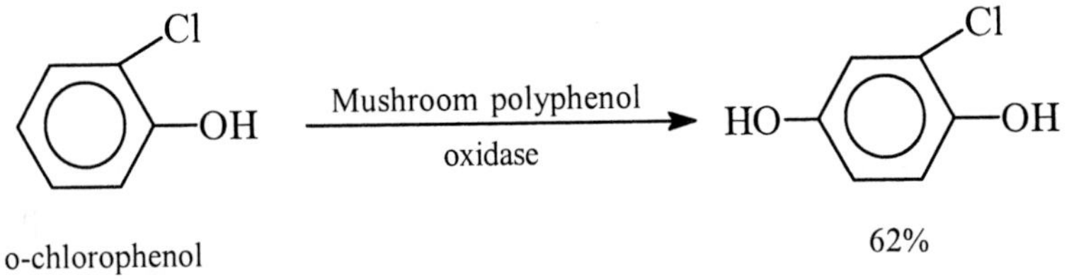
Cl
OH
o-chlorophenol
Mushroom polyphenol oxidase
Cl
HO
OH
62%

Scheme 21

Table 1

Substrate	Product	ee (%)	Ref.
		100	28
		100	28
		100	28
		100	29
		96	30

11.3 Biochemical (Microbial) Reductions

Like enzymatic oxidations, the enzymatic reductions are straight forward and highly stereoselective. Prelog was the first to study the reduction of carbonyl compounds with a number of enzymatic systems. For example, reduction of ketones with curvularia fulcata gave predictable stereochemical induction based on the groups present (large and small) in the keto group. This is known as Prelog's rule.[31] According to this rule, if the steric difference between large (L) and small (S) groups attached to the carbonyl group is large enough, the enzyme delivers hydrogen from the less hindered face to give the corresponding alcohol.

Two most common enzymatic systems are yeast alcohol dehydrogenase (YAD) and horse liver alcohol dehydrogenase (HLADH). The selectivity observed with these enzymes is determined by non-bonded interaction of substrate and enzyme in the hydrogen transfer transition state.[31]

Baker's yeast (*Saccharomyces cerevisiae*) is a very common 'reagent' and it selectively reduces β-ketoesters and β-diketones. Thus, reduction of ethylacetoacetate with Baker's yeast gave the (S)-alcohol. On the other hand, reduction of ethyl β-ketovalerate gave the (R)-alcohol (Scheme 22).

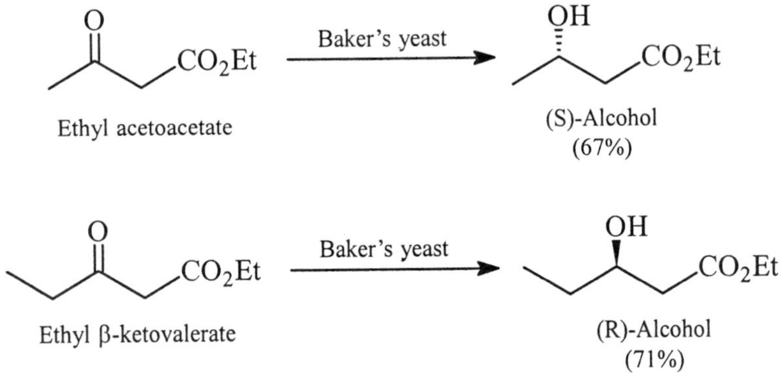

Ethyl acetoacetate

Baker's yeast

(S)-Alcohol
(67%)

Ethyl β-ketovalerate

Baker's yeast

(R)-Alcohol
(71%)

Scheme 22

It was shown that the selectivity of reduction changed from (S) selectivity with small chain esters to (R) selectivity with long chain esters.[32]

The (S)-alcohol obtained above (Scheme 22) is used in the Mori's synthesis of (S)-(+)-sulcatol.[33]

The selectivity of reduction is also illustrated by the observation that 2-butanone is reduced by *Thermoanaerobium brockii* which gave the (R)-alcohol (2-butnaol) in 12% yield and 48% ee, R, but the large ketones are reduced to the (S)-alcohol (85% yield and 96% ee, S)[34] (Scheme 23).

The above examples illustrate the enantioselectivity of the reduction and that selectivity depends on the size and nature of the groups around the carbonyl.

The (S)-alcohol obtained above (Scheme 23) is used in the Mori's synthesis (S)-(+)-sulcatol.[33]

There are a number of synthetic applications of the use of Baker's yeast. Thus, reduction of the β-ketoester (Scheme 24) gave 71% yield of the alcohol, which was used in the Hoffmann's synthesis of the cigarette beetle.[34a]

The selectivity of all these reductions is in consistent with the (S)-selectivity as predicted by Prelog's rule. It is found that 1,3-diketones are normally reduced to β-ketoalcohol. Thus, 2,4-hexanedione gave quantitatively (S)-5-hydroxy-3-hexanone (90% ee).[35]

Reduction of ethyl acetoacetate with *Aspergillus niger* gives 98% of a 75:21 mixture favouring (R)-alcohol. This is in contrast to the formation of (S)-alcohol with Baker's yeast or geotrichum candidum.[36]

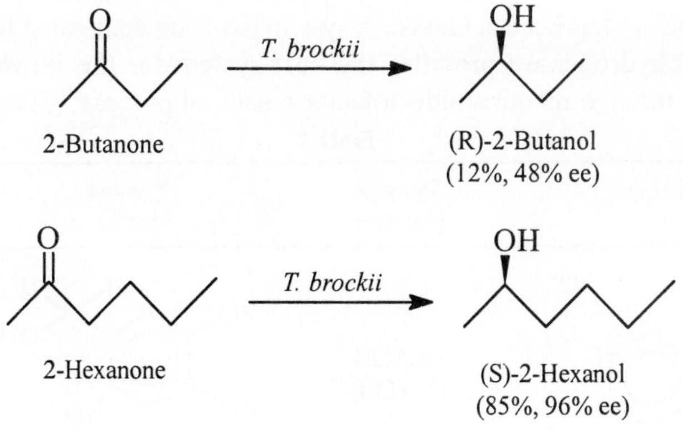
O
T. brockii
OH
2-Butanone
(R)-2-Butanol
(12%, 48% ee)
O
T. brockii
OH
2-Hexanone
(S)-2-Hexanol
(85%, 96% ee)

Scheme 23

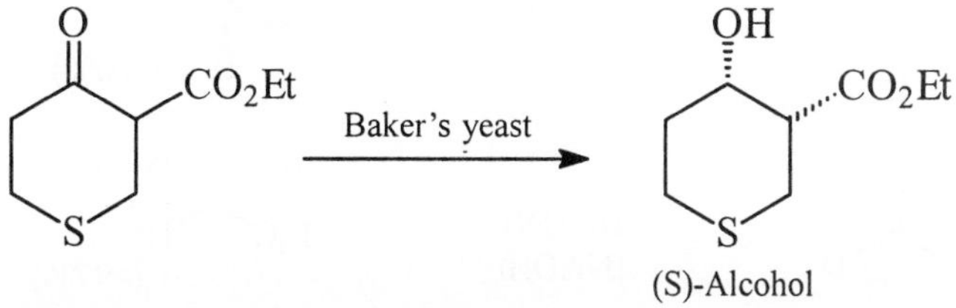
O
CO2Et
Baker's yeast
OH
CO2Et
S
S
(S)-Alcohol

Scheme 24

Baker's yeast reduces simple ketones[37] as shown by the selective reduction of the ketonic moiety on the side chain of the cyclopentadione to the (R)-alcohol (Scheme 25). The formed (R)-alcohol is used in the synthesis is norgestral.[38]

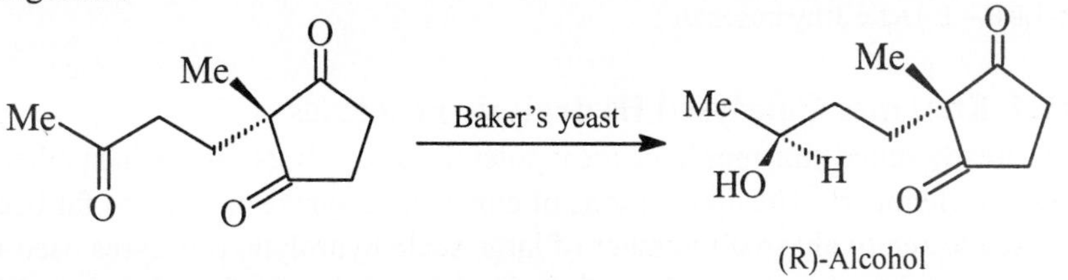
Me
Me
O
O
O
Baker's yeast
Me
Me
HO
H
O
O
(R)-Alcohol

Scheme 25

Geranial on reduction with Baker's yeast gave (R)-citronellol. However, reduction of the Z isomer (neral) gave a 6:4 R:S mixture probably due to isomerisation of the double bond in neral prior to the delivery of hydrogen.[39]

Miscellaneous Reductions

Asymmetric reduction of carbonyl compounds and production of isotopically labelled species has been achieved. A system based on deuterated formate and formate dehydrogenase provides the best system for the introduction of deuterium through nicotinamide-cofactor catalysed process[40] (Table 2).

Table 2

Substrate	Enzyme (cofactor)	Product (ee%)	Ref.
	HLADH (NADH)		41
	HLADH (NADH)		42
	HLADH (NADH)		43
	L–LDH (NaDH)		44

HLADH = Horse liver alcohol dehydrogenase
L–LDH = L-lactic dehydrogenase

11.4 Enzymes Catalysed Hydrolytic Processes

As already stated enzymes have great potential as catalysts for use in synthetic organic chemistry. The applications of enzymes in synthesis have so far been limited to relatively small number of large scale hydrolytic processes used in industry and to a large number of small scale synthesis of products used in research. Following are given some of the applications of enzymes in hydrolytic processes.

11.4.1 Enantioselective Hydrolysis of Meso Diesters

Pig liver esterase has been used for the enantioselective hydrolysis of the following meso substrates (Table 3).

Table 3

Substrate	Product	ee (%)	Ref.
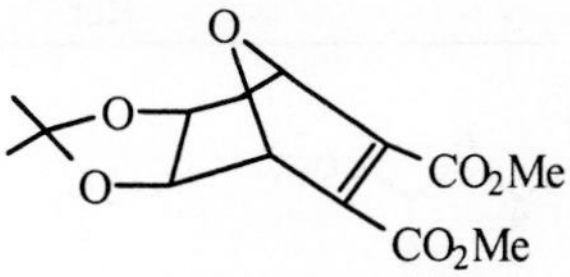	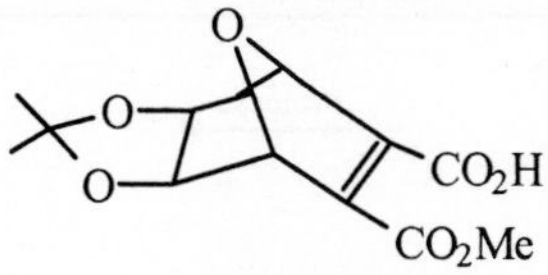	77	45
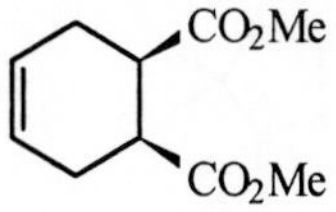	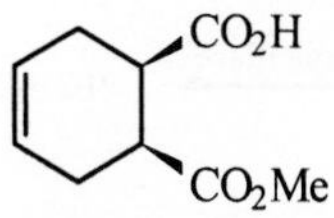	96	46, 47
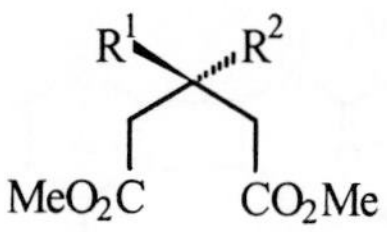	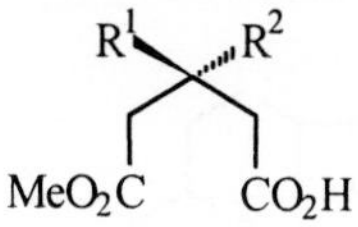	100	48, 49, 50

R^1=OH, CH$_3$, H, PhCH$_2$OCONH
R^2=CH$_3$, H, CH$_2$Ph, NO$_2$, C$_6$H$_5$, CHMe$_2$, cyclohexyl

The enantioselective hydrolysis of the following have been achieved by hog pancreatic lipase (Scheme 26).[51]

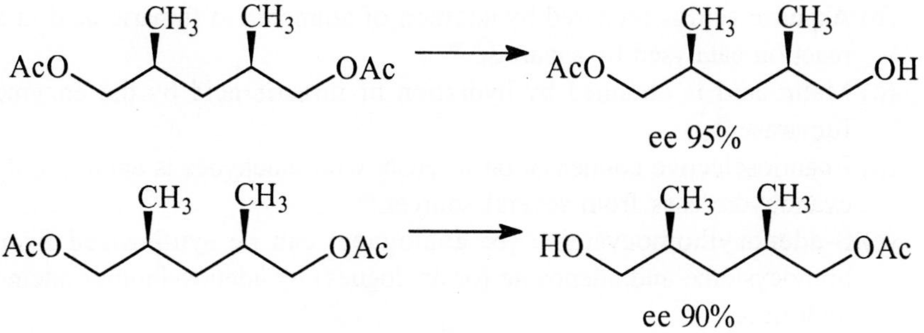

Scheme 26

11.4.2 Hydrolysis of N-acylamino Acids
The hydrolytic enzymes 'amidases' are useful for the hydrolysis of N-acylamino acids for the synthesis of amino acids and in the formation of amide bonds in polypeptides and proteins. In fact, this method is the resolution of amino acids.[52]

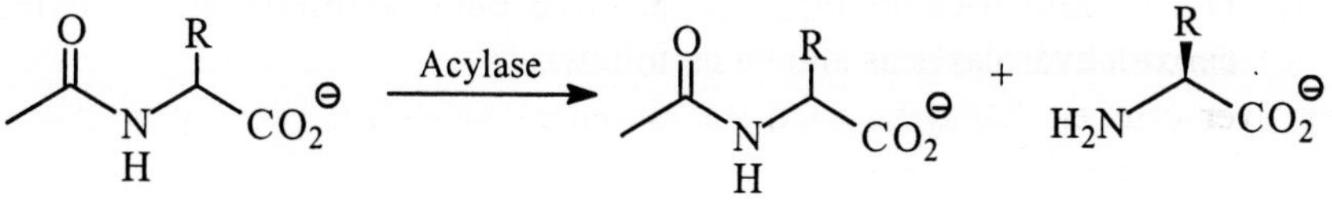

Some other applications in use of amidases are also given (Table 4):

Table 4

Reaction	Ref.
	53
	54, 55

11.4.3 Miscellaneous Applications of Enzymes

(a) A number of commercial applications of Isomerases and Lyases are recorded. For example, glycosidases are used in large quantity in conversion of corn starch to glucose[56] and glucose isomerase catalyses the equilibration of glucose and fructose.[57]

(b) Aspartic acid is prepared by addition of ammonia to fumaric acid in a reaction catalysed by aspartase.[58]

(c) Malic acid is obtained by hydration of fumaric acid by the enzyme fumarase.[59]

(d) Enantioselective condensation of HCN with aldehydes is catalysed by cyanohydrolases from several sources.[60]

(e) S-adenosylhomocysteine (or analogues) can be synthesised from homocysteine and adenosine (or analogues) by adenosylhomocysteine hydrolase.[61]

(f) Ester groups at Sn-1 and Sn-2 positions of glycerol moiety can be hydrolysed by phospholipases A_1 and A_2 respectively.[62]

(g) L-Phenylalanine synthesised by addition of isotopically labelled ammonia to cinnamic acid catalysed by phenylalanine ammonia lyase.[63]

(h) Using hydrolytic deaminations, L-citraline, L-arginine has been prepared on a large scale.[64]

(i) Acrylamine has been synthesised from acrylonitrile by nitrile hydratases.[65]

(j) D-, L- and mesotastaric acids have been synthesised by using epoxidehydrolases as shown as follows: [66]

E_1 = D-tartarate epoxidase
E_2 = L=tartarate epoxidase
E_3 = epoxide hydrolysate from rabit liver microsomes

(k) The synthesis of 5-phospho-D-ribosyl-1-pyrophosphate, a key intermediate in the biosynthesis of nucleotide has been achieved.[67]

References

1. A. Kiener, CHEMTECH, September 1995, pp. 31-35.
2. I.M. Shirley and S.C. Taylor, *J. Chem. Soc. Chem. Commun.*, 1983, 954.
3. S.V. Ley and A.J. Redgrave, *J. Synlett.*, 1990, 393.
4. S.V. Ley, F. Sternferd and S. Taylor, *Tetrahedron Lett.*, 1987, **28**, 225.
5. A. Kleinzeller and Z. Fenel, *Chem. Listy*, 1952, **46**, 300; *Chem. Abstr.*, 1953, **47**, 4290.
6. D.T. Gibson, J.R. Koch, C.L. Schuld and R.E. Kallio, *Biochemistry*, 1968, **7**, 3795; D.T. Gibson, M. Hansley, H. Yoshioka and T.J. Mabry, *Biochemistry*, 1970, **9**, 1626.
7. T. Hudlicky and J.D. Price, *Synlett.*, 1990, 159; T. Hudlicky, H. Lund, J.D. Price and F. Rulin, *Tetrahedron Lett.*, 1989, **30**, 4053.
8. C.J. Sih and J.P. Rosazza, in Applications of Biochemical Systems in Organic Chemistry; J.B. Jones, C.J. Sih and D. Perlman, Eds., Wiley, New York, 1976; Part II, pp. 100-102; G.S. Fanken and R.A. Johnson, Chemical Oxidations with Microorganisms, Marcel Dekker, New York, 1972, pp. 157-164.
9. C.C. Ryerson, D.P. Ballou and C. Walsh, *Biochemistry*, 1982, **21**, 2644; N.A. Donoghu, D.B. Norris and P.W. Trudgill, *Eur. J. Biochem.*, 1976, **63**, 175.
10. B.P. Branchaud and C.T. Walsh, *J. Am. Chem. Soc.*, 1985, **107**, 2153.
11. J.D. Blck and M.J. Taschner, *J. Am. Chem. Soc.*, 1988, **110**, 6892.

12. Ch. Tamm, *Angew. Chem.*, 1962, **74**, 225; *Angew. Chem. Int.*, Ed. 1962, **1**, 78; D. Perlan (ed.), Fermentation Advances, Academic, New York, 1969; K. Kieslich, *Synthesis*, 1969, 120; W. Charney and H.L. Herzog, Microbial Transformations of Steroids, Academic, New York, 1967; A. Capek, O. Hanc and M. Tadra, Microbial Transformations of Steroids, Academia, Prague, 1966; M. Raynaud, Ph. Daste, F. Grossin, J.F. Biellmann and R. Wennig, *Ann. Inst.*, Pasteur, 1960, **115**, 731; H. Tizuka and A. Naqito, Microbial Transformation of Steroids and Alkaloids, University Park Press, State College, Pennsylvania, 1967; J.B. Davis, Petroleum Microbiology, Elsevier, Amsterdam, 1967; C. Ralledge, *Chem. Ind.*, 1970, 843; L. Wallen, F.H. Stodola and R.W. Jacksom, Type Reactions in Fermentation Chemistry, U.S. Department of Agriclture, 1959, pp. 185-189; D.W. Ribbons, *Ann. Rept. Chem. Soc.*, London, 1965, **62**, 445; W.C. Evans, *Ann. Rept.Chem. Soc.*, London, 1956, **53**, 279; O. Hayashi and M. Noyaki, *Science*, 1969, **164**, 338; D.T. Gibson, *Science*, 1968, **161**, 1093; Grunther S. Fonken and Roy A. Johnson, Chemical Oxidations with Microorganism, Mercel Dekker, New York, 1972.
13. D.H. Peterson and H.C. Murray, *J. Am. Chem. Soc.*, 1952, **174**, 1871; H.C. Murray and D.H. Peterson, *U.S. Patent, 2, 602, 769* (July 8, 1952).
14. W.F. Vander Waard, D. Vander Sijde and J. de Flines, *Trans. Chim.*, 1966, **85**, 712.
15. P. Crabbe and C. Cassas Campillo, *U.S. Patent, 3, 375, 175* (March 26, 1968).
16. I.I. Zaretskaya, L.M. Kogan, O.B. Tikhomirova, Jr., D. Sis, N.S. Wulfon, V.I. Zareksu, V.G. Zaikin, G.K. Skrybin and I.V. Torgov, *Tetrahedron*, 1968, **24**, 1595.
17. J. Ureaht, E. Vischer and A. Wettstein, *Held. Chim. Acta*, 1996, **43**, 1077.
18. J.T. McCurdy and R.D. Garrett, *J. Org. Chem.*, 1968, **33**, 660.
19. F.J. Fried, R.W. Thoma and A. Klingsberg, *J. Am. Chem. Soc.*, 1953, **75**, 5764.
20. R.L. Prairie and P. Talalay, *Biochemistry*, 1963, **2**, 203.
21. B.P. Branchaud and C.T. Walsh, *J. Am. Chem. Soc.*, 1985, **107**, 2153.
22. J.B. Jones and I.J. Jokovac, *Org. Synth.*, 1984, **63**, 10.
23. J. Grunwald, B. Wirz, M.P. Scollar and A.M. Klibanov, *J. Am. Chem. Soc.*, 1986, **108**, 6732.
24. A. Brossi, A. Ramel, J. O'Brien and S. Teitel, *Chem. Pharm. Bull.*, 1973, **21**, 1839.
25. B.C. Saunders and B.P. Stark, *Tetrahedron*, 1967, **23**, 1867.
26. R.Z. Kazandjian and A.M. Klibanov, *J. Am. Chem. Soc.*, 1985, **107**, 5448.
27. Milos Hudlicky, Oxidations in Organic Chemistry, ACS Monograph 186, American Chemical Society, Washington DC, 1990.
28. G.S.Y. Ng., L.C. Yuan, I.J. Jakovac and J.B. Jones, *Tetrahedron*, 1984, **40**, 1235.
29. J.B. Jones and I.J. Jakovac, *Can. J. Chem.*, 1982, **60**, 19.
30. J.B. Jones, *Methods Enzymol.*, 1976, **44**, 831.
31. V. Prelog, *Pure Appl. Chem.*, 1964, **9**, 119.
32. B. Zhou, A.S. Gopalan, F. van Middlesworth, W.R. Shieh and C.J. Sih, *J. Am. Chem. Soc.*, 1983, **105**, 5925.
33. K. Mori, *Tetrahedron*, 1981, **37**, 1341.
34. E. Kienam, E.K. Hafeli, K.K. Seth and R. Lamed, *J. Am. Chem. Soc.*, 1986, **108**, 162.
34a. R.W. Hoffman, W. Helbig and W. Landner, *Tetrahedron Letters*, 1982, **23**, 3479.
35. J. Bolte, J.G. Gourey and H. Veschambre, *Tetrahedron Lett.*, 1986, **27**, 4051.
36. R. Bernardi, R. Cardillo and D. Ghiringhelli, *J. Chem. Soc. Chem. Commun.*, 1984, 460.
37. J.K. Lieser, *Synth. Commun.*, 1982, **13**, 765.
38. W.H. Zhou, D.Z. Hung, O.C. Deng, Z.P. Zhuang and Z.O. Wang, *Nat. Prd. Proc. Sino-Am. Symp.*, 1980, 299; *Chem. Abstr.*, 1983, **88**, 198545w.
39. M. Bostmembrum-Desrut, G. Douphin, A. Kergomard, M.F. Renard and H. Veschambre, *Tetrahedron*, 1985, **41**, 3679.
40. C.H. Wong and G.M. Whitesides, *J. Am. Chem. Soc.*, 1983, **105**, 5012.

41. A.R. Battershy, P.W. Sheldrake, J. Staunton and D.C. Williams, *J. Chem. Soc. Perkin Trans.*, 1976, **1**, 1056.
42. D.R. Dodds and J.B. Jones, *J. Chem. Soc. Chem. Commun.*, 1982, 1080.
43. C.H. Wong and G.M. Whitesides, *J. Am. Chem. Soc.*, 1983, **105**, 5012.
44. B.C. Hirschbein and G.M. Whitesides, *J. Am. Chem. Soc.*, 1982, **104**, 4458.
45. Y. Ito, T. Shibata, M. Arita, H. Sawai and M. Ohno, *J. Am. Chem. Soc.*, 1981, **103**, 6739.
46. H.J. Gais and K.L. Lukas, *Angew. Chem.*, 1984, **96**, 140; *Angew. Chem. Int. Ed. Engl.*, 1984, **23**, 142.
47. S. Kobayashi, K. Kamiyama, T. Limori and M. Ohno, *Tetrahedron Lett.*, 1984, **23**, 2557.
48. F.C. Huang, L.F.H. Lee, R.S.D. Mittal, P.R. Ravi Kumar, J.A. Chan and C.J. Sih, *J. Am. Chem. Soc.*, 1975, **97**, 4144; C.H. Chervenka and P.E. Wilson, *J. Biol. Chem.*, 1956, **222**, 635.
49. Y.F. Wang, T. Izawa, S. Kabayaski and M. Ohno, *J. Am. Chem. Soc.*, 1982, **104**, 6465.
50. C.J. Francis, J.B. Jones, *J. Chem. Soc. Chem. Commun.*, 1984, 579.
51. Y.F. Wang, C.S. Chen, G. Girdaukas and C.J. Sih, *J. Am. Chem. Soc.*, 1984, **106**, 3695.
52. I. Chibata, Immobilized Enzymes – Research and Development, Halsted Press, New York, 1978; Y. Izumi, I. Chibata and T. Itoh, *Angew. Chem.*, 1978, **90**, 187; *Angew. Chem. Int. Ed., Engl.*, 1978, **17**, 176.
53. H.D. Jakubki, P. Kuhl and A. Könnecke, *Angew. Chem.*, 1985 (97); *Angew. Chem. Int. Ed. Engl.*, 1985, **24**, 85.
54. B.J. Abbott, *Adv. Appl. Microbiol.*, 1976, **20**, 203.
55. D.L. Regan, M.D. Dunnill and M.D. Lilly, *Biotechnol. Bioeng.*, 1974, **16**, 333.
56. H.M. Walton, J.E. Eastman and A.E. Staly, *Biotechnol. Bioeng.*, 1973, 447; J.H. Wilson and M.D. Lilly, *Biotechnol. Biology*, 1969, **11**, 349; J.J. Marshall and W.J. Whelan, *Chem. Ind.*, London, 1971, **25**, 701; C. Gruesbeck and H.F. Rase, *Ind. End. Chem. Proc. Res. Dev.*, 1972, **11**, 74.
57. H.H. Weetall, *Process Biochem.*, 1975, **10**, 3; H.H. Weetall, W.P. Vann, W.H. Pitcher, Jr., D.D. Lee, Y.Y. Lee et al., *Methods Enzymol*, 1976, **44**, 776; G.W. Strandberg and K.L. Similey, *Appl. Microbiol.*, 1971, **21**, 588; N.B. Havewala and W.H. Pitcher, Jr., *Enzyme Eng.*, 1974, **2**, 315; N.H. Mermelstein, *Food Technol.*, Chicago, 1975, **29**, 20.
58. T. Tosa, T. Sato, T. Mori, Y. Matuo and I. Chibata, *Biotechnol. Biology*, 1973, **15**, 69.
59. K. Yamamoto, T. Tosa, K. Yamashita and I. Chibata, *Eur. J. Appl. Microbiol.*, 1976, **3**, 169.
60. W. Becker and E. Pteil, *J. Am. Chem. Soc.*, 1966, **88**, 4299.
61. B. Chabannes, A. Garib, L. Cronenberger and H. Pacheco, *Prep. Biochem.*, 1983, **12**, 395; R.C. Knudsen and I. Yall, *J. Bacteriol.*, 1972, **112**, 569; S.K. Shapiro and D.J. Ehninger, *Anal. Biochem.*, 1966, **15**, 323.
62. G. Rao, H.O.O. Schmid, K.R. Reddy and J.G. White, *Biochim. Biphys. Acta*, 1982, **715**, 205; H. Eibi, *Angew. Chem. Int. Ed. Eng.*, 1984, **23**, 257 (a review).
63. A.R. Battersby, *Chem. Ber.*, 1984, **20**, 611.
64. Y. Izumi, I. Chibata and T. Itoh, *Ang. Chem. Int. Ed. Engl.*, 1978, **17**, 176.
65. Y. Asano, T. Yasuda, Y. Tani and H. Yamada, *Agric. Biol. Chem.*, 1982, **46**, 1183.
66. M. Ohno, *Ferment. Ind. Tokyo*, 1979, **37**, 836; H. Sato, Jap. Patent 75 140 684, Japan Kokai; *Chem. Abstr.*, 1975, **84**, 149212; R.H. Allen, W.B. Jakoby, *J. Biol. Chem.*, 1969, **244**, 2078.
67. A. Gross, O. Abril, J.M. Lewis, S. Geresh and G.M. Whitesides, *J. Am. Chem. Soc.*, 1983, **205**, 7428.

12. Aqueous Phase Reactions

12.1 Introduction

The use of water as a solvent for carrying out organic reactions was non-existent till about the middle of the 20th century. In view of the environmental concerns caused by pollution of organic solvents, chemists all over the world have been trying to carry out organic reactions in aqueous phase. The advantage of using water as a solvent is its cost, safety (it is non-inflammable, and is devoid of any carcinogenic effects) and simple operation. Water has the highest value for specific heat of all substances. It's unique enthalpic and entropic properties has led the chemists to use it as a solvent in organic reactions. Water has an abnormally low volatility because its molecules are associated with each other by means of hydrogen bonds. In fact, the H bonding is the main reason why covalent compounds have low solubility in water. Ionic material become hydrated and polar materials take part in the hydrogen bonding, so they are soluble.

Under high pressure and temperature, ordinary water behaves very differently.[1] The electrolytic conductance of aqueous solutions increases with increase in pressure. However, for all other solvents the electrical conductivity of solutions decrease with increase in pressure. This unusual behaviour of water is due to its peculiar associative properties.[2]

Water becomes less dense due to thermal expansion with increase in temperature. The density of water is 1.0 g/cm³ at room temperature, which changes to 0.7 g/cm³ at 306 °C. At critical point, the densities of the two phases become identical and they become a single fluid, which is called supercritical fluid. The density of water at this point is ~ 0.3 g/cm³. In the supercritical region, most of the properties of water vary widely. The most important of these is the heat capacity at constant pressure, which approach infinity at the critical point. Also, the dielectric constant of dense, supercritical water ranges from 5 to 20 on variation of applied pressure.

As the temperature of water increases to the critical point, its electrolytic conductance rises sharply independent of the pressure. This is attributed to decrease in its viscosity over this range. However, near the supercritical point, the conductance begin to drop off.

Following are given some of the reactions which have been carried out in aqueous medium.

12.2 Diels-Alder Reaction

The most important method[3] to form cyclic structures is the well-known Diels-Alder reaction. However, the first Diels-Alder reaction in aqueous media was carried out in the beginning of the 19th century.[4] Thus, furan reacted with maleic anhydride in hot water to give the adduct (Scheme 1).

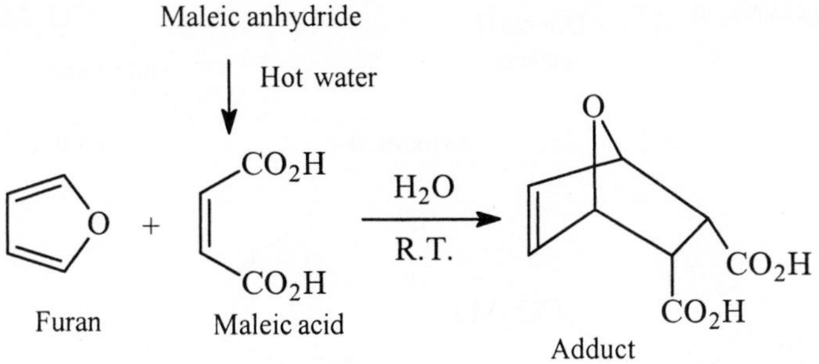

Scheme 1

The product obtained was a diacid (Scheme 1) showing that the reaction occurred via the formation of maleic acid from maleic anhydride.

A typical reaction of cyclopentadiene with N-sec. butylmaleimide gave quantitative yield of the adduct (Scheme 2).[5]

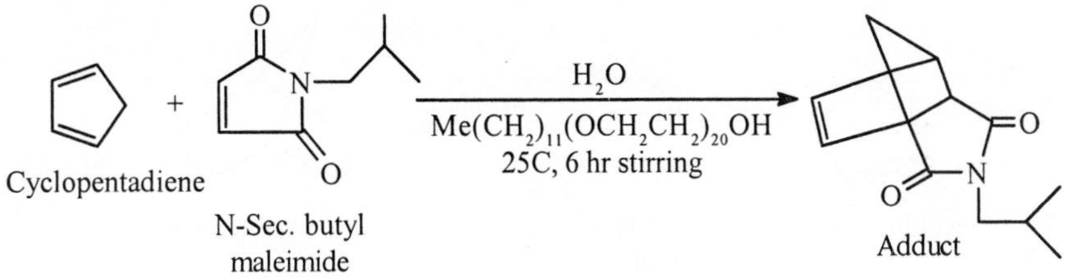

Scheme 2

However, it was only in 1980 that Breslow[6] observed that the Diels-Alder reaction of cyclopentadiene with butenone in water (Scheme 3) was more than 700 times faster than the same reaction in isooctane.

Similarly, cyclopentadiene reacted with dimethyl maleate or methyl methacrytate in aqueous medium (Schemes 4 and 5).

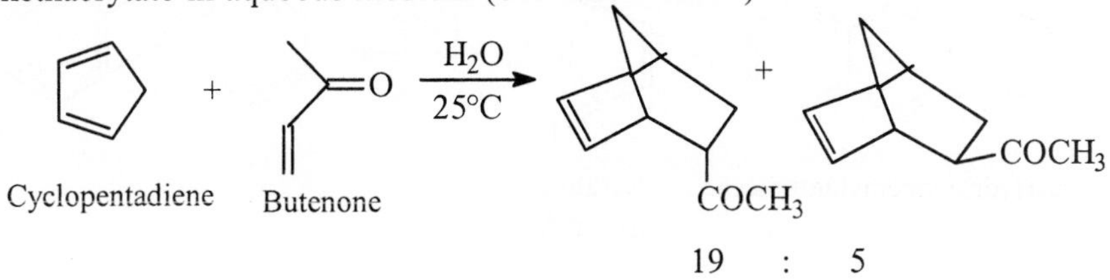

Scheme 3

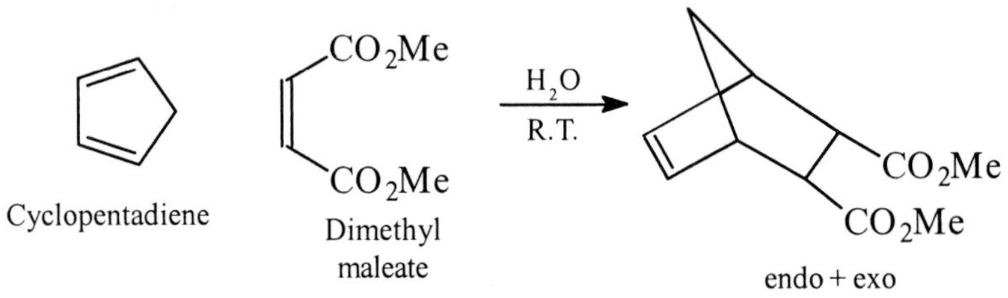

Scheme 4

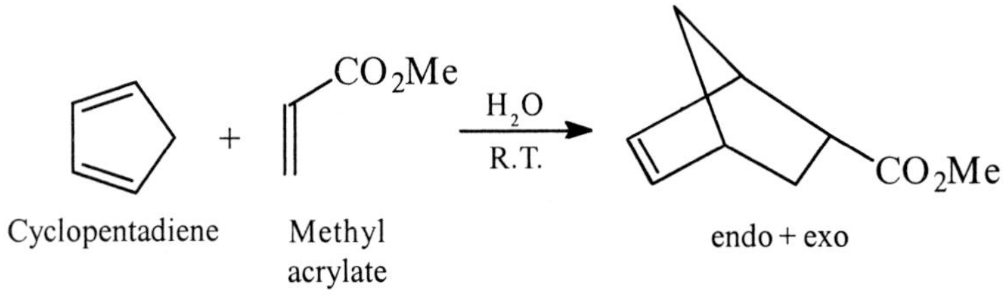

Scheme 5

The following two reactions (Schemes 6 and 7) have also been found to proceed in quantitative yield.[6]

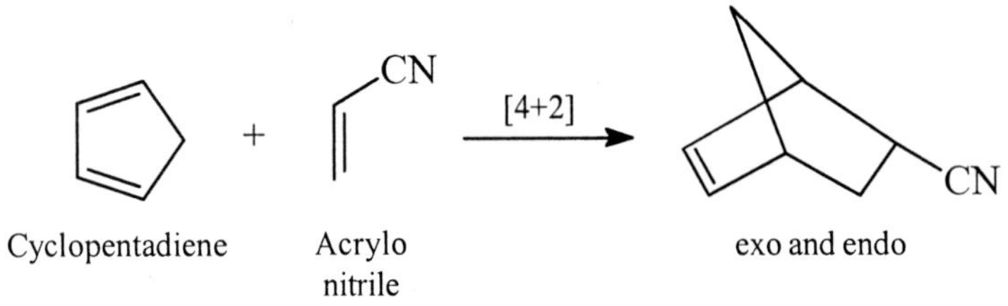

Scheme 6

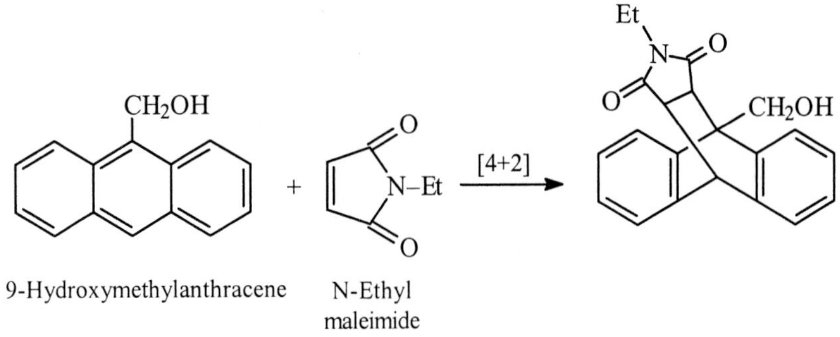

Scheme 7

It has been found that in the Diels-Alder reaction (Scheme 7), the rate increased 2.5 times if the reaction was carried out in a 4.86 M aqueous solution of LiCl. In this case LiCl is a prohydrophobic ('salting out') agent. However, the rate decreased considerably if the reaction was carried out in presence of 2.0 M aqueous solution of guanidinum perchlorate. In this case, guanidinium perchlorate is an antihydrophobic ('salting-in') agent.

It has been demonstrated[7] that the use of 'salting-out' prohydrophobic agents considerably increased the yields of cyclo-adduction of the diene carboxylates with a variety of dienophiles in water at ambient temperature. Similar results were obtained by using corresponding sodium or ammonium carboxylates.

It is well known that the conventional Diels-Alder reactions in aprotic organic solvents are catalysed by Lewis acids. In view of this, the use of Lewis acids in aqueous Diels-Alder reactions has been investigated and the reaction was found to occur much faster (Scheme 8).[8]

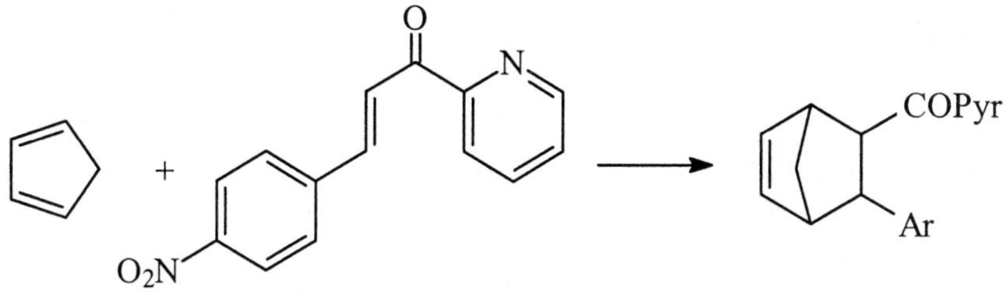

Scheme 8

The above reaction (Scheme 8) in aqueous $Cu(NO_3)_2$ proceeded about 800 times faster than in water alone and 2,50,000 times faster than in acetonitrile.

It is appropriate to state that the reaction between cyclopentadiene and methyl acrylate (Scheme 5) proceed about four times faster in formamide and six times faster in ethylene glycol than it does in methanol.[9] Similar results were reported by Liotta et al.[10] These solvents, viz. formamide and ethylene glycol are referred to as 'water-like' solvent systems. Also addition of LiCl in 'water-like' solvents led to a further rate increase. In this case, it is noteworthy that addition of traditional 'antihydrophobic' additives like urea also increased the rate (rather than retarding the rate) of the Diel-Alder reaction.[11]

The Diels-Alder reaction in aqueous medium has tremendous synthetic potentialities.[11a] This technique has been used in the field of terpenes, steroids and alkaloids.

Due to the convenience of conducting Diels-Alder reactions, in aqueous phase, this methodology has found a number of applications in pharmaceutical industry. Some interesting applications are:

(i) Synthesis of antifungals based on aqueous Diels-Alder reaction.[12]
(ii) Intramolecular version of Diels-Alder reaction with a dienecarboxylate

was used in synthetic study of the antibiotic ilicicolin H.[13]

(iii) Synthesis of α,ω-alkane dicarboxylic acid by the reaction of cyclooctene with ozone in H_2O at 10° followed by treatment with H_2O_2 in presence of an emulsifier (Scheme 9).[14]

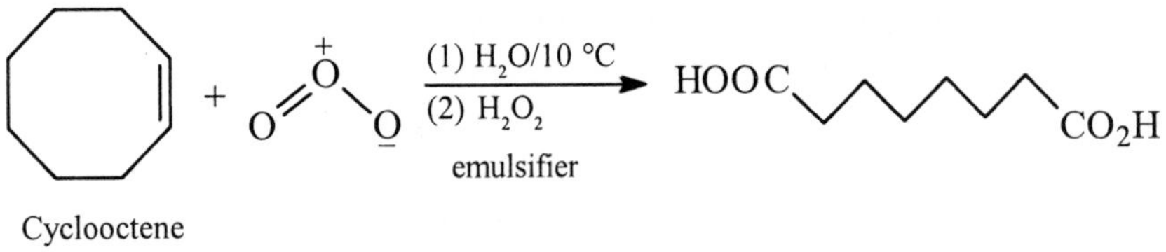

Cyclooctene

Scheme 9

(iv) A hetero-Diels-Alder reaction for the synthesis of heterocyclic compounds with nitrogen- or oxygen-containing dienophiles are particularly useful.[15] The first example of hetero-Diels Alder reaction was reported in 1985 by Grieco. In this reaction an iminium salt (generated *in situ* under Mannich-like conditions) reacted with dienes in water to give aza-Diels-Alder reaction products (Scheme 10) which are useful in the synthesis of alkaloids.[16]

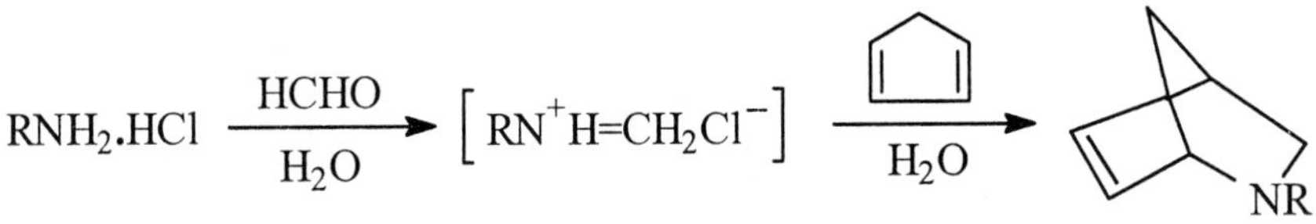

Scheme 10

The intramolecular aza-Diels-Alder reaction[17] also occurs in aqueous media. This reaction gave fused ring system (Scheme 11).

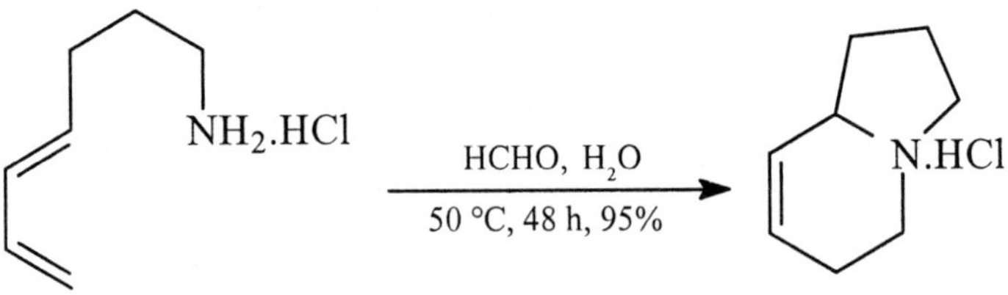

Scheme 11

For more details, refer to a review on hetero Diels-Alder reactions.[18]

(v) A convenient one-pot synthesis of a variety of heterocyclic products by the reaction of an appropriate oxime with NaOCl in H_2O/CH_2Cl_2 (Scheme 12) has been achieved.[19]

Also see Diels-Alder reaction in ionic liquids, which is considered to be better than in aqueous medium (Chapter 14).

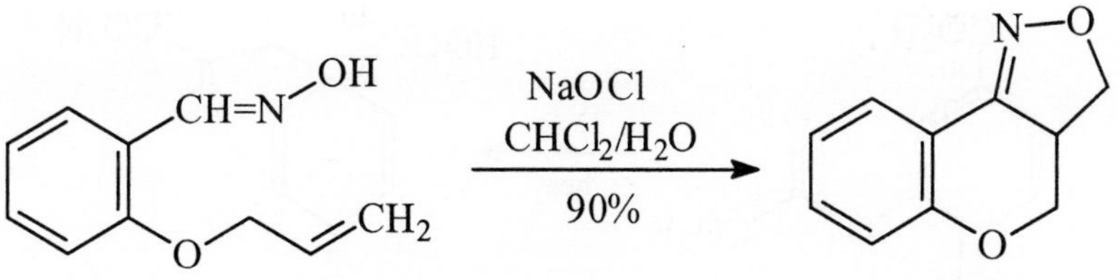
CH=N
OH
CH2
O
NaOCl
CHCl2/H2O
90%
N—O
O

Scheme 12

Diels-Alder reaction can also be performed in the solid phase. Thus, phenylpropiolic acid derivatives upon heating at 80 °C give the products (anhydrides) in 20-50% yields (Scheme 13).[20]

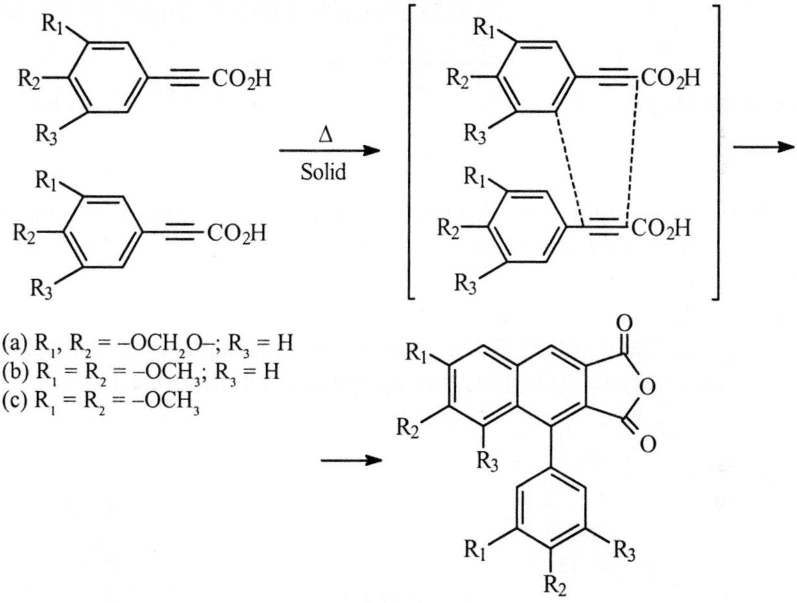
R1
R2
R3
CO2H
Δ
Solid
(a) R1, R2 = –OCH2O–; R3 = H
(b) R1 = R2 = –OCH3; R3 = H
(c) R1 = R2 = –OCH3
O

Scheme 13

12.3 Claisen Rearrangement

The thermal rearrangement of allyl phenyl ethers to o-allyl phenol and its mechanism is very well known to organic chemists.[21] Both the aliphatic and aromatic Claisen rearrangements involve a 3,3-sigmatropic shift.[22] There are reviews providing usefulness of this rearrangement reaction.[23]

The first reported use of water in promoting Claisen rearrangements was in 1970.[24] The first example of the use of pure water for Claisen rearrangement of chorismic acid is given (Scheme 14).[25]

A simple aliphatic Claisen rearrangement, a [3,3]-sigmatropic rearrangement of an allyl vinyl ether in water gave the aldehyde (Scheme 15).[26]

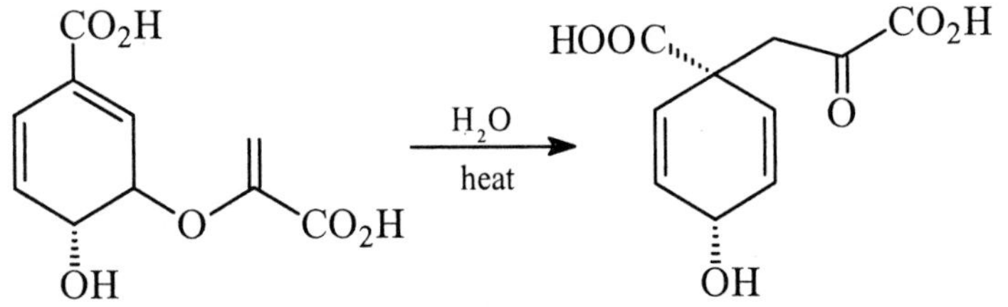

Scheme 14

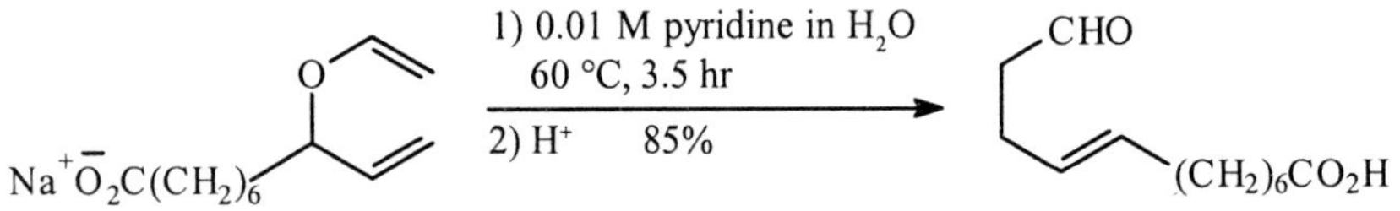

Scheme 15

The corresponding ester also underwent similar rearrangement. Similarly both allyl vinyl ether and 2-hepta-3,5-dienyl vinyl-ether underwent 3,3-shift. The best results were obtained in 2/1 methanol-water; the rates were about 40 times than those in acetone solvent.[27]

A special feature of the Claisen rearrangement in aqueous medium is that it is not necessary to protect the free hydroxyl group (Scheme 16).[28]

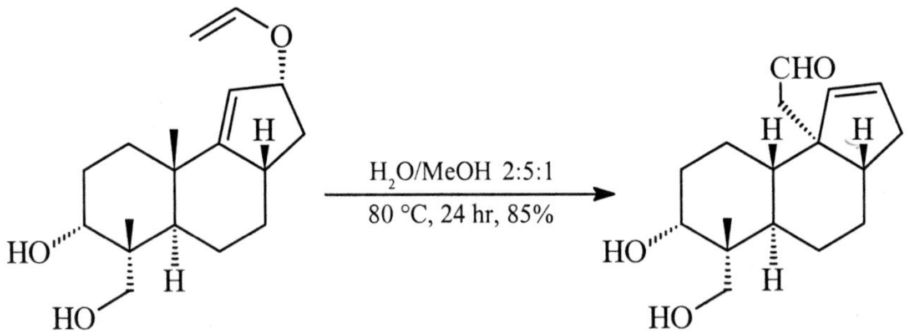

Scheme 16

The above rearrangement for protected analog under usual claisen condition resulted in elimination of acetaldehyde.[29]

Following are given some of the important applications of Claisen rearrangement in aqueous solution.

(i) Synthesis of fenestrene aldehyde having trans ring fusion between the two 5-membered rings (Scheme 17).[30]

(ii) The Claisen rearrangement of the allyl vinyl ether (Scheme 18) gave the aldehyde in 82% yield.[30]

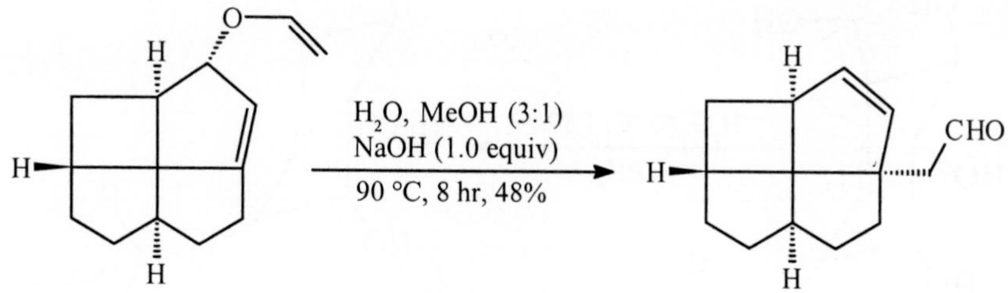

H₂O, MeOH (3:1)
NaOH (1.0 equiv)
90 °C, 8 hr, 48%

Scheme 17

1N aq. NaOH
95 °C, 5 hr, 82%

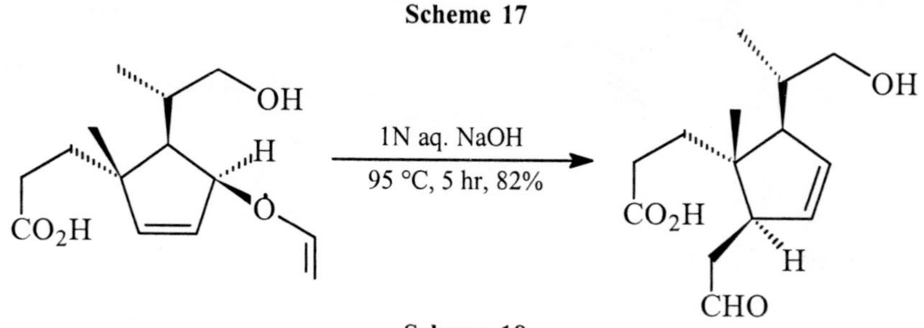

Scheme 18

(iii) Claisen rearrangements of 6-β-glycosylallyl vinyl ether (Scheme 19) and of 6-α-glycosylalkyl vinyl ether (Scheme 20) has been successful in aqueous medium.[31] In both these reactions NaBH₄ was added so that the formed aldehyde gets converted into the corresponding alcohol.

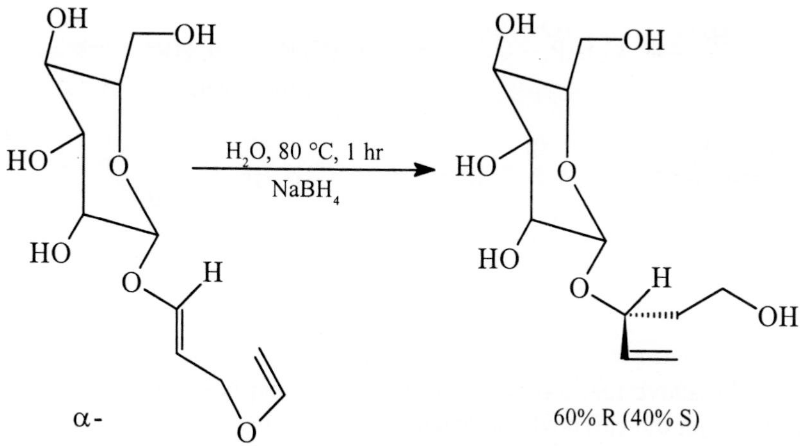

H₂O, 80 °C, 1 hr
NaBH₄

α-

60% R (40% S)

Scheme 19

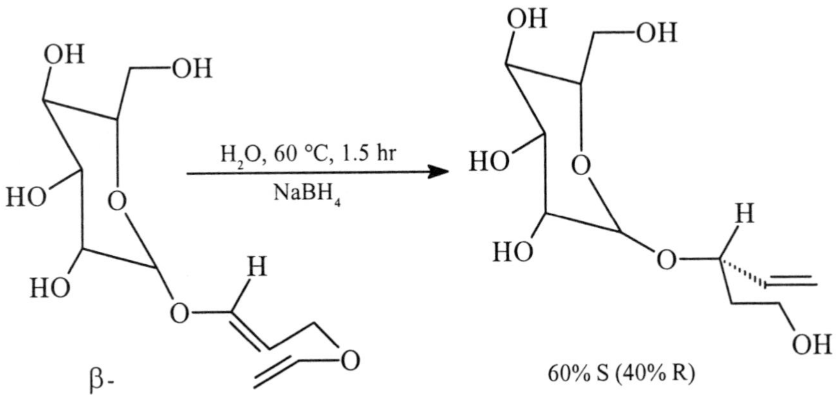

Scheme 20

12.4 Wittig-Horner Reaction

The original wittig reaction[32] has been extensively used for the preparation of olefins from alkylidene phosphoranes (ylids) and carbonyl compounds (Scheme 21).

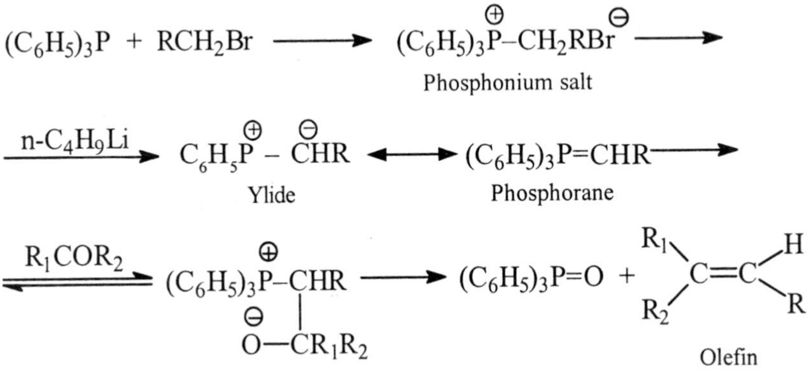

Scheme 21

In the above reaction the ylide is unstable and is generated *in situ* for reaction with the carbonyl compound.

A modification of the above reaction, known as the Wittig-Horner reaction or Horner-Wadsworth-Emmons reaction uses phosphonate esters. Thus, the reaction of ethyl bromoacetate with triphenylphosphite gives the phosphonate ester, which on treatment with base (NaH) and reaction with cyclohexanone gives α,β-unsaturated ester, ethyl cyclohexylidineacetate in 70% yield (Scheme 22).

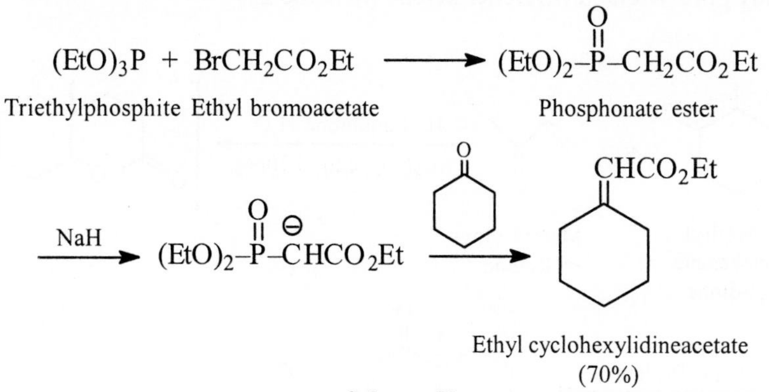

$$(EtO)_3P \ + \ BrCH_2CO_2Et \longrightarrow (EtO)_2{-}\overset{\overset{\displaystyle O}{\|}}{P}{-}CH_2CO_2Et$$

Triethylphosphite   Ethyl bromoacetate   Phosphonate ester

Ethyl cyclohexylidineacetate
(70%)

Scheme 22

The above reaction is sometimes performed in an organic/water biphase system.[33] It is now reported[34] that in the above reaction (Scheme 22) in place of strong base like NaH, a PTC can be used in aq. NaOH with good results. In the above reaction (Scheme 22) the base used is NaH or any other strong base. It has been found that the reaction proceeds with a much weaker base, such as K_2CO_3 or $KHCO_3$. Even compounds with base and acid sensitive functional groups can be used directly. In a typical example, under such conditions, β-dimethylhydrazoneacetaldehyde can be obtained efficiently.[35]

12.5 Michael Reaction

It is an addition reaction[36] between an α,β-unsaturated carbonyl compound and a compound with an active methylene group (e.g., malonic ester, acetoacetic ester, cyanoacetic ester, nitroparaffins etc.) in presence of a base, e.g., sodium ethoxide or a secondary amine (usually piperidine).

The first successful report of Michael reaction in aqueous medium was in the 1970s. 2-Methylcyclopentane-1,3-dione when reacted with vinyl ketone in water gave an adduct without the use of a basic catalyst (pH > 7). The adduct further cyclises to give a 5-6 fused ring system (Scheme 23).[37]

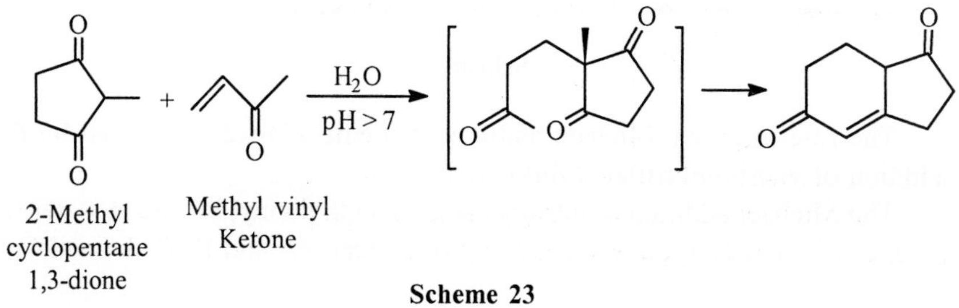

2-Methyl
cyclopentane
1,3-dione

Methyl vinyl
Ketone

Scheme 23

In this reaction (Scheme 23), use of water as solvent gave better yields and pure compound compared to reaction with methanol in presence of a base.

Michael reaction of 2-methyl-cyclohexane-1,3-dione with vinylketone give optically pure Wieland-Miescher ketone (Scheme 24).[38]

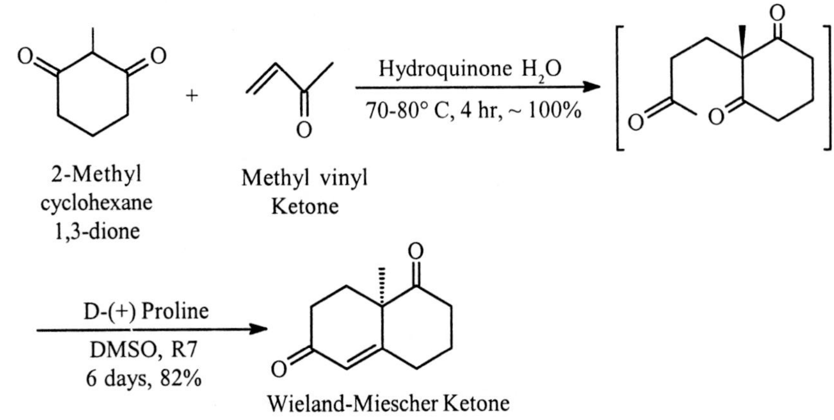

2-Methyl
cyclohexane
1,3-dione

Methyl vinyl
Ketone

Wieland-Miescher Ketone

Scheme 24

The Michael addition of 2-methyl-cyclopentane 1,3-dione to acrolein in water gave an adduct (Scheme 25) which was used for the synthesis of 13-α-methyl-14α-hydroxysteroid.[39]

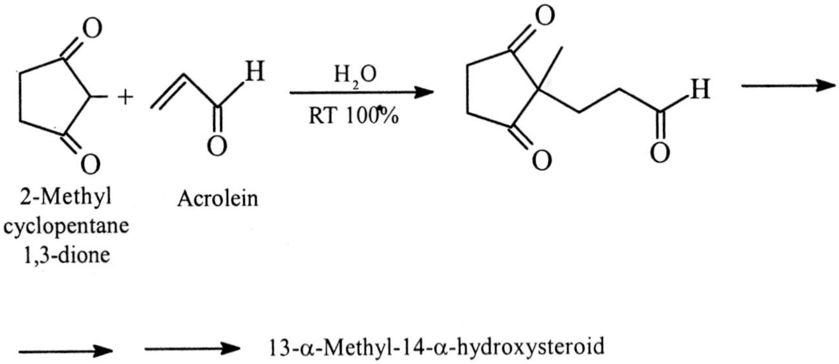

2-Methyl
cyclopentane
1,3-dione

Acrolein

13-α-Methyl-14-α-hydroxysteroid

Scheme 25

The rate of above Michael addition (Scheme 25) was enhanced by the addition of ytterbium triflate [yb(OTf)$_3$].

The Michael addition of nitromethane to methyl vinyl ketone in water (in absence of a catalyst) gave 4:1 mixture of adducts (A and B) (Scheme 26).[40]

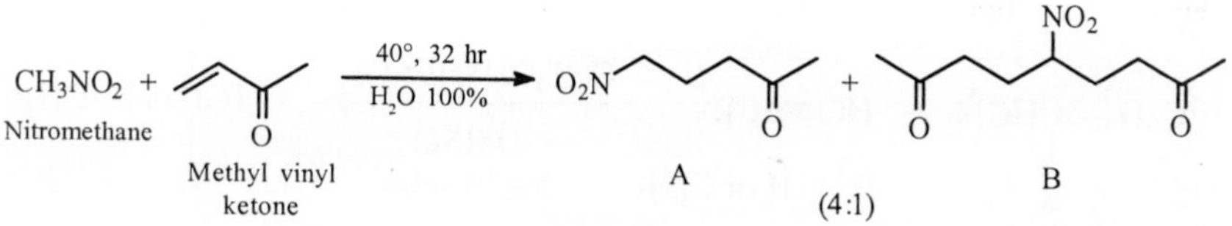

Scheme 26

Use of methyl alcohol as a solvent (in place of H_2O) gave 1:1 mixture of A and B. The above reaction does not occur in neat conditions or in solvents like THF, PhMe etc. in the absence of a catalyst.

A typical synthesis of allylrethrone, an important component of an insecticidal pyrthroid has been carried out by a combination of michael reaction of 5-nitro-1-pentene and methyl vinyl ketone in presence of Al_2O_3 followed by an intramolecular aldol type condensation (Scheme 27).[41]

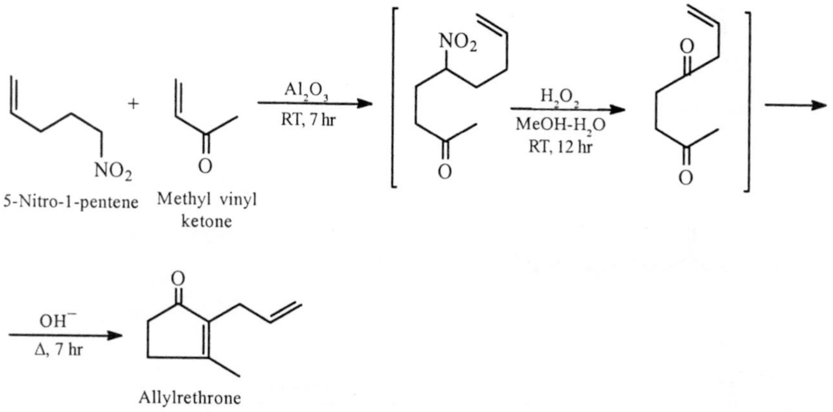

Scheme 27

The Michael addition of cyclohexenone to ascorbic acid was carried out in water in presence of an inorganic acid (rather than a base) (Scheme 28).[42]

The reaction[43] of active nitriles with acetylenes can be catalysed by quaternary ammonium salts (PTC) (Scheme 29).

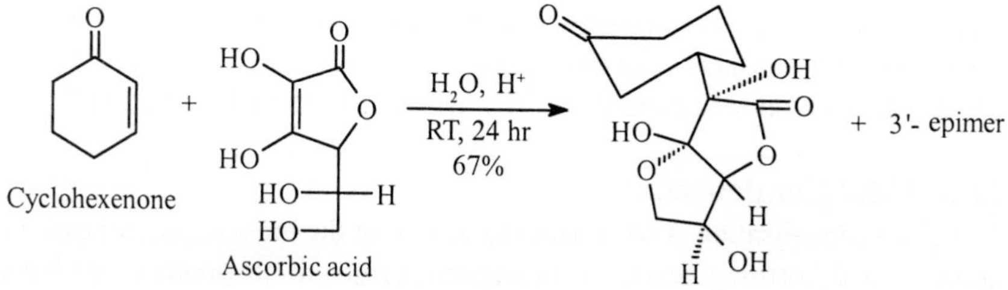

Scheme 28

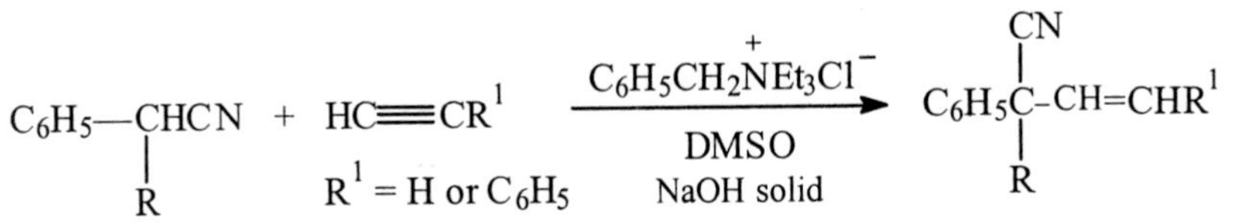

$$C_6H_5-\underset{R}{\overset{|}{CH}}CN + HC\equiv CR^1 \quad \xrightarrow[\substack{DMSO \\ NaOH\ solid}]{C_6H_5CH_2\overset{+}{N}Et_3Cl^-} \quad C_6H_5\underset{R}{\overset{CN}{\underset{|}{\overset{|}{C}}}}-CH=CHR^1$$

$R^1 = H$ or C_6H_5

R = CH$_3$, isopropyl, benzyl

Scheme 29

Very efficient Michael addition reactions of amines, thiphenol and methyl acetoacetate to chalcone in water suspension have been developed.[44] For example, stirring a suspension of powdered chalcone in a small amount of water containing n-Bu$_2$NH and a surfactant, hexadecyltrimethylammonium bromide for 2 hr gave the adduct in 98% yield. Similarly adducts were obtained with thiophenol and methyl acetoacetate (Scheme 30).

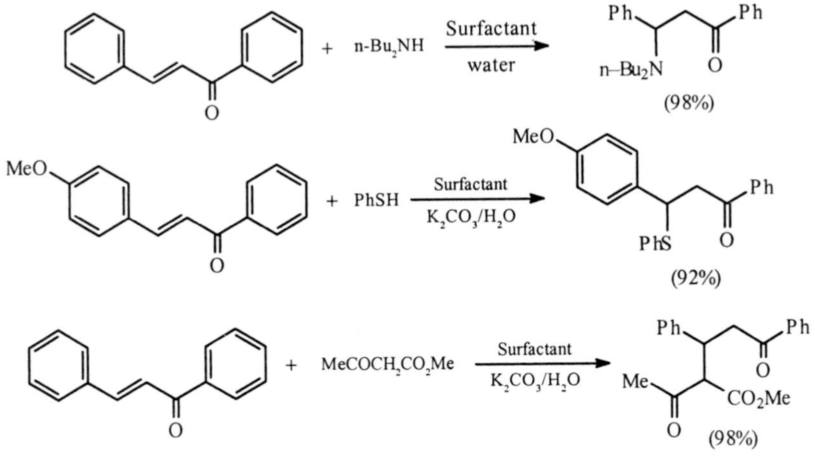

Scheme 30

See also Michael addition in solid state (Sec. 13.2.3).

Asymmetric Michael addition of benzenethiol to 2-cyclohexenone and maleic acid esters proceeds enantioselectively in their crystalline cyclodextrin complexes. The adducts were obtained in 38 and 30% ee respectively. In both cases, the reaction was carried out in water suspension (Scheme 31).[45]

12.6 Aldol Condensation

The aldol condensation is considered to be one of the most important carbon-carbon bond forming reactions in organic synthesis in presence of basic reagents. The conventional aldol condensation involve reversible self-addition of aldehydes containing a α-hydrogen atom. The formed β-hydroxy aldehydes

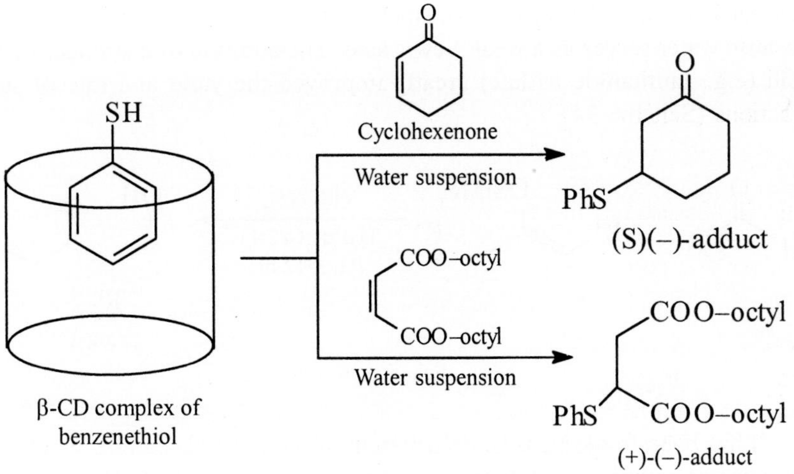

Scheme 31

undergo dehydration to give α,β-unsaturated aldehydes. This has been extensively reviewed.[46] The reaction can occur either between two identical or different aldehydes, two identical or different ketones and an aldehyde and a ketone.

A stereoselective aldol condensation is known as Mukaiyama reaction.[47] It consists in the reaction of an silyl enol ether of 3-pentanone with an aldehyde (2-methyl-butanal) in presence of $TiCl_4$ to yield an aldol product, Manicone, an alarm pheromone (Scheme 32).[48]

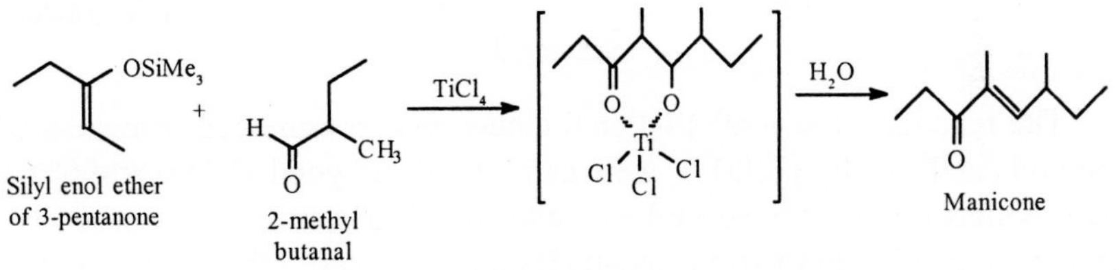

Scheme 32

The above reactions are carried out under non-aqueous conditions.

The first water-promoted aldol reaction of silyl enol ethers with aldehydes was first reported in 1986 (Scheme 33).[49]

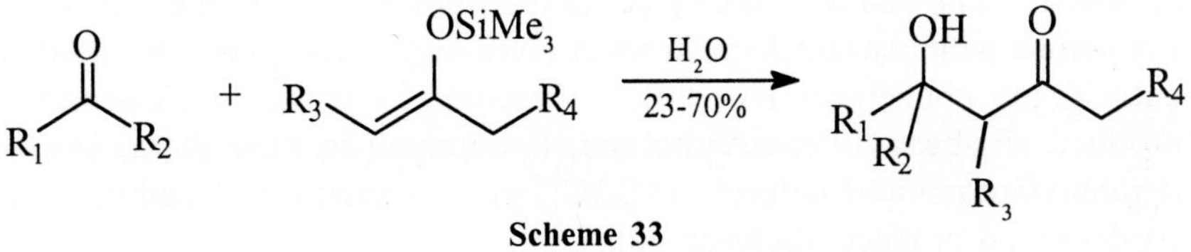

Scheme 33

The above reactions were carried out in aqueous medium without any acid catalyst. The reaction, however, took several days for completion, probably because water serves as a weak Lewis acid. The addition of a stronger Lewis acid (e.g., lanthanide triflate) greatly improved the yield and rate of such reactions (Scheme 34).[50]

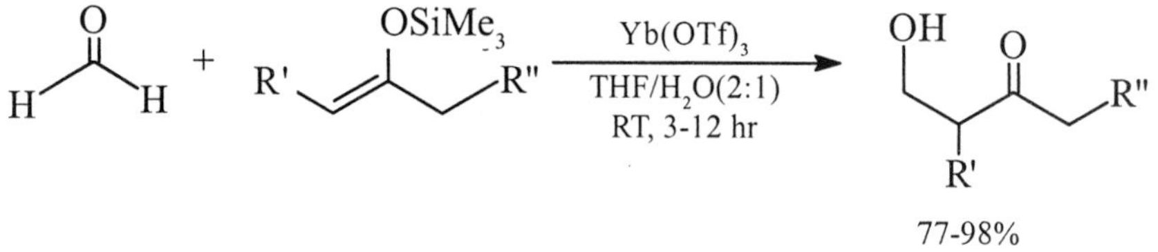

77-98%

Scheme 34

It has been found that the dehydration of the alcohols can be avoided in presence of complexes of Zn with aminoesters or aminoalcohols.[51]

Using the above methodology, vinyl ketones (Scheme 35) can be obtained by the reaction of 2-alkyl-1,3-diketones with aqueous formaldehyde (formalin) using 6-10 M aqueous potassium carbonate as base; the final step involved cleavage of the intermediate with base.[52]

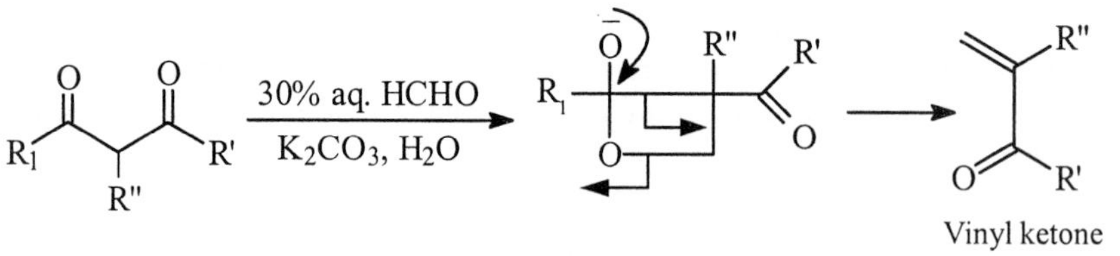

Vinyl ketone

Scheme 35

The reaction of several silyl enol ethers with commercial formaldehyde solution catalysed by yb(OTf)$_3$ were carried out and good yields (80-90%) of the products obtained.[53] Several examples of the aldol condenstion in water have been cited using various aldehydes and silyl enol ethers. The products were obtained in good yield (80-90%).[54] In all the above reactions the catalyst could be recovered and used again and again. The above methodology has been extensively reviewed.[55]

An interesting case of aldol condensation is vinylogous aldol reaction. The γ-hydrogen of α,β-unsuturated ketones, nitriles and esters is 'active' and the electrophilic addition taken place at the γ-position. Thus, the reaction of isophorone with benzaldehyde in water gives only vinylogous aldol addition but with low conversion. However, in presence of CTACI, the condensation product, (E)-benzylideneisophorone, is obtained in 80% yield. Use of tetrabutylammonium chloride (TBACl) gives a mixture of addition and condensation products (Scheme 36).[56]

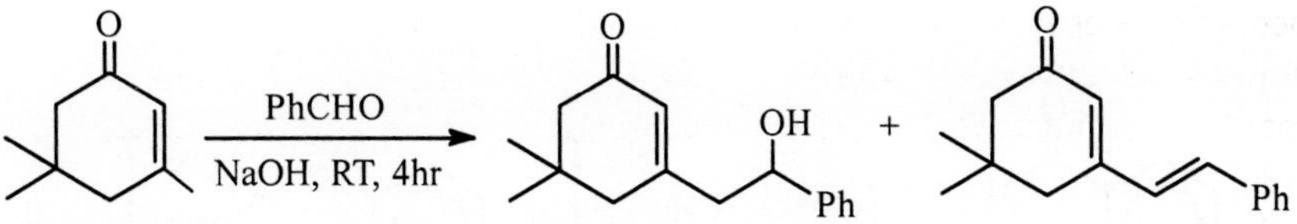

Isophorone		
Water only	24%	–
CTACI	–	80%
TBACI	27%	58%

Scheme 36

Certain aldol condensations have been also carried out in solid state (Sec. 13.2.5).

12.7 Knoevenagel Reaction

The condensation of aldehydes or ketones, with active methylene compounds (especially malonic ester) in presence of a weak base like ammonia or amine (primary or secondary) is known as Knoevenagel reaction.[57,58] However, when condensation is carried out in presence of pyridine as a base, decarboxylation usually occurs during the condensation. This is known as Doebner modification.[59] Some examples are given (Scheme 37).

$$CH_3CHO \; + \; CH_2(COOH)_2 \xrightarrow{\text{Base}} CH_3CH{=}C(COOH)_2$$

$$\downarrow {-}CO_2$$

$$CH_3CH{=}CHCOOH$$

Acetaldehyde   Malonic acid   Crotonic acid

$$C_6H_5CHO \; + \; CH_2(COOC_2H_5)_2 \xrightarrow[\text{benzene}]{\text{Pyridine}} C_6H_5CH{=}C(COOC_2H_5)_2$$

$$\xrightarrow[\text{2) H}_3\text{O}^+]{\text{1) hydrolysis}} C_6H_5CH{=}C(COOH)_2 \xrightarrow[{-}CO_2]{\Delta} C_6H_5CH{=}CHCOOH$$

Cinnamic acid

Scheme 37

The Knoevenagel reaction has been carried out between aldehydes and acetonitrile in water. Thus, salicylaldehydes react with malononitrile at room temperature in the heterogeneous aqueous alkaline medium to give α-hydroxybenzylidene malononitriles, which are converted directly to 3-cyanocoumarins by acidification and heating (Scheme 38).[60]

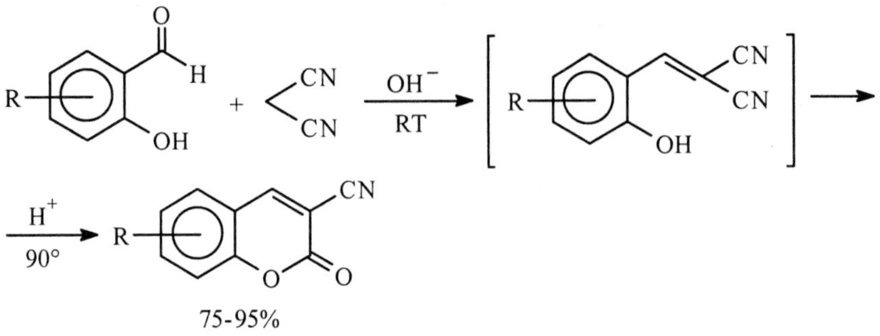

75-95%

R = H, OH, OMe

Scheme 38

In a similar way, use of substituted acetonitriles in the above procedure (Scheme 38) give the corresponding 3-substituted coumarins in 66-98% yields (Scheme 39).

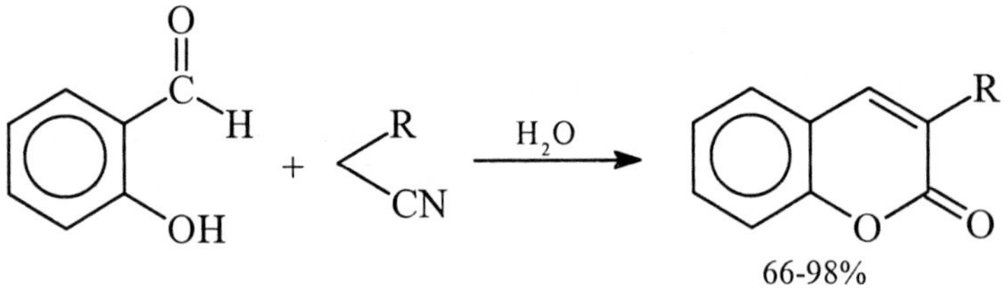

66-98%

R = CN, CO$_2$Et, NO$_2$, Ph, 2-Py

Scheme 39

In case of phenylacetonitrile, a catalytic amount of CTABr (0.1 mole/equiv) is used. The above reaction gives better yields in water compared to in ethanol.

The reaction of benzaldehyde with acetonitrile does not occur in water only but requires the presence of catalytic amount of CTACl or TBACl to give high yields of the corresponding arylcinnamonitriles (Scheme 40).[57]

Knoevenagel-type addition product can be obtained by the reaction of acrylic derivatives in presence of a 1,4-diazabicyclo[2.2.2]octane (DABCO) (Scheme 41).[61]

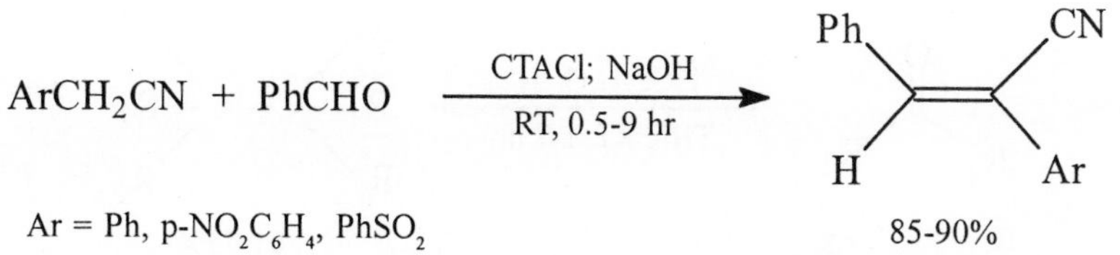

$$ArCH_2CN + PhCHO \xrightarrow[\text{RT, 0.5-9 hr}]{\text{CTACl; NaOH}}$$

Ph, C=C, CN / H, Ar

85-90%

Ar = Ph, p-NO$_2$C$_6$H$_4$, PhSO$_2$

Scheme 40

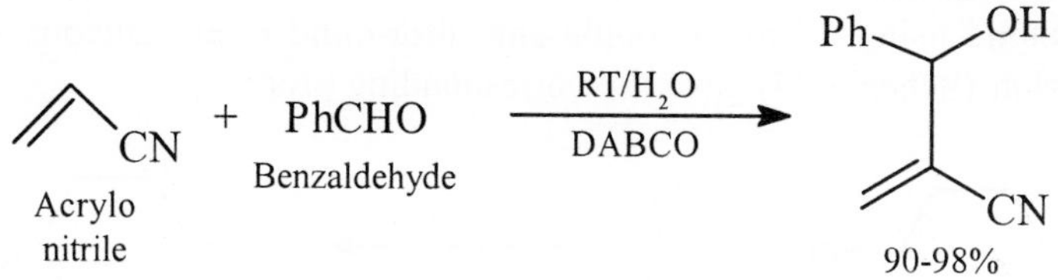

Acrylo nitrile (CH$_2$=CHCN) + PhCHO (Benzaldehyde) $\xrightarrow[\text{DABCO}]{\text{RT/H}_2\text{O}}$ Ph–CH(OH)–C(=CH$_2$)–CN

90-98%

Scheme 41

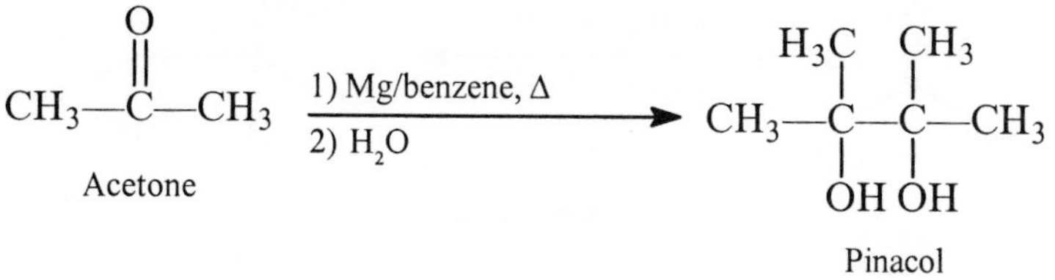

12.8 Pinacol Coupling

Ketones are known to react with Mg/benzene to give 1,2-diols by heating with magnesium in benzene followed by treatment with water. Thus, under these conditions aceton**e** give pinacol (Scheme 42)

$$CH_3-\overset{O}{\underset{}{C}}-CH_3 \xrightarrow[\text{2) H}_2\text{O}]{\text{1) Mg/benzene, }\Delta}$$

Acetone

$$CH_3-\overset{H_3C}{\underset{OH}{C}}-\overset{CH_3}{\underset{OH}{C}}-CH_3$$

Pinacol

Scheme 42

This is known as pinacol coupling. The use of Zn-Cu couple to couple unsaturated aldehydes to pinacols was recorded as early as 1892.[62] Subsequently chromium and vanadium[63] and some ammonical-TiCl$_3$[64] based reducing agents were used.

It has now been found[65] that pinacol coupling takes place in aromatic ketones and aldehydes in aqueous media in presence of Ti(III), under alkaline conditions. However, in presence of acids, only the substrates (aromatic ketones and aldehydes) having electron withdrawing group like CN, CHO, COMe, COOH, COOMe, pyridyl (activating groups) only underwent pinacol coupling[66] (Scheme 43). In case of nonactivated carbonyl compounds, it was necessary to use excess of the substrate as a solvent.[67]

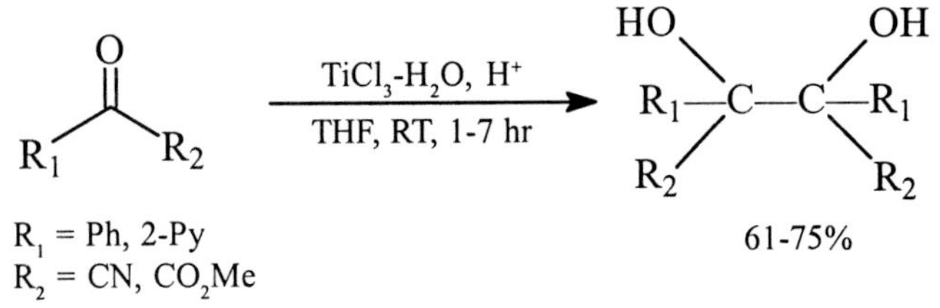

R$_1$ = Ph, 2-Py
R$_2$ = CN, CO$_2$Me

61-75%

Scheme 43

The coupling reaction between an α,β-unsaturated carbonyl compound and acetone using a Zn-Cu couple and ultrasound in an aqueous acetone suspension (Scheme 44) gave the corresponding product.[68]

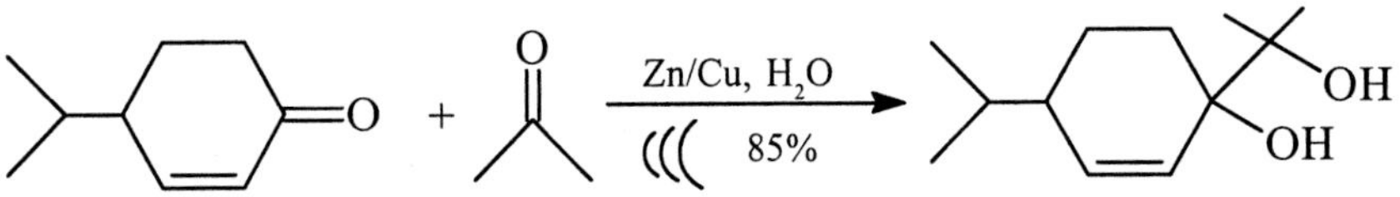

Scheme 44

An interesting example is the coupling of aldimines to give vicinal diamines (Scheme 45) by indium in aqueous ethanol in presence of small amount of ammonium chloride, which accelerates the reaction.[69]

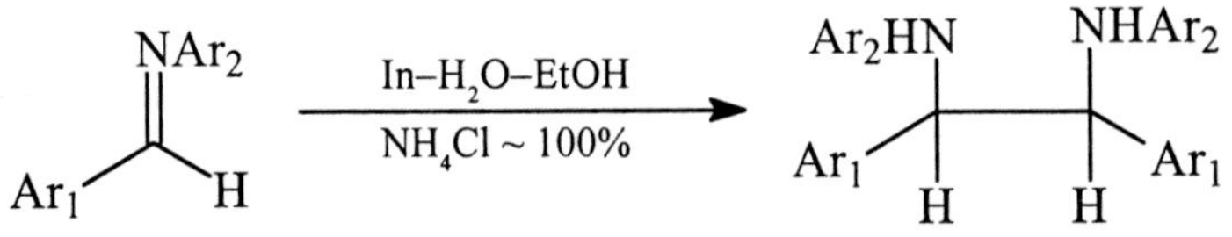

Scheme 45

12.9 Benzoin Condensation

It consists in the treatment of aromatic aldehydes with sodium or potassium cyanide, usually in an aqueous ethanolic solution to give α-hydroxy ketones (benzoins) (Scheme 46).[70]

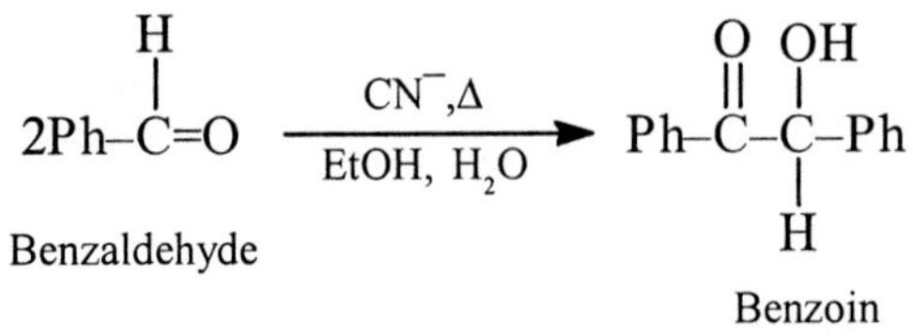

Scheme 46

Benzoin condensation can be considered to occur through a formal Knoevenagel type addition (Scheme 47). The key step of the reaction is the loss of the aldehydic proton, which gives rise to the cyanohydrin anion. In this case the acidity of the proton is increased by the electron-withdrawing power of the cyano group.

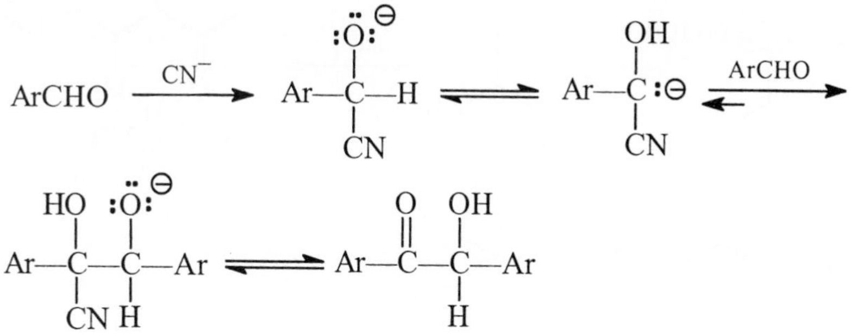

Scheme 47

It is found that benzoin condensation of aldehydes are strongly catalysed by a PTC (quaternary ammonium cyanide in a two phase system).[71] In a similar way, acyloin condensations are easily effected by stirring aliphatic or aromatic aldehydes with a quaternary catalyst (PTC), N-laurylthiazolium bromide in aqueous phosphate buffer at room temperature.[72] The aromatic aldehydes reacted in a short time (about 5 min). However, aliphatic aldehydes require longer time (5-10 hr) for completion. Mixtures of aliphatic and aryl aromatic aldehydes give mixed α-hydroxy ketones.[73]

On the basis of extensive work, Breslow found that the benzoin condensation in aqueous media using inorganic salts (e.g., LiCl) is about 200 times faster than in ethanol (without any salt).[74] The addition of γ-cyclodextrin also accelerates the reaction, whereas the addition of β-cyclodextrin inhibits the condensation.

12.10 Claisen-Schmidt Condensation

Normally Claisen-Schmidt condensation involves the condensation of aromatic aldehydes (without α-hydrogen) with an aliphatic aldehyde or ketone (having α-hydrogen) in presence of a relatively strong base (hydroxide or alkoxide) to form α,β-unsaturated aldehyde or a ketone (Scheme 48).[75]

$$C_6H_5CHO + CH_3CHO \xrightarrow{\text{NaOH}} C_6H_5CH=CHCHO$$

Cinnamaldehyde

$$C_6H_5COCH_3 + C_6H_5CHO \xrightarrow[\text{EtOH}]{\text{NaOH}} C_6H_5COCH=CHC_6H_5$$

Benzalacetophenone
(chalcone)

Scheme 48

A related reaction, known as Mukaiyama reaction[76] involves the reaction of silyl enol ether of the ketone with an aldehyde in an organic solvent in presence of TiCl$_4$ (Scheme 49) (see also Scheme 32).

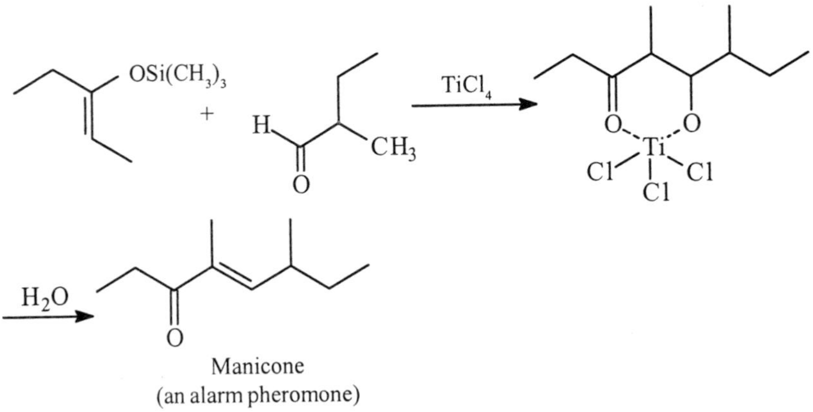

Manicone
(an alarm pheromone)

Scheme 49

It has been shown[77] that trimethyl silyl enol ether of cyclohexanone with benzaldehyde occurs in water in presence of TiCl$_4$ in heterogeneous phase at room temperature and atmospheric pressure (Scheme 50).

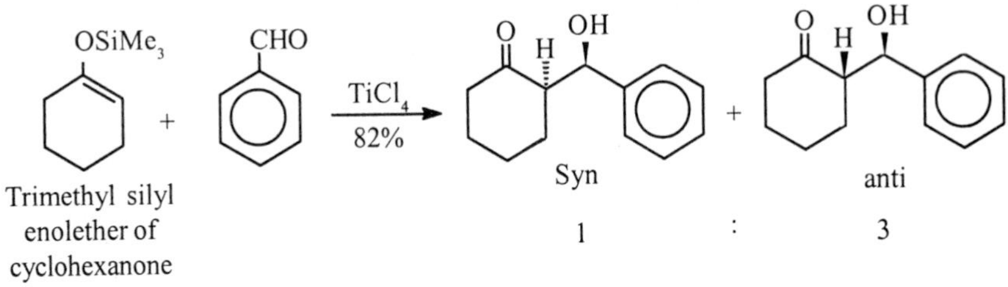

Trimethyl silyl
enolether of
cyclohexanone

Scheme 50

Better yields are obtained under sonication conditions. The reaction is favoured by an electron-withdrawing substituent in the para position of the phenyl ring in benzaldehyde.

Claisen-Schmidt reaction of acetophenones with aromatic aldehydes in presence of cationic surfactants such as cetylammonium compounds, CTACl, CTABr, (CTA)$_2$SO$_4$ and CTAOH in mild alkaline conditions give chalcones, which on cyclisation give flavanols (Scheme 51).

The reaction of cyclohexanone with benzaldehyde in water gives high yield of a 1:1 threo-erythro mixture of the ketol. However in presence of CTACl, the bis-condensation product in obtained quantitatively (Scheme 52).[78]

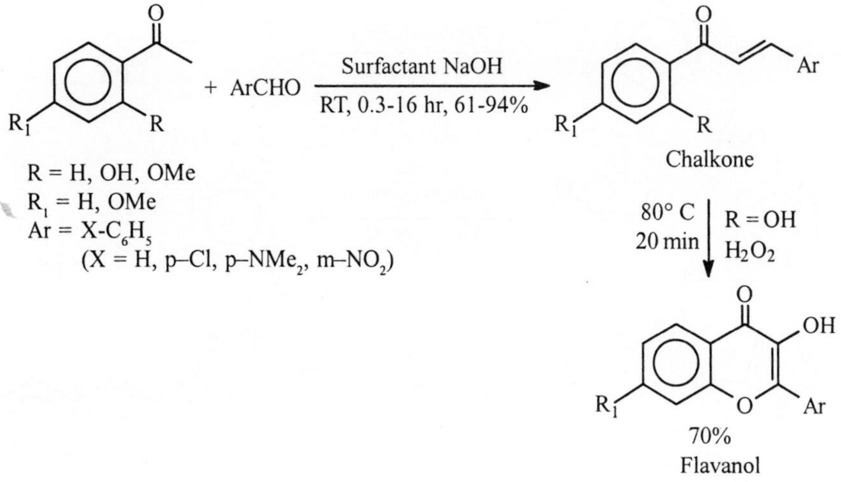

R = H, OH, OMe
R$_1$ = H, OMe
Ar = X-C$_6$H$_5$
    (X = H, p–Cl, p–NMe$_2$, m–NO$_2$)

Chalkone

80° C   R = OH
20 min   H$_2$O$_2$

70%
Flavanol

Scheme 51

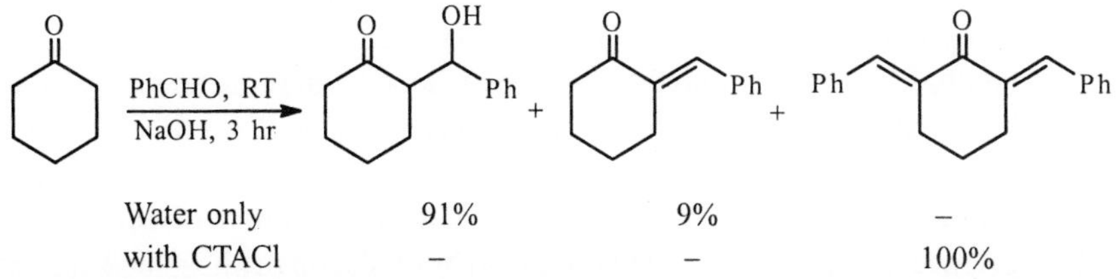

PhCHO, RT
NaOH, 3 hr

Water only   91%   9%   –
with CTACl   –   –   100%

Scheme 52

12.11 Heck Reaction

The Heck reaction, a synthetically used palladium catalysed reaction involves the coupling of an alkene with a halide or triflate in presence of Pd(O) catalyst to form a new alkene (Scheme 53).

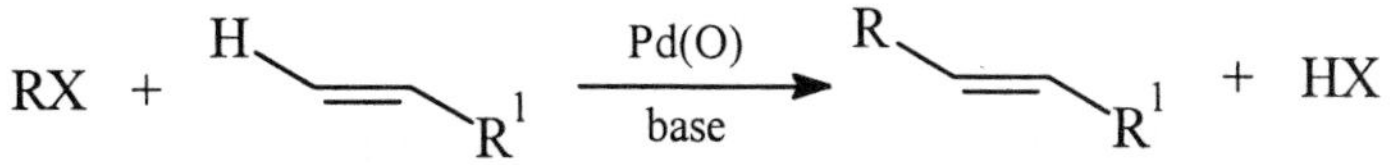

R = aryl, vinyl or alkyl group without β-hydrogens on a sp^3 carbon atom
X = halide or triflate (OSO$_2$CF$_3$)

Scheme 53

Heck reaction uses mild base such as Et$_3$N or anions like OH⁻, ⁻OCOCH$_3$, CO$_3$$^{2-}$ etc. Some other applications of Heck reaction are also given (Scheme 54).

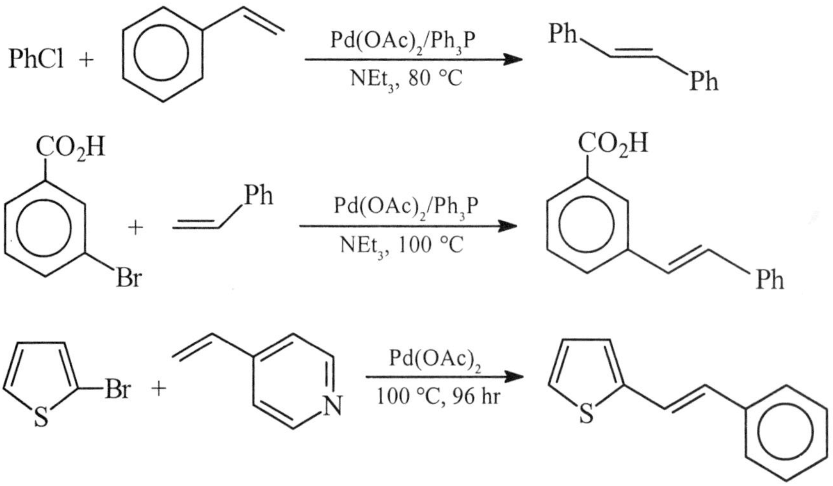

Scheme 54

The traditional technique for carrying out the Heck reaction is to use anhydrous polar solvents (eg., DMF and MeCN) and tert. amines as bases.

Recently it has been found that the Heck reaction can proceed very well in water. In fact, the role of water in the Heck reaction, as well as other reactions catalysed by Pd(O) in presence of phosphine ligands is: (i) transformation of catalyst precursor into Pd(O) species and (ii) the generation of zero-valent palladium species capable of oxidative addition by oxidation of phosphine ligands by the Pd(II) catalyst precursor can be affected by water content of the reaction mixture.

It has been found that the Heck reaction can be accomplished under PTC conditions[79] with inorganic carbonates as bases under mild conditions at room temperature. Such conditions can be used in case of substrates, like methyl vinyl ketone, which do not survive the usual conditions of Heck arylation (action of base at high temperature). Subsequently, it has been shown that the Heck reaction can be carried out in water and aqueous organic solvents, catalysed by simple palladium salts in presence of inorganic bases like K_2CO_3, Na_2CO_3, $NaHCO_3$, KOH etc.[80]

An interesting application of the Heck reaction is the synthesis of cinnamic acid by the reaction of aryl halides with acrylic acid (Scheme 55).

Use of acrylo nitrile in place of acrylic acid in this method (Scheme 55) yield the corresponding cinnamonitriles. Most of the products obtained in Heck-reactions are almost exclusively (E) isomers. However, the reaction of acrylonitrile give a mixture of (E) and (Z) isomers with ratio 3:1, close to that observed under conventional anhydrous conditions.[81]

The Heck reaction can also be performed under milder condition by addition of acetate ion as given in (Scheme 56).

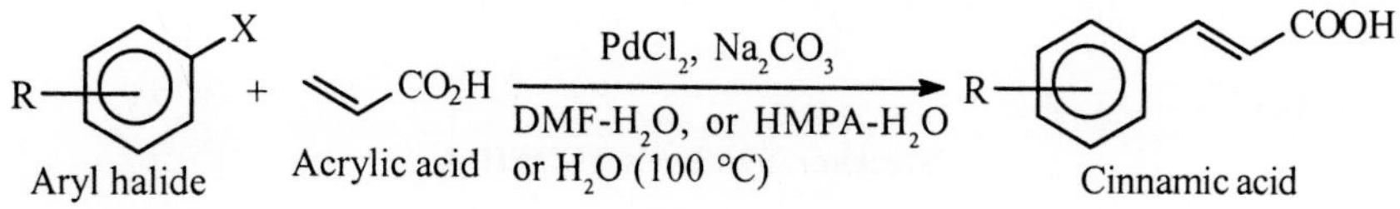

X = I, Br

R = H, p-Cl, p-OMe, p-Me-, p-Ac, p-NO$_2$, p-CHO, p-OH, m-COOH etc.

Scheme 55

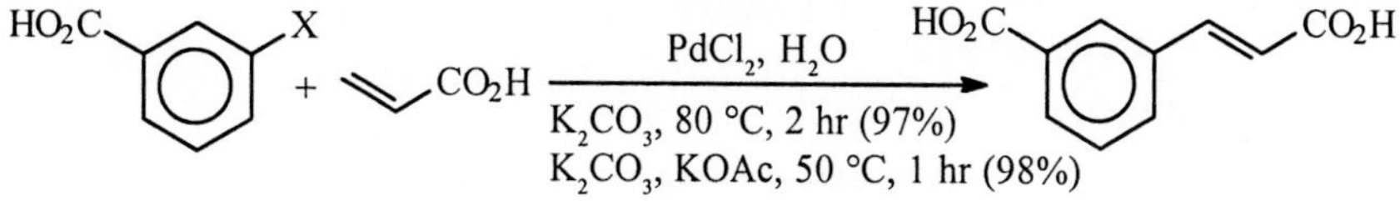

Scheme 56

A number of other application of the Heck reaction have been described in literature.[82]

The Heck reaction has also been performed in ionic liquids (for details see Sec. 14.3.3).

12.12 Strecker Synthesis

This method is used for the synthesis of amino acids by the reaction of an aldehyde with ammonia followed by reaction with HCN to give α–aminonitrile, which on hydrolysis give the corresponding amino acid (Scheme 57).[83]

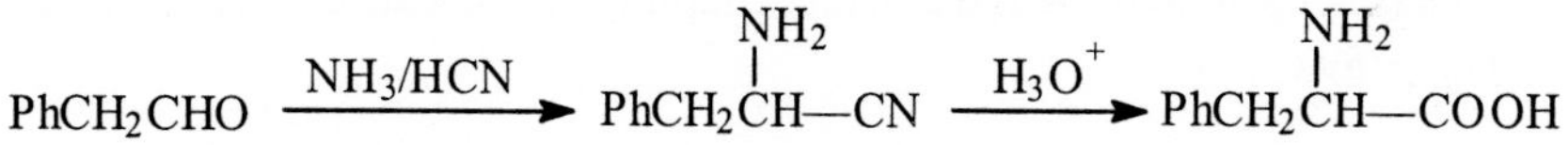

Scheme 57

The α-aminonitrile can also be obtained by the treatment of the aldehyde with HCN followed by reaction of the formed cyanohydrin with ammonia. This method is known as Erlenmeyer modification.[84] A more convenient route is to treat the aldehyde in one step with ammonium chloride and sodium cyanide (this mixture is equivalent to ammonium cyanide, which in turn dissociate into ammonia and HCN). This procedure is referred to as the Zelinsky-Stadnikott modification.[85] The final step is the hydrolysis of the intermediate α-aminonitrile under acidic or basic conditions.

Using strecker synthesis, disodium iminodiacetate (DSIDA) an intermediate for the manufacture is Monsantos' Roundup (herbicide) was synthesised.[86]

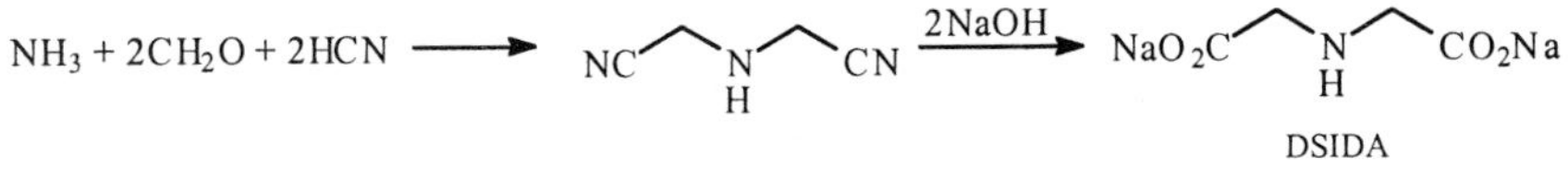

$$NH_3 + 2CH_2O + 2HCN \longrightarrow$$

Strecker Synthesis of DSIDA

In the above synthesis hydrogen cyanide, a hazardous chemical is used and this requires special handling to minimise the risk to workers and the environment. An alternative green synthesis of DSIDA was developed by Monsanto.

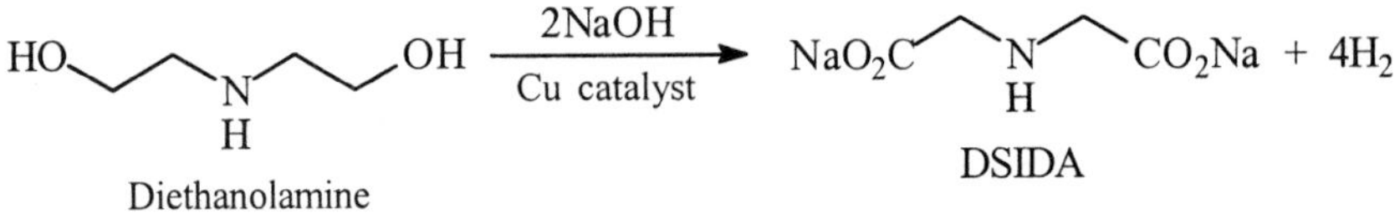

Diethanolamine

Alternative Synthesis of DSIDA

The new method avoids the use of HCN and CH_2O and is safer to operate.

12.13 Wurtz Reaction

It involves coupling of alkyl halides with sodium in dry ether to give hydrocarbons.[87] An example is the synthesis of hexane (Scheme 58).

$$CH_3CH_2CH_2Br \xrightarrow[\text{ether}]{\text{Na}} CH_3(CH_2)_4CH_3$$

Propyl bromide   Hexane

Scheme 58

It has been shown[88] that the Wurtz coupling can be carried out by Zn/H_2O (Scheme 59).

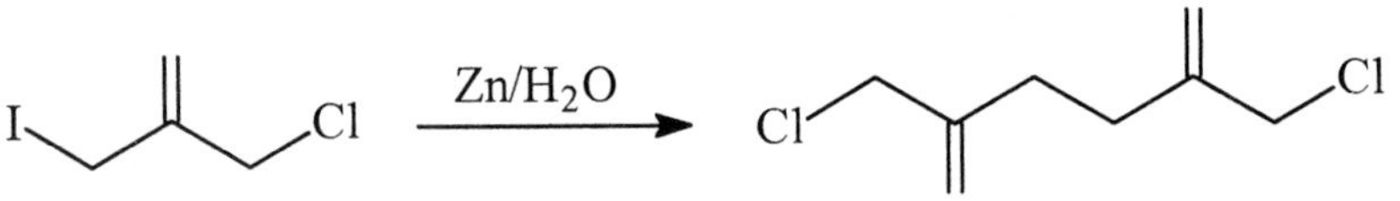

Scheme 59

12.14 Oxidations

One of the most widely investigated process in organic chemistry is oxidation. A large number of reactions involving oxidation are used in industries besides being of interest in the laboratory. A number of oxidizing agents with different substrates have been described.[89]

Oxidations have been known to be carried out in aqueous medium for a long time. The well known oxidation of arenes with $KMnO_4$ in aqueous alkaline medium is very well known.[90] However, the yields are considerably increased by using KMnO$_4$ in presence of a phase transfer catalyst particularly in the oxidation of toluene. Also, H$_2$O$_2$ in water has been used quite frequently in many organic solvents. It is environment-friendly since it forms water as a secondary product.

In the present unit, some innovative and recent oxidations by chemical reagents in aqueous media are given. Enzymatic oxidations have also been known to occur in water. However, this subject will be discussed in a separate section. Following are given some of the important reactions in aqueous medium.

12.14.1 Epoxidation

Peracids are known to react with alkenes to give stable three-membered rings containing oxygen atom, called epoxides or oxiranes (Scheme 60).

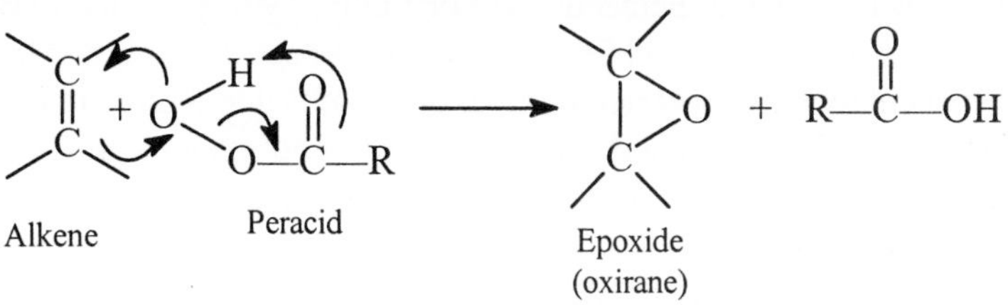

Alkene   Peracid   Epoxide
(oxirane)

Scheme 60

A number of peracids can be used. The reaction takes place in nonpolar solvents such as dichloromethane and benzene. The above epoxidations are stereoselective and take place by syn addition to the double bond, as established by x-ray analysis of the products obtained. As the cis alkene gives only cis epoxide and trans alkene gives trans epoxide, the reaction must be concerted, i.e., the one step mechanism retains the stereochemistry of the starting alkenes.

It has now been found[91] that epoxidation of simple alkenes with m-chloroperoxybenzoic acid in aqueous solution of NaHCO$_3$ (pH ~ 8.3) at room temperature proceeds well and gives good yields of the epoxides. Using this procedure some of the alkenes like cyclopentene, cyclohexene, cycloheptene, cyclooctene, methyl cyclohexene, (+)-3-carene have been reacted with m-chloroperbenzoic acid at 20 °C for 30 min to give 90-95% of the epoxide. Styrene could be epoxidized at 20 °C (1 hr) giving 95% yield. α-Methylstyrene and trans-β-methylstyrene could be epoxidized at 0 °C in 1 hr giving 63% and 93% respectively yields of the epoxides. In aqueous medium, the reaction occurs in heterogeneous phase, but this does not effect the reactivity, which sometimes is higher than in homogenous organic phase.

For direct epoxidation of simple alkenes by H$_2$O$_2$, the peroxide must be activated (Scheme 61). This is done in buffered aqueous tetrahydrofuran (THF), 50% H$_2$O$_2$ activated by stoichiometic amounts of organophosphorus anhydride.

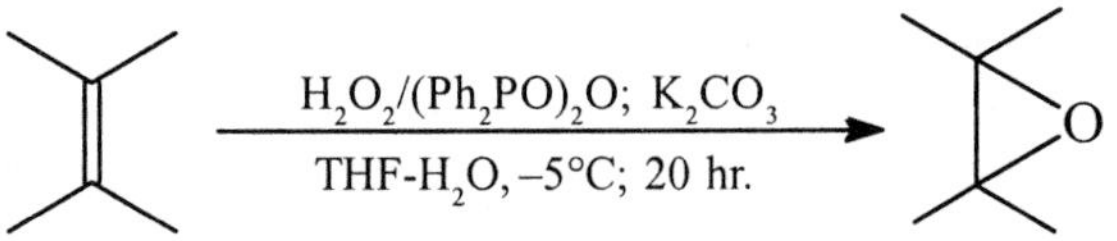

Alkene

Epoxide
60-100%

Scheme 61

Using this method a variety of alkenes could be epoxidized.[92] The epoxidation of electron-deficient olefins can be achieved with H_2O_2 in presence of sodium tungstate as a catalyst.[93] The epoxidation of alkene has also been effected with a variety of oxidizing reagents such as $PhIO_4$, NaClO, O_2, H_2O_2, ROOH, $KHSO_5$ etc. in aqueous medium in presence of metalloporphyrins.[94]

On an industrial scale, epoxidation of alkenes is generally carried out by using hydrogen peroxide, peracetic acid or t-butyl hydroperoxide (TBHP).[95] A novel safe and cheap method for epoxidation has been developed.[96] It consist in using nacent oxygen generated by electrolysis of water at room temperature by using Pd black as an anode. Using this method cyclohexene could be epoxidized in good yield. In this illustration, water is used as a reaction medium as well as a reagent.

Regioselective epoxidation of allyl alcohols in presence of other C=C bonds by using monoperphthalic acid (MPPA) in presence of cetyltrimethyl ammonium hydroxide (CTAOH) (which controls the pH of the aqueous medium) (Scheme 62).[97]

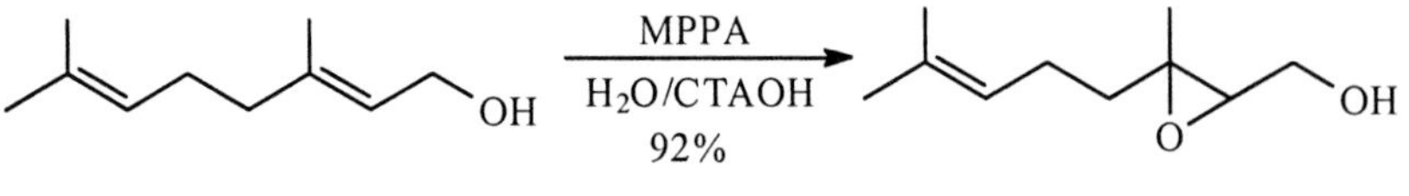

Scheme 62

It is interesting to note how the epoxidation takes place at different double bonds in the terpenoids viz. geraniol (I), nerol (II), farnesol (III) and linalool (IV) with MPPA by carrying out the reaction at different pH.

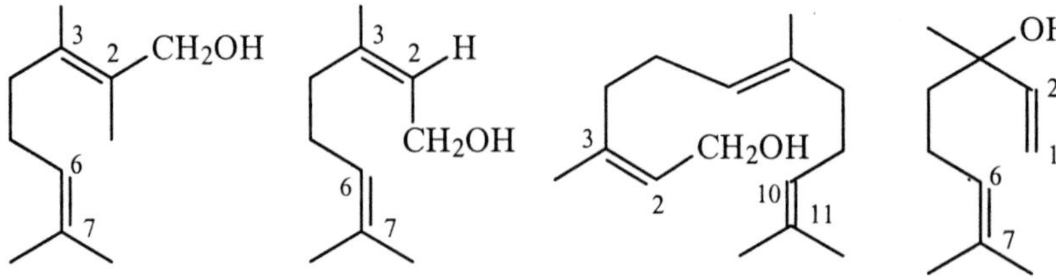

Geraniol
I

Nerol
II

Farnesol
III

Linalool
IV

It has been found[98] that 2,3-epoxidation takes place in I, II and III with MPPA in aqueous medium at pH 12.5 and in about 90% yields. In I, II and IV

6,7-epoxidation takes place in aqueous medium at pH 8.3 in about 60-90% yields. The 10,11-epoxidation takes place at pH 12 in 88% yield. In case of linalool (IV), 1,2-epoxidation does not take place. It is appropriate to state that 6,7-epoxidation of geraniol (I) has been reported earlier with t-C_4H_9OOH/VO (acac)$_2$ in benzene (refluxing) and 2,3-epoxidation achieved by using m-chloroperbenzoic acid.[99]

It is interesting to note that attempts to epoxidise linalool and its analogues with MPPA in organic solvents give the tetrahydrofuran and tetrahydropyran derivatives.

Epoxidation of α,β-unsaturated carbonyl compounds in aqueous media can be achieved traditionally by Michael reaction with alkaline hydroperoxides.[100] It has been reported that the epoxidation of α,β-unsaturated carbonyl compound can be conveniently accomplished[101] by using sodium perborate (SPB) in aqueous media at pH 8 to give α,β-epoxyketones (Scheme 63).

The epoxidation of α,β-unsaturated carbonyl compounds with hydrogen peroxide under basic biphase condition, known as the Weitz-Scheffer epoxidation (Scheme 64)[102] is a very convenient and efficient method for giving the epoxides.

The reaction (Scheme 64) has been used for the epoxidation of a number of α,β-unsaturated aldehydes, ketones, nitriles, esters and sulfones etc.

The epoxidation of α,β-unsaturated carboxylic acid is difficult. However, the epoxidation can be achieved with H_2O_2 in presence of Na_2WO_4 at pH 5.8-6.8 (Scheme 65).

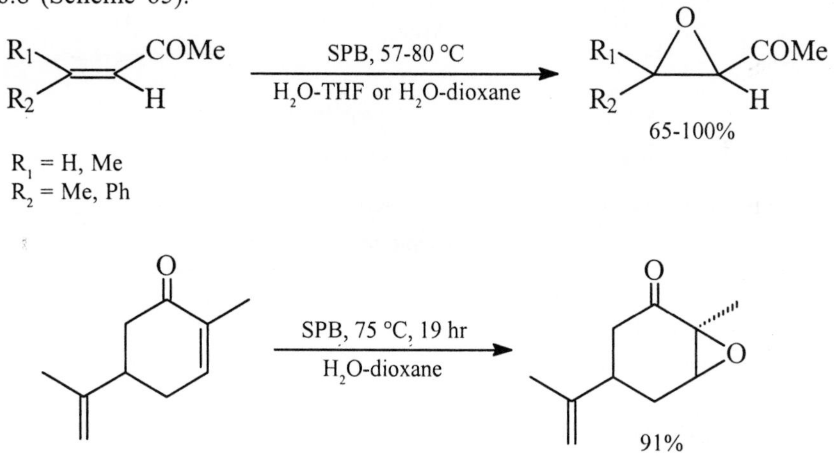

R$_1$ = H, Me
R$_2$ = Me, Ph

Scheme 63

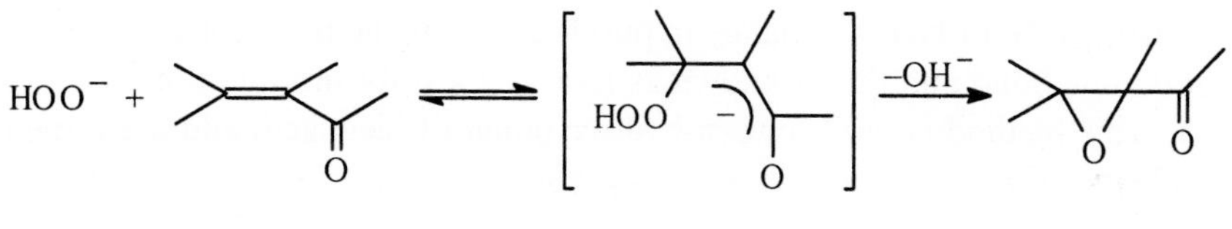

Scheme 64

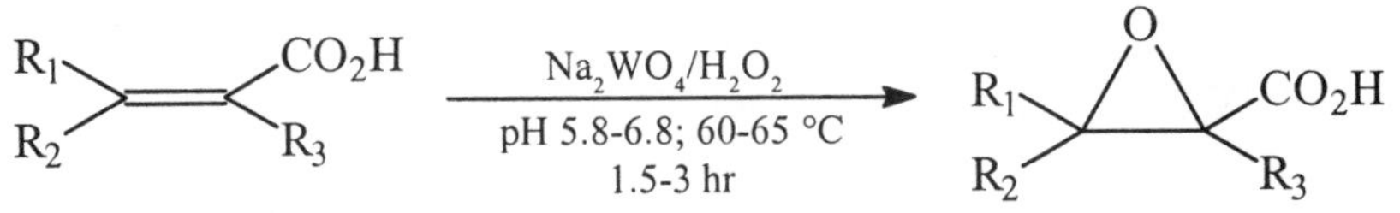

R_1, R_2, R_3 = H, Me, Br

Scheme 65

The above reaction is known as Payne's reaction[103] using the modified procedure of sharpless.[104]

Alternatively the above epoxidation of α,β-unsaturated carboxylic acids can be achieved by using ozone-acetone system and buffering the reaction with $NaHCO_3$[105] in 75-80% yield.

Epoxidation of fumaric acid can be achieved by using ozone in water at neutral pH in quantitative yield.[106]

The epoxidation of chalkones with NaOCl (commercially available) in water suspension proceeded very efficiently[107] in presence of a PTC hexadecyltrimethylammonium bromide in excellent yields (50-100%) (Scheme 66).

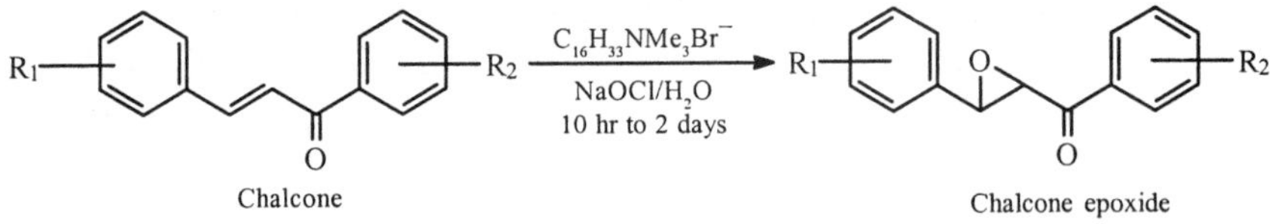

Chalcone   Chalcone epoxide

$R_1 = R_2 = H$ R_1 = p-MeO; R_2 = H
R_1 = p-Br; R_2 = H R_1 = p-Me; R_2 = H
R_1 = H; R_2 = p-Br R_1 = H; R_2 = p-Me
R_1 = m-Me; R_2 = H $R_1 = R_2$ = p-Cl
R_1 = p-Cl; R_2 = H $R_1 = R_2$ = p-Me

Scheme 66

12.14.2 Dihydroxylation
In case of alkens, one can get either syn- or anti-dihydroxylation.

12.14.2.1 Syn-Dihydroxylation
One of the earliest known method of syn-dihydroxylation of alkenes is by treatment with dilute $KMnO_4$ solution in presence of sodium hydroxide (Scheme 67). In fact the change in purple colour is the basis for the presence of double bond and this is known as Baeyer's test for unsaturation.

This method is used for syn-hydroxylation of oleic acid and norbornene (Scheme 68).

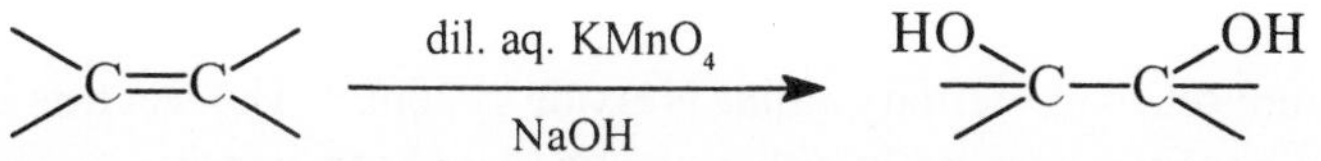

Scheme 67

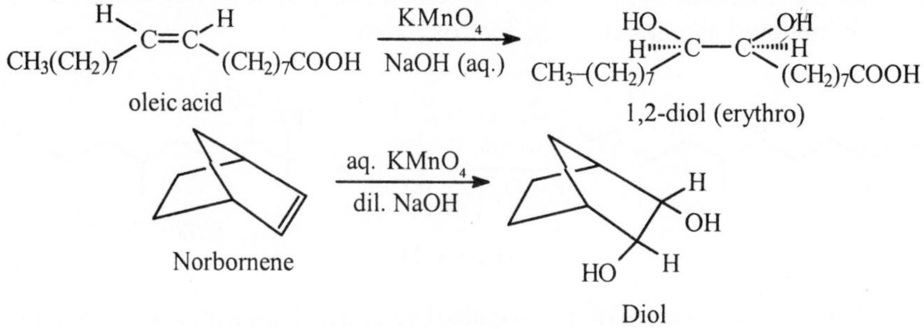

oleic acid

1,2-diol (erythro)

Norbornene

Diol

Scheme 68

There are numerous other examples of syn-dihydroxylation of alkenes in the literature.

Osmium tetroxide in dry organic solvent[108] was subsequently used for syn-dihydroxylations of alkenes. In fact, only a catalytic amount of osmium tetraoxide was needed; the reaction was done in presence of chlorate salt as primary oxidant. The reaction is normally carried out in water-tetrahydrofuran solvent mixture (Scheme 69). Silver or barium chlorate gave better yields.

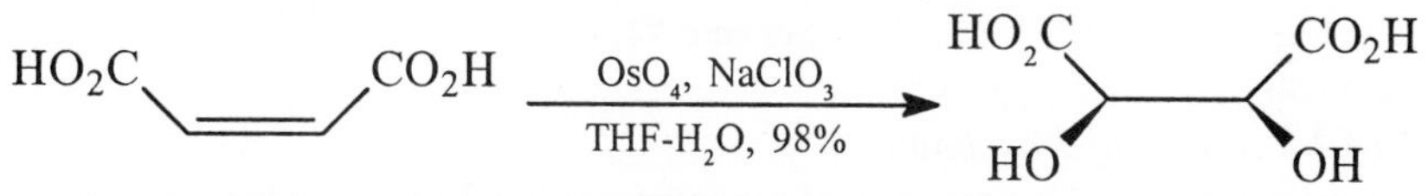

Scheme 69

The syn-hydroxylation of alkenes can also be effected by hydrogen peroxide in presence of catalytic amount of OsO_4. This procedure was used earlier in solvents such as acetone or diethyl ether.[109] By this method allyl alcohol is quantitatively hydroxylated in water (Scheme-70).[110]

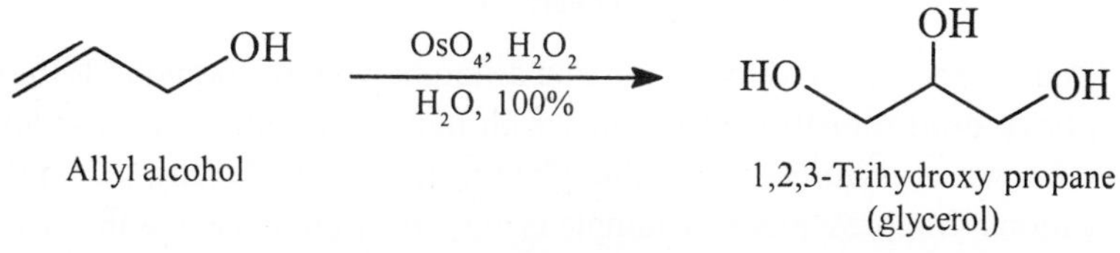

Allyl alcohol

1,2,3-Trihydroxy propane (glycerol)

Scheme 70

Following are given some other methods used for syn-dihydroxylation of alkenes:

(i) Osmium-tetroxide-tertiary amine N-oxide system.[111] The reaction is carried out in aqueous acetone in either one or two phase system.

(ii) $K_3Fe(CN)_6$ in presence of K_2CO_3 in aqueous or tertiary butyl alcohol provides a powerful system for the osmium-catalysed dihydroxylation of alkenes (Scheme 71).[112] Using this method even alkene having low reactivity or hindered alkenes could be hydroxylated.

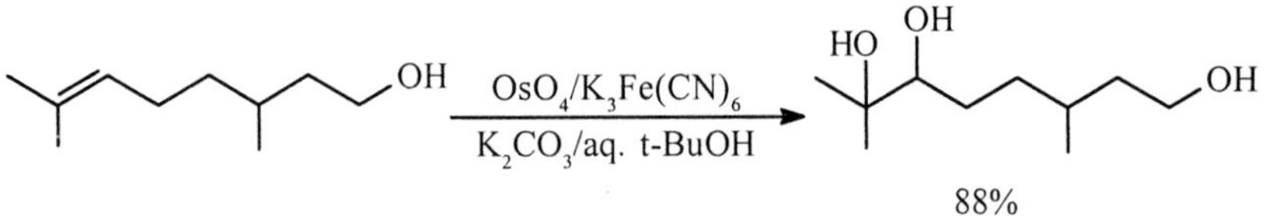

Scheme 71

Syn-hydroxylation of olefins has also been carried out with $KMnO_4$ solution using a PTC catalyst under alkaline conditions. Thus, under alkaline conditions, cyclooctene gives[113] 50% yield of cis 1,2-cyclooctane diol compared to an yield of about 7% by the classical technique (Scheme 72).

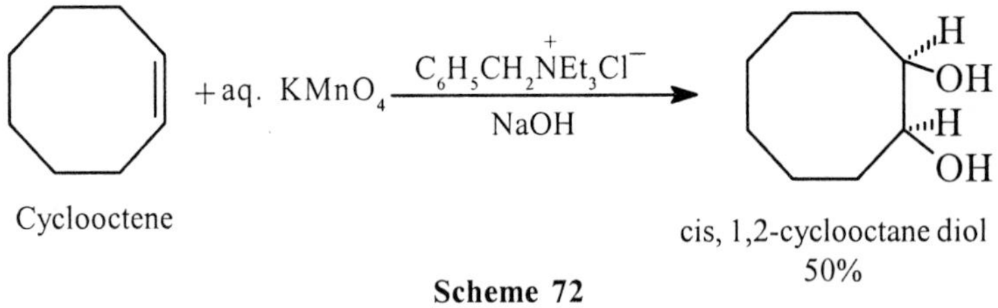

Cyclooctene

cis, 1,2-cyclooctane diol
50%

Scheme 72

12.14.2.2 Anti Dihydroxylation

Hydrogen peroxide in presence of tungsten oxide (WO_3) or selenium dioxide (SeO_2) react with alkene to give anti-dihydroxylation products (Scheme 73).[114]

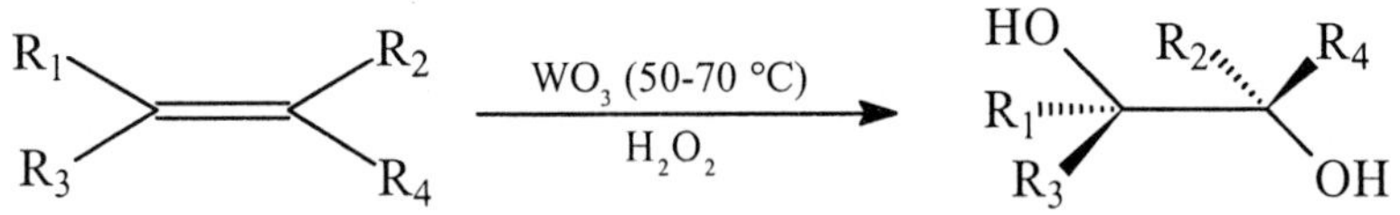

Scheme 73

Using the Sharpless dihydroxylation different types of compounds (having C=C) have been transformed to diols with high enantiomeric-excess levels. This is known as asymmetric dihydroxylation and has a wide range of synthetic applications. A representative example is dihydroxylation used as the key step for the synthesis of squalestatin (Scheme 74).[115]

A one pot procedure for the antihydroxylation of the carbon-carbon double bond can be achieved as shown below (Scheme 75).[116]

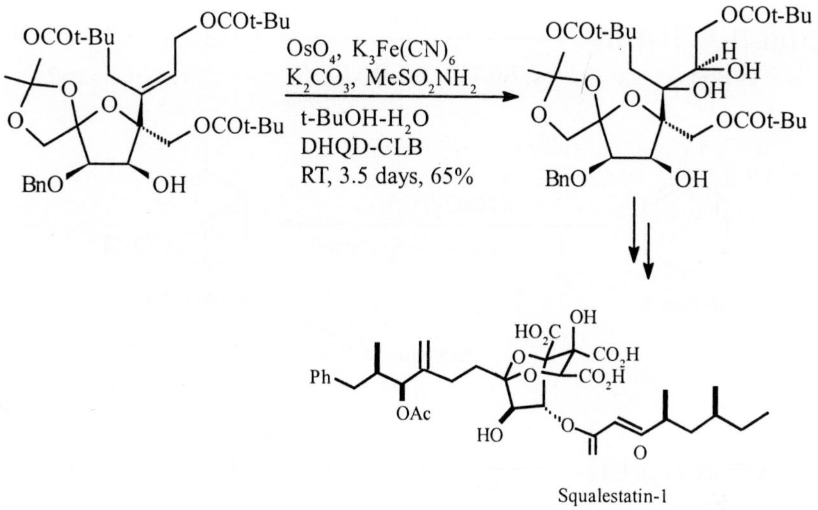

Scheme 74

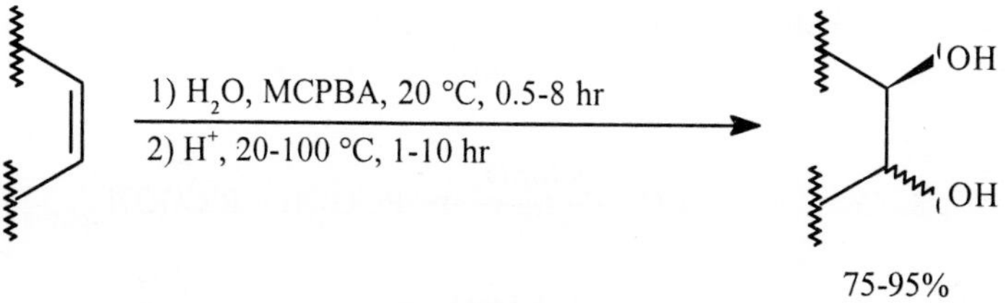

1) H$_2$O, MCPBA, 20 °C, 0.5-8 hr

2) H$^+$, 20-100 °C, 1-10 hr

75-95%

Scheme 75

12.14.3 Miscellaneous Oxidations in Aqueous Medium

12.14.3.1 Alkenes

The oxidation of alkenes with aqueous solution of KMnO$_4$ in presence of a phase transfer catalyst (e.g. CH$_3$(CH$_2$)$_{15}$N$^+$(CH$_3$)$_3$Cl$^-$) or a crown ether (e.g. 18-crown-6) gives 79% yield of the carboxylic acid. Some examples are given (Scheme 76).

An interesting example is the oxidation of n-octane to 1-octanol (Scheme 77) using *Psuedomonas oleovorans*.[117] This procedure is used for the commercial production of 1-octanol (>98% pure).[118]

$$CH_3(CH_2)_5CH=CH_2 \xrightarrow[\substack{CH_3(CH_2)_{15}\overset{+}{N}(CH_3)_3Cl^-}]{aq.\ KMnO_4} CH_3(CH_2)_4CH_2COOH$$

1-octene   Heptanoic acid

KMnO₄/H₂O

Dicyclohexano-18-crown-6

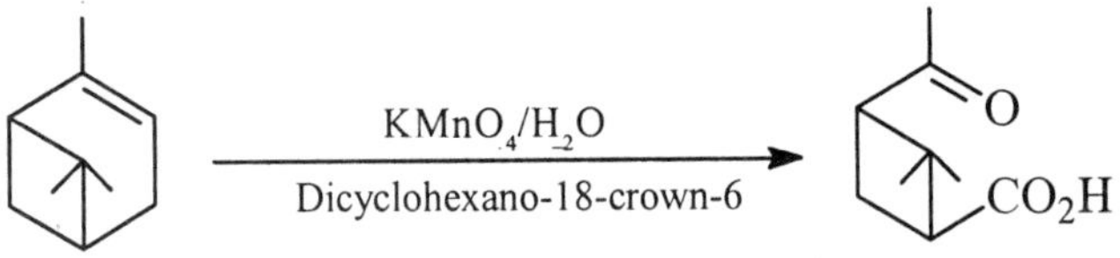

α-Pinene   cis-pinonic acid

Scheme 76

$$CH_3(CH_2)_6CH_3 \xrightarrow[\substack{oleovorans\ H_2O}]{Pseudomonas} CH_3(CH_2)_6CH_2OH$$

n-Octane   1-Octanol

Scheme 77

12.14.3.2 Alkynes

Alkynes can be oxidised with KMnO₄ in aqueous medium to give a mixture of carboxylic acids. Some examples are given below (Scheme 78).

$$R-C\equiv C-R' + 4[O] \xrightarrow{KMnO_4} RCOOH + R'COOH$$

$$CH_3(CH_2)_7C\equiv C(CH_2)_7COOH \xrightarrow[pH\ 7.5]{aq.\ KMnO_4} CH_3(CH_2)_7-\overset{\overset{O}{\|}}{C}-\overset{\overset{O}{\|}}{C}-(CH_2)_7-COOH$$

$$\xrightarrow{alk.\ KMnO_4} CH_3(CH_2)_7COOH + HOOC(CH_2)_7COOH$$

Scheme 78

12.14.3.3 Oxidation of Aromatic Side Chains and Aromatic Ring System

Some examples are given (Scheme 79).

12.14.3.4 Oxidation of Aldehydes and Ketones

A number of procedures are available for the oxidation of aldehydes to the corresponding carboxylic acids in aqueous and organic media.[119]

Aromatic aldehydes have been conveniently oxidized by aqueous performic acid obtained by addition of H₂O₂ to HCOOH at low temperature (0-4 °C).[120]

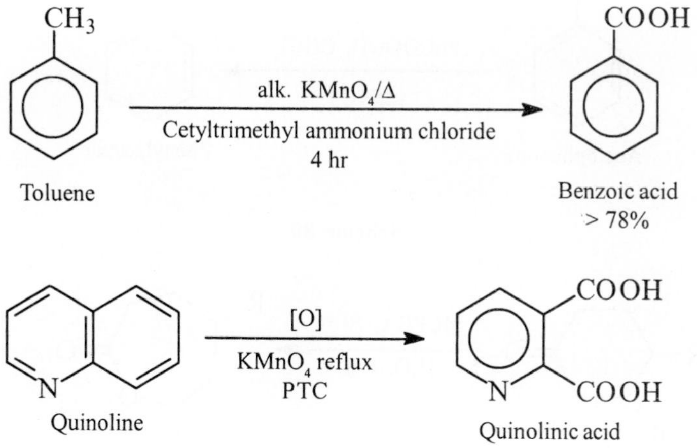

Scheme 79

The carboxylic acids precipitate out of the reaction mixture and can be isolated by filtration. Even the hetroaromatic aldehydes like formyl pyridines, formyl quinolines and formylazaindoles can be oxidised by the above procedure to the corresponding carboxylic acids; in this procedure, the formation of N-oxides is avoided.

Chemoselective oxidation of formyl group in presence of other oxidizable groups can be carried out in aqueous media in presence of a surfactant. For example, 4-(methylthio)benzaldehyde is quantitatively oxidised to 4-(methylthio)benzoic acid with TBHP in a basic aqueous medium in presence of cetyltrimethyl ammonium sulphate.[121]

Aromatic aldehydes having hydroxyl group in ortho or para position to the formyl groups can be oxidised with alkaline H_2O_2 (Dakin reaction) in low yields.[122] This reaction has been recently carried out in high yields using sodium percarbonate (SPC; Na_2CO_3, 1.5 H_2O_2) in H_2O-THF under ultrasonic irradiation.[123] Using this procedure following aldehydes have been oxidised in 85-95% yields: o-hydroxybenzaldehyde; p-hydroxybenzaldehyde; 2-hydroxy-4-methoxybenzaldehyde, 2-hydroxy-3-methoxybenzaldehyde and 3-methoxy-4-hydroxybenzaldehyde.

The Baeyer-Villiger oxidation[124] is well known for the conversion of aromatic ketones (e.g., acetophenone) into the corresponding esters (Scheme 80).

The common peracids used are perbenzoic acid, performic acid and m-chloroperbenzoic acid. The reaction is generally carried out in organic solvents. The Baeyer-Villiger oxidation of ketones has been satisfactorily carried out in aqueous heterogeneous medium with MCPBA at room temperature.[125] Some examples using the above methodology are given (Scheme 81).

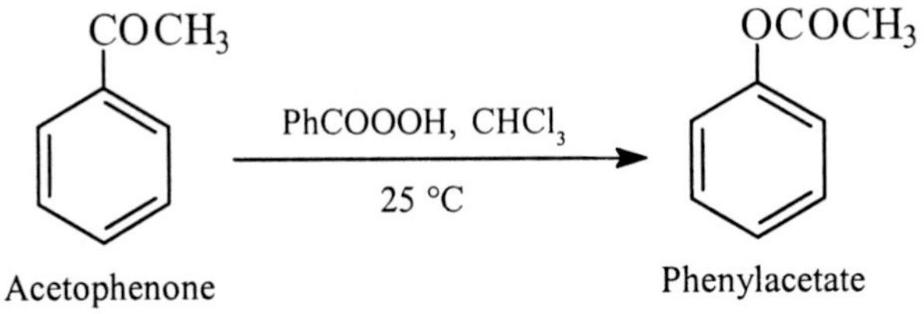

Scheme 80

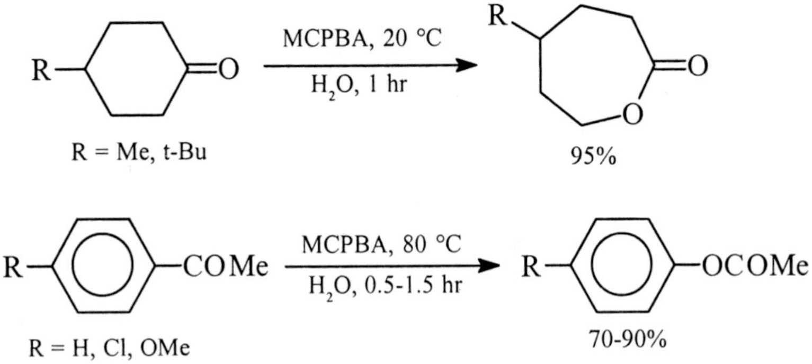

R = Me, t-Bu   95%

R = H, Cl, OMe   70-90%

Scheme 81

Using the above procedure, ketones, which are reactive (e.g. anthrone which usually gives anthraquinone) and ketones, which are unreactive or give the expected lactones in organic solvents with difficulty[126] can also be oxidised (Scheme 82).

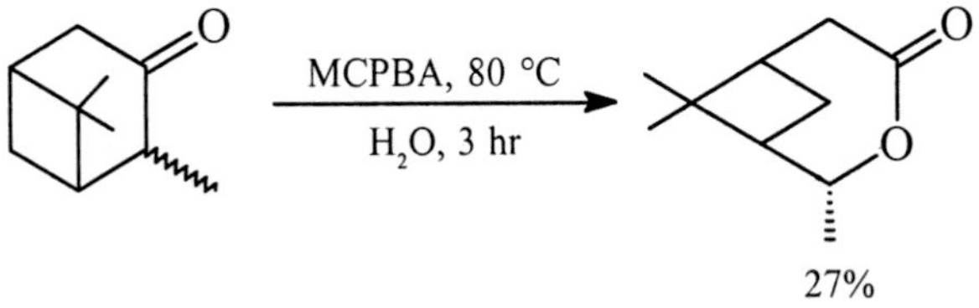

27%

Scheme 82

Some Baeyer-Villiger oxidation of ketones with m-chloroperbenzoic acid proceed much faster in the solid state than in solution. In this method, a mixture of powdered ketone and 2-mole equivalent of m-chloroperbenzoic acid is kept at room temperature to give the product.[127] The yields obtained in solid state are much better than in $CHCl_3$. Some representative examples are given as follows (Scheme 83).

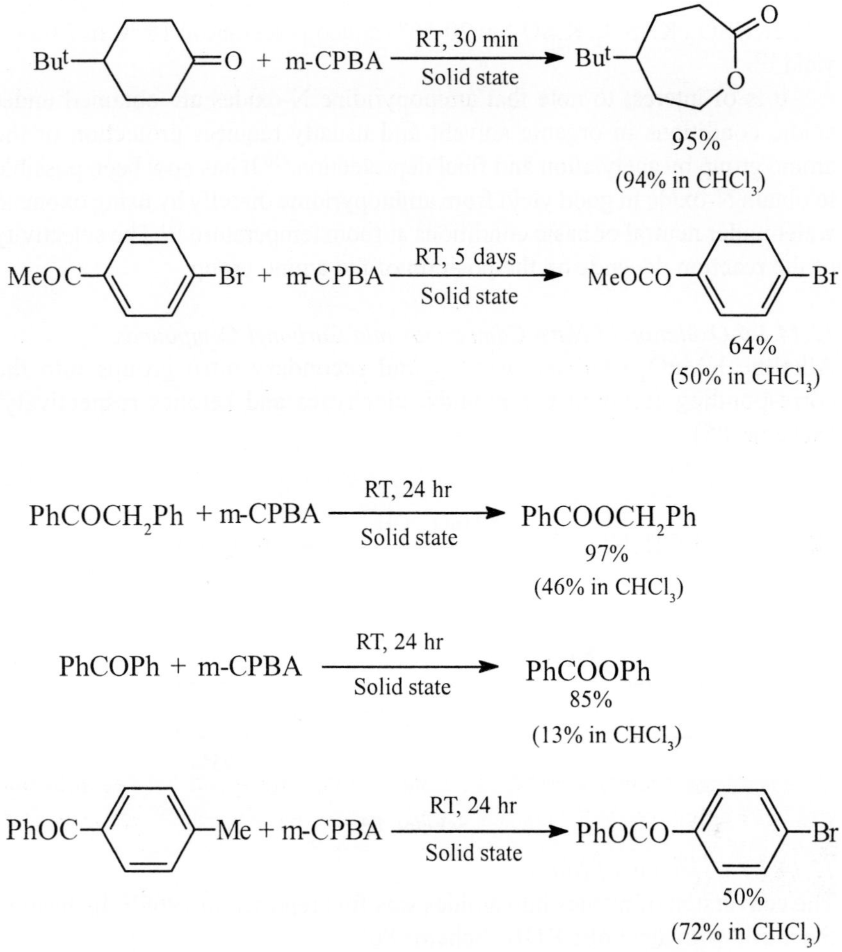

$$\text{But}-\!\!\bigcirc\!\!=\!O + \text{m-CPBA} \xrightarrow[\text{Solid state}]{\text{RT, 30 min}} \text{But}-\!\!\bigcirc\!\!\!\!\bigcirc$$

95%

(94% in $CHCl_3$)

$$\text{MeOC}-\!\!\bigcirc\!\!-\text{Br} + \text{m-CPBA} \xrightarrow[\text{Solid state}]{\text{RT, 5 days}} \text{MeOCO}-\!\!\bigcirc\!\!-\text{Br}$$

64%

(50% in $CHCl_3$)

$$\text{PhCOCH}_2\text{Ph} + \text{m-CPBA} \xrightarrow[\text{Solid state}]{\text{RT, 24 hr}} \text{PhCOOCH}_2\text{Ph}$$

97%

(46% in $CHCl_3$)

$$\text{PhCOPh} + \text{m-CPBA} \xrightarrow[\text{Solid state}]{\text{RT, 24 hr}} \text{PhCOOPh}$$

85%

(13% in $CHCl_3$)

$$\text{PhOC}-\!\!\bigcirc\!\!-\text{Me} + \text{m-CPBA} \xrightarrow[\text{Solid state}]{\text{RT, 24 hr}} \text{PhOCO}-\!\!\bigcirc\!\!-\text{Br}$$

50%

(72% in $CHCl_3$)

Scheme 83

12.14.3.5 Oxidation of Amines into Nitro Compounds

Alkaline $KMnO_4$ oxidises tertiary alkyl amines into nitro compounds (Scheme 84).[128]

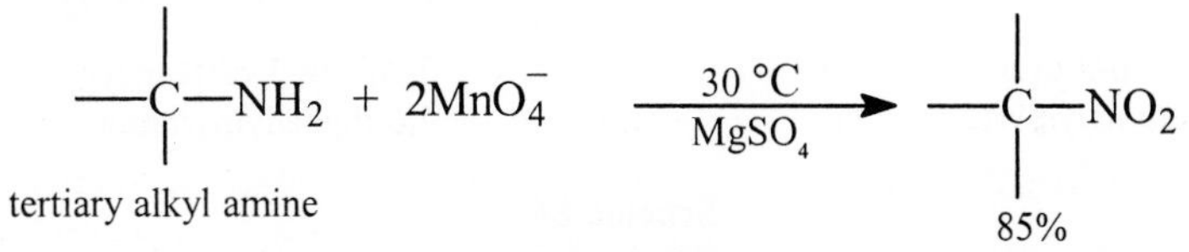

$$-\!\!\overset{|}{\underset{|}{C}}\!-\text{NH}_2 + 2\text{MnO}_4^- \xrightarrow[\text{MgSO}_4]{30\ ^\circ C} -\!\!\overset{|}{\underset{|}{C}}\!-\text{NO}_2$$

tertiary alkyl amine

85%

Scheme 84

Primary and secondary alkyl amines remain uneffected under above conditions.
Aromatic amines containing a carboxylic or alcoholic groups can be oxidised

to nitro compounds by oxone (potassium hydrogen peroxymonosulfate triple salt, $2KHSO_3$, $KHSO_4$, K_2SO_4) in 20-50% aqueous acetone at 18 °C in 73-84% yield.[129]

It is of interest to note that aminopyridine N-oxides are obtained under acidic conditions in organic solvent and usually requires protection of the amino group by acetylation and final deprotection.[130] It has now been possible to obtain N-oxide in good yield from aminopyridine directly by using oxone in water under neutral or basic conditions at room temperature.[131] The selectivity of the reaction depends on the position of the amino group.

12.14.3.6 Oxidation of Nitro Compounds into Carbonyl Compounds
Alkaline $KMnO_4$ oxidises primary and secondary nitro groups into the corresponding carbonyl compounds (aldehydes and ketones respectively) (Scheme 85).

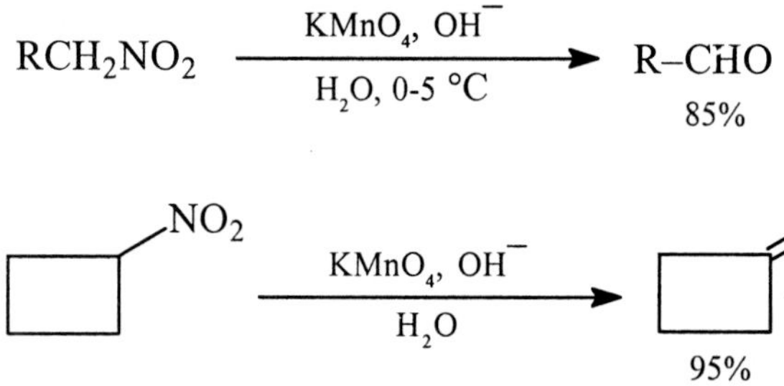

$$RCH_2NO_2 \xrightarrow[\text{H}_2\text{O, 0-5 °C}]{\text{KMnO}_4, \text{ OH}^-} R{-}CHO$$
85%

Scheme 85

12.14.3.7 Oxidation of Nitriles
The conversion of nitriles into amides was first reported in 1968[132] by heating the nitrile with alcoholic KOH (Scheme 86).

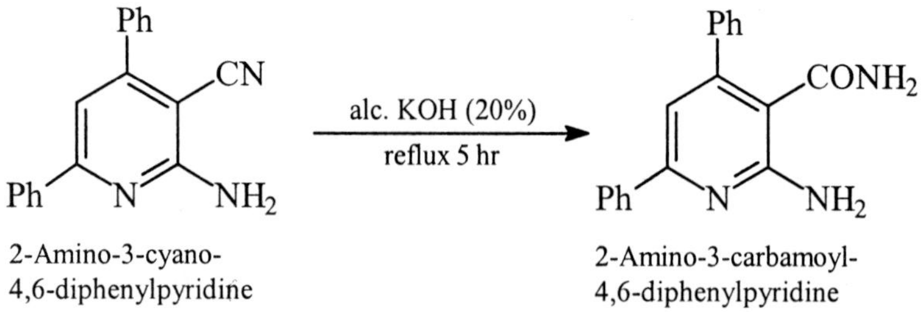

2-Amino-3-cyano-
4,6-diphenylpyridine

2-Amino-3-carbamoyl-
4,6-diphenylpyridine

Scheme 86

It is now well known that the conversion of nitriles into amides can be carried out under a variety of conditions in presence of metal catalyst.[133]

A convenient method[134] has been developed. It involves in using urea-hydrogen peroxide adduct (UHP, H_2NCONH_2, H_2O_2) in presence of catalytic amount of K_2CO_3 in water-acetone at room temperature (Scheme 87).

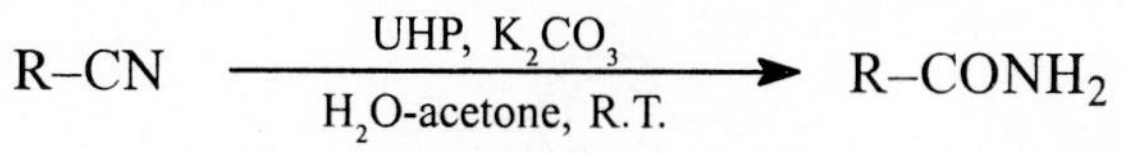

$$R-CN \quad \xrightarrow[\text{H}_2\text{O-acetone, R.T.}]{\text{UHP, K}_2\text{CO}_3} \quad R-CONH_2$$

Scheme 87

Using the above method following nitriles have been converted into corresponding amides in 85-95% yield: benzonitrile, methyl cyanide and chloromethyl cyanide.

Nitriles can also be converted into amides by using sodium perborate (SPB; $NaBO_3$, nH_2O, n = 1 to 4) in aqueous media such as H_2O-MeOH[135]; H_2O-acetone[136] and H_2O-dioxan[137]. An interesting application of this reaction is the synthesis of quinazolin-4-(3H)-ones[138] (Scheme-88).

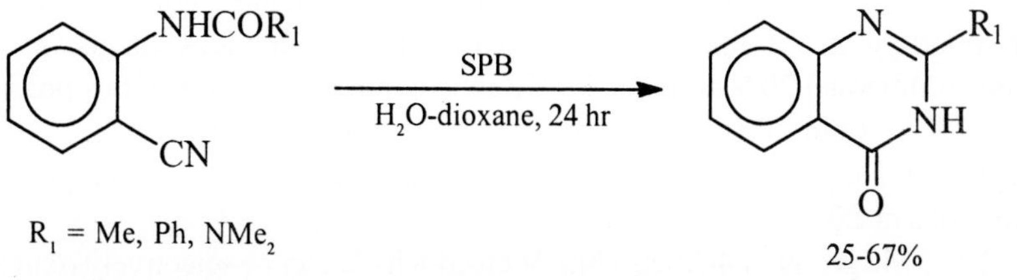

R_1 = Me, Ph, NMe$_2$

25-67%

Scheme 88

The quinazolin-4-(3H)-ones are interesting systems to build pharmaceutical compounds.

It has been reported[139] that the CN group of 4-(methylthio)benzonitrile is quantitatively and selectively oxidised to amide by tertiary butyl hydroperoxide (TBHP) in strong alkaline aqueous medium in presence of cetyltrimethyl ammonium sulfate [$(CTA)_2SO_4$] (Scheme 89). TBHP does not oxidise the CN group at pH 7 (even at 100 °C), however in the absence of $(CTA)_2SO_4$, only the methylsulfenyl group is oxidised to methylsulfinyl. But under basic conditions, TBHP converts both groups into amide and sulfonyl groups respectively (Scheme 89).

12.14.3.8 Oxidation of Sulphides

A number of reagents (e.g. H_2O_2/acetic acid) are available for the oxidation of sulphides to sulfoxides and sulphones. These methods are useful only for small scale preparations. On a large scale, an oxidant like oxone in aqueous acetone, buffered to pH 7.8-8.0 with sodium bicarbonate is used.[140] This procedure is economical and environmentally benin. The formation of the oxidation products, viz. sulfoxides or sulphones depend on the equivalent of oxone used, temperature and reaction time. In aqueous medium at pH 6-7 (buffered with phosphate), the reaction is very fast and excellent conversions to sulfoxides and sulphones are obtained.[141]

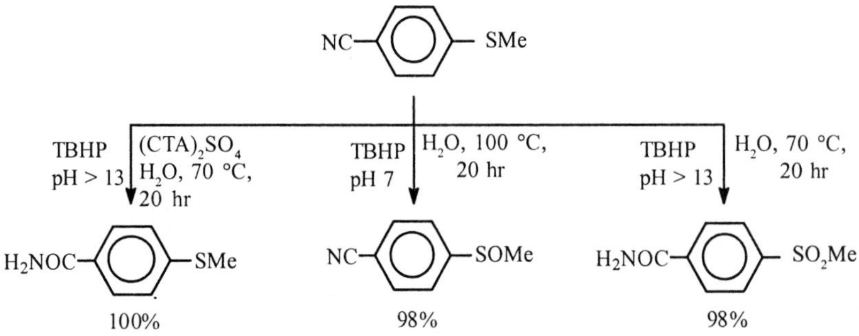

Scheme 89

Another cheaply available industrial chemical, sodium perborate (SPB) in aqueous methanolic sodium hydroxide oxidises sulphides into sulphones in very good yield.[142] Sulphides can be oxidised to sulphoxides exclusively by using commercial 70% aqueous TBHP in water in the heterogeneous phase at 20-70 °C.[143] Using this procedure some of the sulphides like Et_2S, PhSMe, PhSPh, p-OHC_6H_4SMe can be oxidised quantitatively into the corresponding sulfoxides at 20 °C.

The SMe group of 4-(methylthio)benzaldehyde can be selectively oxidised to the sulfinyl group in water at 70 °C at pH 7 with tertiary butyl hydroperoxide (TBHP) (Scheme-90).[144] However, under strong basic condition both the CHO and SMe groups are oxidised to –COOH and –SO_2Me respectively. By using MCPBA under basic conditions, oxidation of –CHO group is prevented and SMe group is oxidised to SO_2Me in good yield.

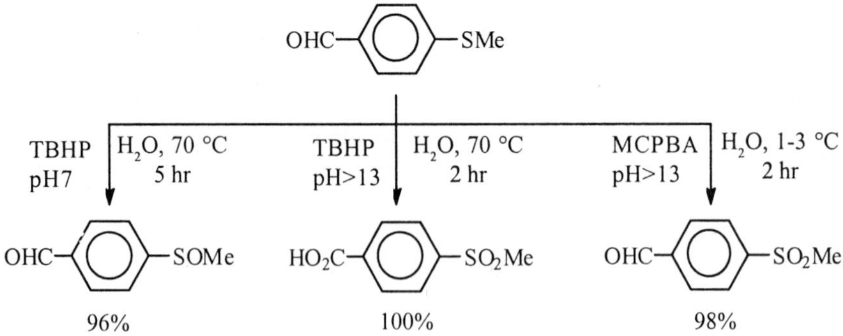

Scheme 90

12.14.3.9 Oxidations with Hypochlorite

Hypochlorite is a well known oxidizing agent in the haloform reaction for the oxidation of methyl ketones to carboxylic acids. It has been shown[145] that the hypochlorite anion can be transferred into organic solutions by PTC (quaternary cations). Some of the applications of this technique are given as follows (Scheme 91).

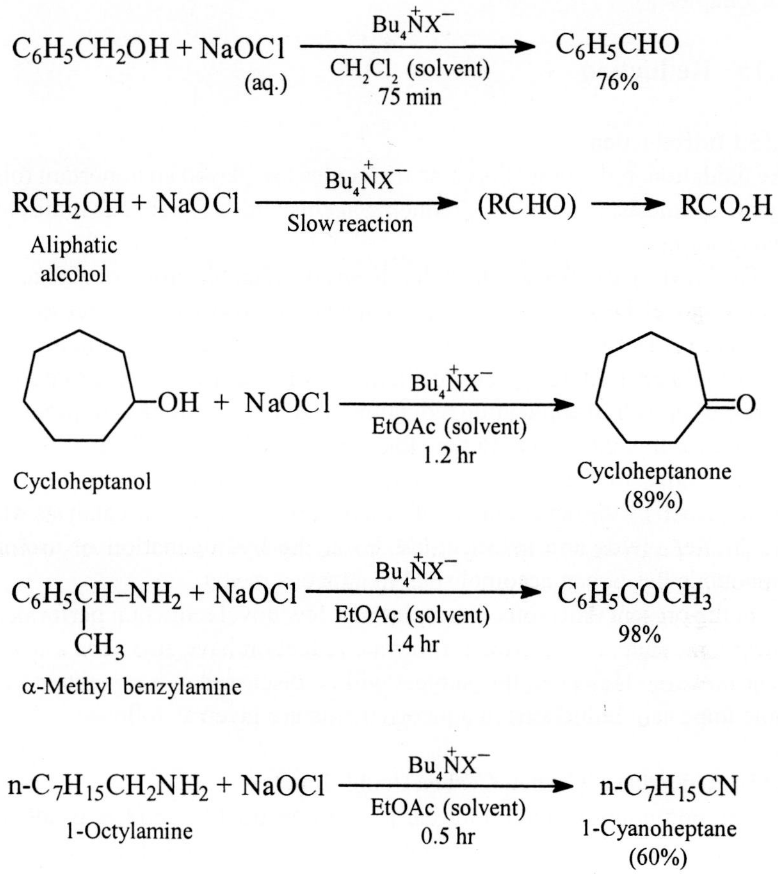

$$C_6H_5CH_2OH + NaOCl \xrightarrow[\substack{CH_2Cl_2 \text{ (solvent)} \\ 75 \text{ min}}]{Bu_4\overset{+}{N}X^-} C_6H_5CHO$$

(aq.)   76%

$$RCH_2OH + NaOCl \xrightarrow[\text{Slow reaction}]{Bu_4\overset{+}{N}X^-} (RCHO) \longrightarrow RCO_2H$$

Aliphatic
alcohol

Cycloheptanol $\xrightarrow[\substack{\text{EtOAc (solvent)} \\ 1.2 \text{ hr}}]{Bu_4\overset{+}{N}X^-}$ Cycloheptanone (89%)

$$C_6H_5\underset{\underset{CH_3}{|}}{CH}-NH_2 + NaOCl \xrightarrow[\substack{\text{EtOAc (solvent)} \\ 1.4 \text{ hr}}]{Bu_4\overset{+}{N}X^-} C_6H_5COCH_3$$

α-Methyl benzylamine   98%

$$n\text{-}C_7H_{15}CH_2NH_2 + NaOCl \xrightarrow[\substack{\text{EtOAc (solvent)} \\ 0.5 \text{ hr}}]{Bu_4\overset{+}{N}X^-} n\text{-}C_7H_{15}CN$$

1-Octylamine   1-Cyanoheptane (60%)

Scheme 91

12.14.3.10 Oxidation with Ferricyanide

Potassium ferri cyanide oxidises 1,2-disubstituted hydrazines in presence of 2,4,6-triphenyl phenol (as PTC) in presence of NaOH[146] to give 1,2-disubstituted azo compounds (Scheme 92).

$$RNHNHR' + K_3Fe(CN)_6 \xrightarrow[\text{NaOH}]{\text{TPP}} RN=NR$$
$$63\text{-}98\%$$

Scheme 92

Besides what has been mentioned above, a number of oxidations can be performed in aqueous phase in the presence of a phase transfer catalyst (see Chapter 8).

12.15　Reduction

12.15.1 Introduction

Like oxidation, reduction of organic molecules has played an important role in organic synthesis. A number of reducing agents with different substrates have been described.[147]

During the past 10 years, there has been considerable progress with respect to the types of bonds which can be reduced and also with respect to regio- and stereo-selectivity of the reduction processes. The only reducing agent which could be used in aqueous medium is sodium borohydride. From a point of view of industrial application reduction in aqueous medium is of paramount importance. It is interesting to note that hydride reductions which at one time seemed impossible to be carried out in aqueous medium have now been accomplished by the development of a number of water-soluble catalysts which give higher yields and selectivities. Even the hydrogenation of aromatic compounds have been accomplished in aqueous media.

In the present unit, some examples of a few novel reduction performed in aqueous medium are described. Enzymic reduction have also been known to occur in water. However, this subject will be discussed in a separate section. Some important reductions in aqueous media are given as follows:

12.15.2 Reduction of Carbon-Carbon Double Bonds

Organic compounds containing carbon-carbon double bond (e.g. alkenes) can be reduced to the corresponding saturated compounds (eg., alkanes) by PtO_2/H_2, Pd/H_2 or Raney Ni/H_2. Even diimide is useful for reducing carbon-carbon double bond in compounds like p-nitrocinnamic acid, cyclohexene 1,2-dicarboxylic acid and oleic acid.

The reduction of carbon-carbon double bonds by the use of water soluble hydrogenation catalysts has received attention.[148] Thus, hydrogenation of 2-acetamidoacrylates with hydrogen at room temperature in water in the presence of water soluble chiral Rh(I) and Ru(II) complexes with (R)-BINAP (SO_3Na) [BINAP is 2,2'-bis(diphenylphosphino-1,1'-binaphthyl] (Scheme 93).[149]

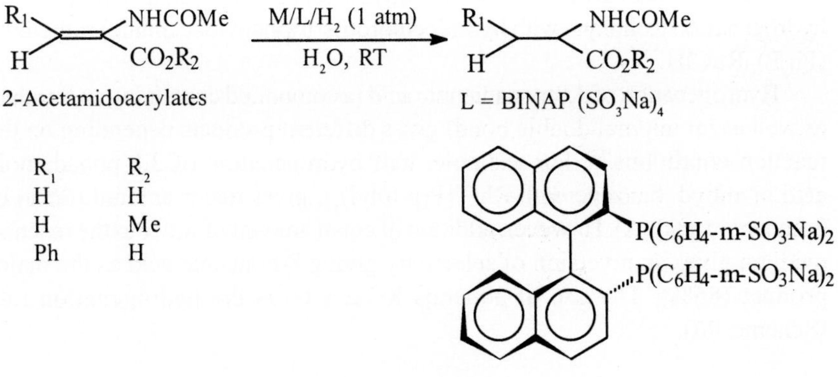

$$R_1 \diagdown \diagup NHCOMe$$
$$H \diagup \diagdown CO_2R_2$$

2-Acetamidoacrylates

$\xrightarrow[\text{H}_2\text{O, RT}]{\text{M/L/H}_2 \text{ (1 atm)}}$

$$R_1 \diagdown \diagup NHCOMe$$
$$H \diagup \diagdown CO_2R_2$$

L = BINAP (SO$_3$Na)$_4$

R$_1$	R$_2$
H	H
H	Me
Ph	H

P(C$_6$H$_4$-m-SO$_3$Na)$_2$
P(C$_6$H$_4$-m-SO$_3$Na)$_2$

Scheme 93

Ruthenium complexes are found to be more stable than the corresponding rhodium analogue; the ee of the final reduced product is found to be 68-88%.

The carbon-carbon double bond of α,β-unsaturated carbonyl compounds is conveniently reduced by using Zn/NiCl$_2$ (9:1) in 2-methoxyethanol (ME)-water system (Scheme 94).[150] Sonication increases the yield.

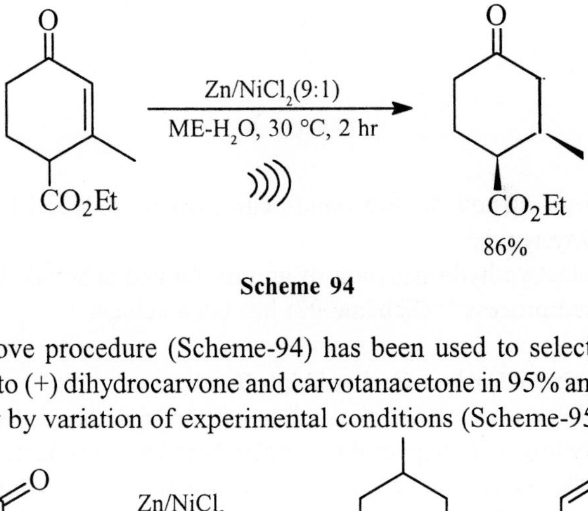

$\xrightarrow[\text{ME-H}_2\text{O, 30 °C, 2 hr}]{\text{Zn/NiCl}_2(9:1)}$

))))

86%

Scheme 94

The above procedure (Scheme-94) has been used to selectively reduce (-)-carvone to (+) dihydrocarvone and carvotanacetone in 95% and 83% yields respectively by variation of experimental conditions (Scheme-95).[151]

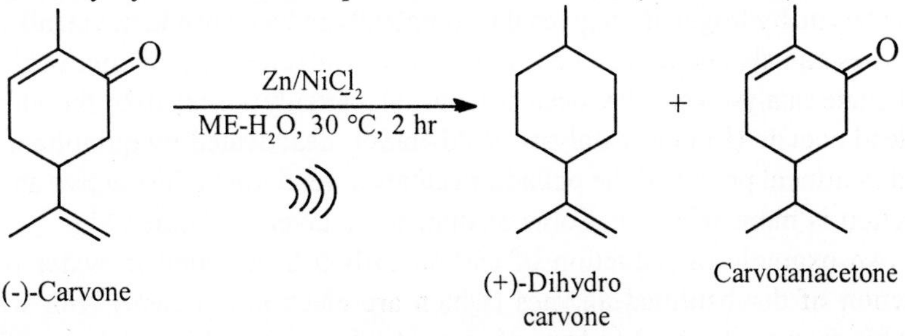

$\xrightarrow[\text{ME-H}_2\text{O, 30 °C, 2 hr}]{\text{Zn/NiCl}_2}$

))))

(-)-Carvone

(+)-Dihydro carvone

+

Carvotanacetone

Scheme 95

Carvone was earlier reduced to dihydrocarvone by using homogeneous hydrogenation technique with hydridochlorotris(triphenylphosphine) ruthenium $(Ph_3P)_3 RuClH$.[152]

Hydrogenation of 3,8-nonadienoic acid (a compound containing an terminal as well as an internal double bond) gives different products depending on the reaction conditions.[153] For example, half hydrogenation of 3,8-nonadienoic acid in anhyd. benzene with $RhCl[P(p-tolyl)_3]_3$ gives major amount (66%) of 3-nonenoic acid (A). However, addition of equal amount of water to the reaction medium gives an inversion of selectivity giving 8-nonenoic acid as the major product (85%). The use of aqueous KOH retards the hydrogenation rate (Scheme 96).

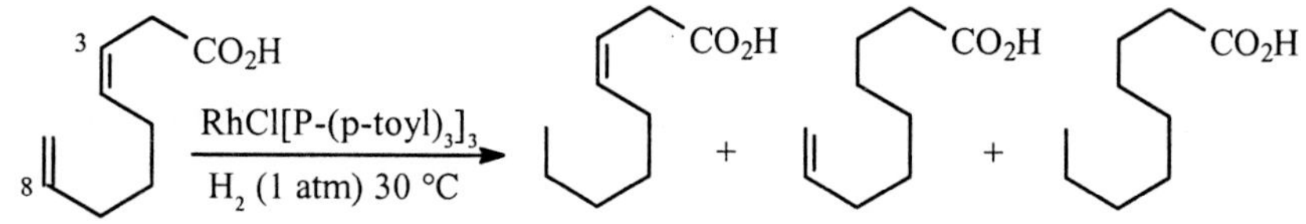

	A	B	C
	3-Nonenoic acid	8-Nonenoic acid	Decanoic acid (capric acid)
C_6H_6 (4 hr)	66%	10%	20%
C_6H_6-H_2O (2 hr)	0.7%	85%	8%
C_6H_6-KOH aq.(20 hr)	6%	18%	39%

Scheme 96

The carbon-carbon double bonds can also be reduced by samarium diiodide-H_2O system.[154]

Chemoselective hydrogenation of an unsaturated aldehyde by transition metal catalysed process[155] (Scheme 97) has been achieved.

12.15.3 Reduction of Carbon-Carbon Triple Bonds

It is very well documented that the carbon-carbon triple bonds (e.g., alkynes) on catalytic hydrogenation gives the completely reduced product, viz. alkanes. Alkynes can also be reduced partially to give z-alkenes by palladium-calcium carbonate catalyst which has been deactivated (partially poisoned) by the addition of lead acetate (Lindlar catalyst) or Pd-BaSO$_4$ deactivated by quinoline. The lead treatment poisoned the palladium catalyst, rendering it less active and the reaction is more selective. Some examples are given (Scheme 98).

An example of reduction of carbon-carbon triple bond in water is the reaction of disubstituted alkynes (which are electron deficient) with water-soluble monosulfonated and trisulfonated triphenylphosphine (Scheme 99).[156]

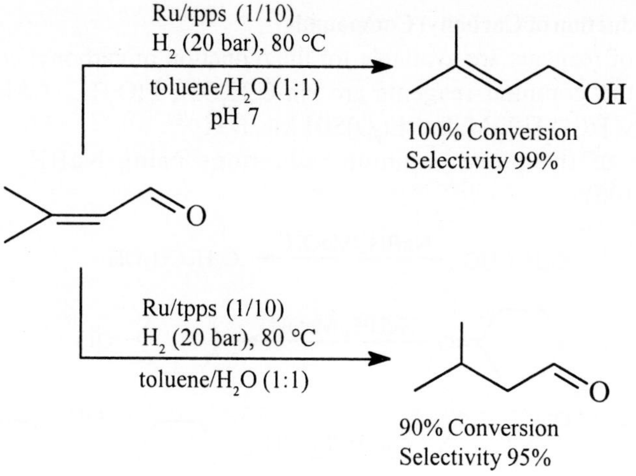

Ru/tpps (1/10)
H$_2$ (20 bar), 80 °C

toluene/H$_2$O (1:1)
pH 7

100% Conversion
Selectivity 99%

Ru/tpps (1/10)
H$_2$ (20 bar), 80 °C

toluene/H$_2$O (1:1)

90% Conversion
Selectivity 95%

Scheme 97

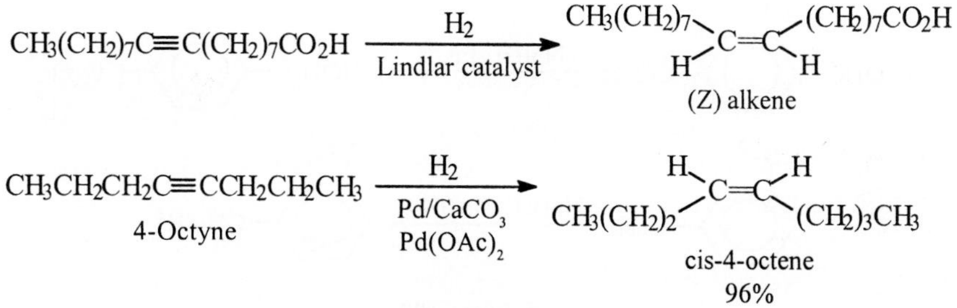

CH$_3$(CH$_2$)$_7$C≡C(CH$_2$)$_7$CO$_2$H $\xrightarrow[\text{Lindlar catalyst}]{\text{H}_2}$ CH$_3$(CH$_2$)$_7$ (CH$_2$)$_7$CO$_2$H

C=C

H   H

(Z) alkene

CH$_3$CH$_2$CH$_2$C≡CCH$_2$CH$_2$CH$_3$ $\xrightarrow[\substack{\text{Pd/CaCO}_3 \\ \text{Pd(OAc)}_2}]{\text{H}_2}$ CH$_3$(CH$_2$)$_2$ (CH$_2$)$_3$CH$_3$

4-Octyne

H   H

C=C

cis-4-octene
96%

Scheme 98

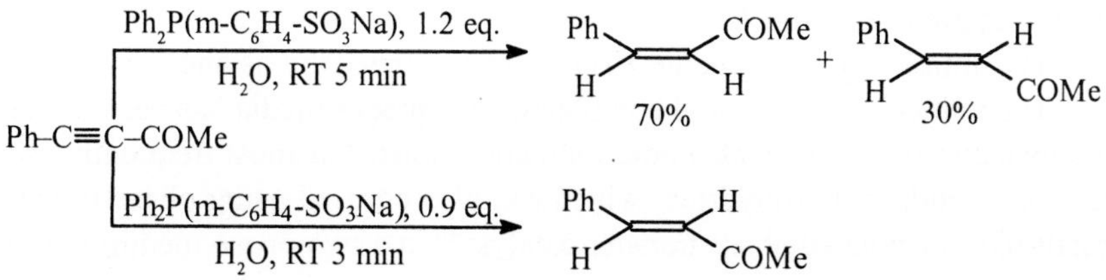

Ph$_2$P(m-C$_6$H$_4$-SO$_3$Na), 1.2 eq.

H$_2$O, RT 5 min

Ph   COMe   Ph   H

C=C   +   C=C

H   H   H   COMe

70%   30%

Ph–C≡C–COMe

Ph$_2$P(m-C$_6$H$_4$-SO$_3$Na), 0.9 eq.

H$_2$O, RT 3 min

Ph   H

C=C

H   COMe

Scheme 99

In the above procedure (Scheme 99), the water acts both as a solvent as well as a reactant, and the amount of phosphine controls the cis/trans ratio of the formed alkenes since it catalyses the cis-trans olefin isomerisation.

12.15.4 Reduction of Carbonyl Compounds

A variety of reagents are available for the reduction of carbonyl compounds. Some of the common reagents are $Na-C_2H_5OH$, PtO_2/H_2, LAH, $NaBH_4$, Na_2BH_3CN, $HCO_2H/EtMgBr$, $(Et_2O)SiH.Me$, B_2H_6.[157]

Some of the more common reductions using $NaBH_4$ are given (Scheme 100).

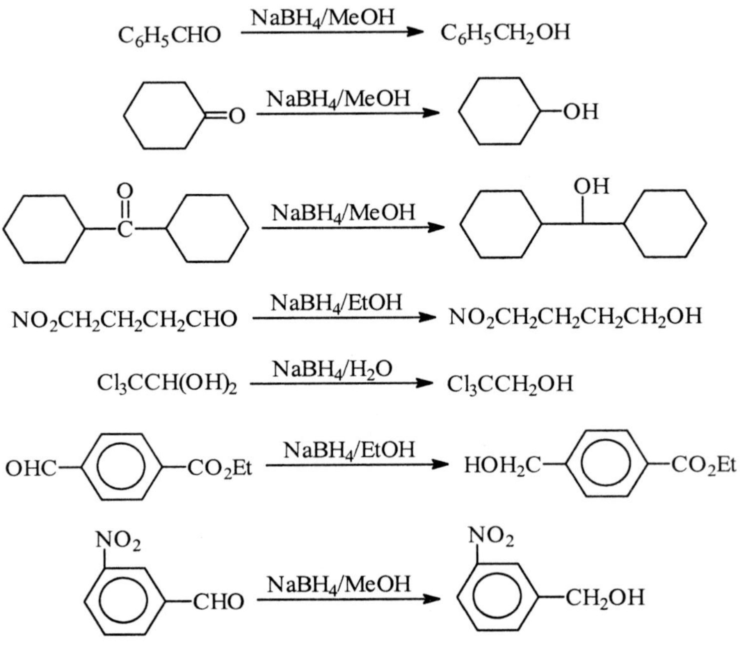

$$C_6H_5CHO \xrightarrow{NaBH_4/MeOH} C_6H_5CH_2OH$$

$$NO_2CH_2CH_2CH_2CHO \xrightarrow{NaBH_4/EtOH} NO_2CH_2CH_2CH_2CH_2OH$$

$$Cl_3CCH(OH)_2 \xrightarrow{NaBH_4/H_2O} Cl_3CCH_2OH$$

Scheme 100

Using sodium borohydride in aqueous medium, 2-alkylresorcinols have been prepared (Scheme 101).[158]

The entire process (Scheme 101) could be carried out as one pot reaction.

The reduction of carbonyl compounds in aqueous media has been carried out by a number of reagents under mild conditions. The most frequently used reagent is sodium borohydride, which can also be used using phase-transfer catalysts[159] or inverse phase transfer catalyst[160] in a two phase medium in the presence of surfactants.

The carbonyl compounds can be quantitatively reduced regio- and stereo-selectively by $NaBH_4$ at room temperature in aqueous solution containing glycosidic amphiphiles like methyl-β-D-glactoside, dodecanoyl-β-D-maltoside, sucrose etc.[161] By using this procedure, α,β-unsaturated ketones give

1,2-reduction product (corresponding allylic alcohols) and cyclohexanones give the more stable alcohol.

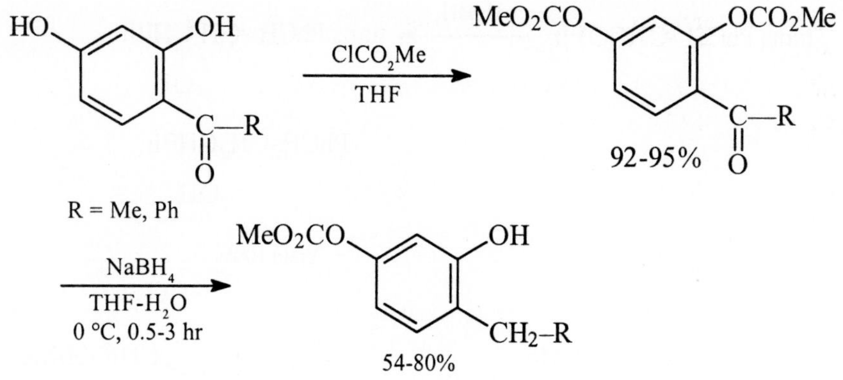

$R = Me, Ph$

Scheme 101

Reduction of ketones with NaBH$_4$ also proceeds in the solid state.[162] In this method a mixture of powdered ketone and 10-fold molar amount of NaBH$_4$ is kept in a dry box at room temperature with occasional mixing and grinding using an agate mortar and pestle for 5 days to give the reduced product. Following ketones were reduced by this procedure (Scheme 102).

Enantioselective hydrogenation of β-ketoesters has been achieved by using a ruthenium catalyst derived from (R,R)-1,2-bis(trans-2,5-diisopropylpholano)ethane [(R,R)-i-Pr-PPE-Ru] to give β-hydroxy esters with high conversion and high ee under mild conditions (Scheme 103).[163]

The reduction of aldehydes like benzaldehyde and p-tolualdehyde with Raney Ni in 10% aqueous NaOH give the corresponding benzyl alcohols in 17-80% yields[164] along with the corresponding carboxylic acids as byproducts obviously by cannizzaro reaction. It has been found that in aqueous NaHCO$_3$ under sonication conditions give the corresponding alcohols in good yields.

Another interesting reagent used for reduction of carbonyl compounds is cadmium chloride-magnisum in H$_2$O-THF system (Scheme 104).[165]

Certain other reagents like samarium iodide in aqueous THF[166], sodium dithionite in aqueous DMF[167], sodium sulfide in presence of polyethylene glycol[168] and metallic zinc along with nickel chloride.[169] Using the latter reagent (Zn/NiCl$_2$), α,β-unsaturated carbonyl compounds can be very readily reduced under ultrasound conditions[170] (Scheme 105).

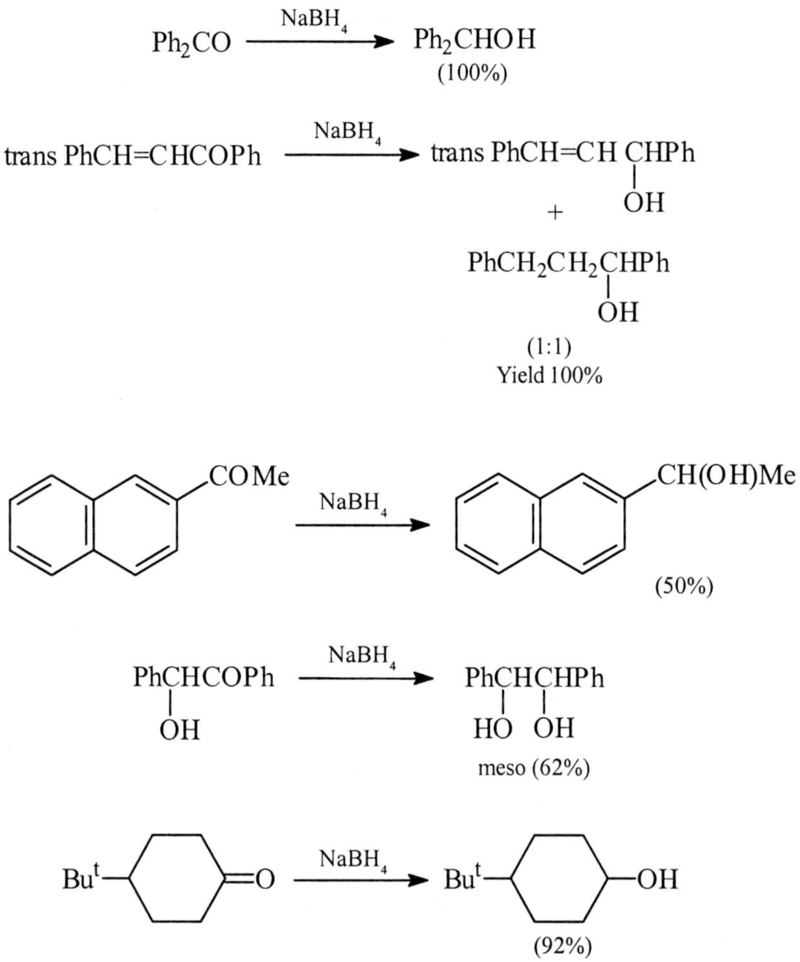
Ph_2CO
$NaBH_4$
Ph_2CHOH
(100%)
trans PhCH=CHCOPh
$NaBH_4$
trans PhCH=CH CHPh
OH
+
$PhCH_2CH_2CHPh$
OH
(1:1)
Yield 100%
COMe
$NaBH_4$
CH(OH)Me
(50%)
PhCHCOPh
OH
$NaBH_4$
PhCHCHPh
HO OH
meso (62%)
Bu^t
O
$NaBH_4$
Bu^t
OH
(92%)

Scheme 102

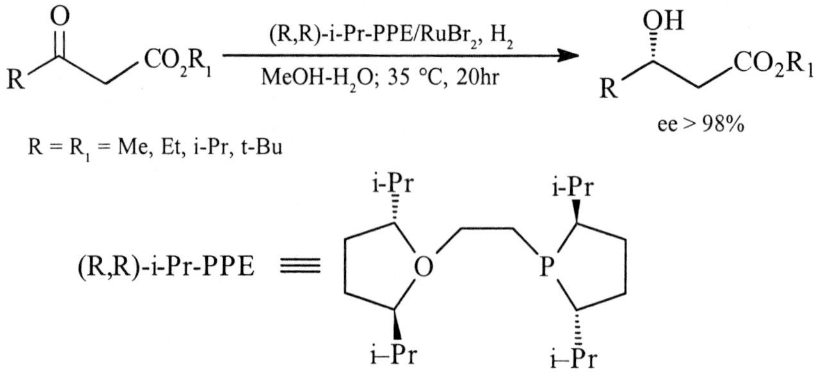
O
R
CO_2R_1
(R,R)-i-Pr-PPE/$RuBr_2$, H_2
MeOH-H_2O; 35 °C, 20hr
OH
R
CO_2R_1
ee > 98%
R = R_1 = Me, Et, i-Pr, t-Bu
(R,R)-i-Pr-PPE ≡
i-Pr
i-Pr
O
P
i-Pr
i-Pr

Scheme 103

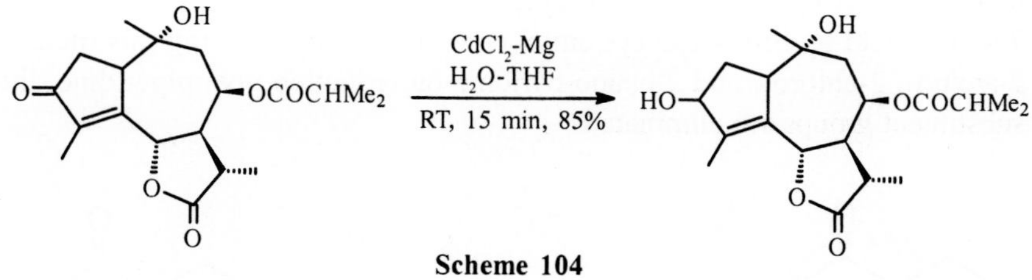

Scheme 104

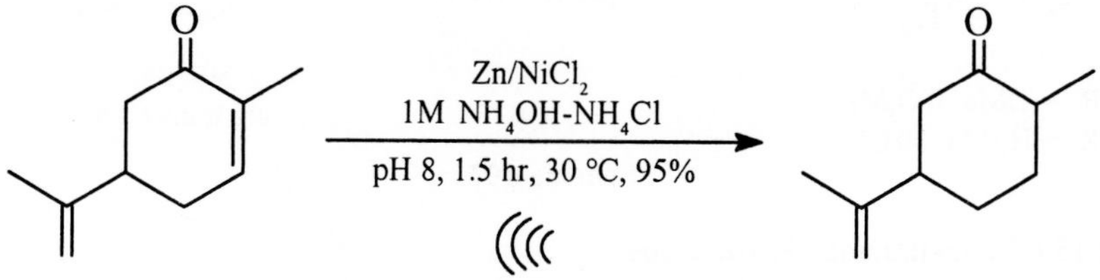

Scheme 105

Ketones can also be reduced in an aqueous medium by $SmI_2\text{-}H_2O$[171] (Scheme 106).

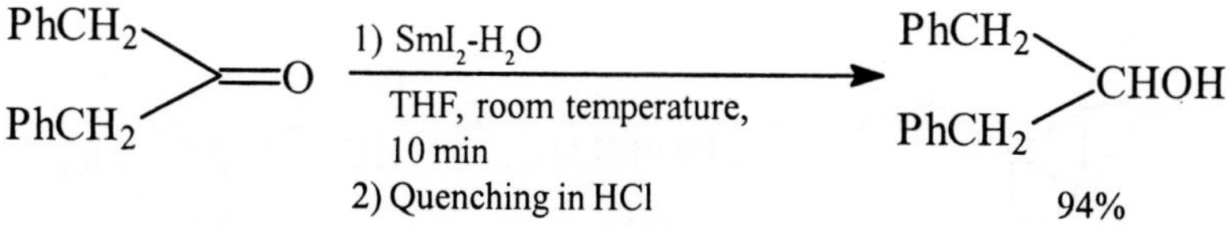

Scheme 106

12.15.5 Reduction of Aromatic Ring

The hydrogenation of benzenoids to cyclohexane derivatives is very useful for various methodologies. Aromatic hydrocarbons require drastic conditions for reduction (for example $PtO_2/H_2/CH_3COOH$; Raney $Ni/H_2/Pr/\Delta$, $Rh\text{-}Al_2O_3/H_2$).

It is now possible to reduce aromatic ring in aqueous medium at 50 atm of H_2 and at room temperature with ruthenium trichloride stabilized by trioctylamine ($RuCl_3/TOA$).[172] One such example is given (Scheme 107).

In the given procedure (Scheme 107) the rate of the reaction is 10-12 times the rate in organic solvent and in the aqueous medium, in comparison to cis, the trans isomer is the major product.

The heterocyclic compounds, viz. pyridine, 2-phenylpyridine and 3-methylpyridine can be reduced to the corresponding hexahydro product by Sm-20% HCl in 90-95% yields.[173] However in the reduction of 4-aminopyridine by the above method, the 4-amino group is eliminated giving piperidine as the major product (60%). The heterocyclic compounds can also be reduced in 70-94% yield by SmI_2-H_2O system at 0 °C for 2.5 hr.[174] Using this method 2-amino-, 2-chloro-, and 2-cyano-pyridine on reduction give piperidine, the substituent groups are eliminated.

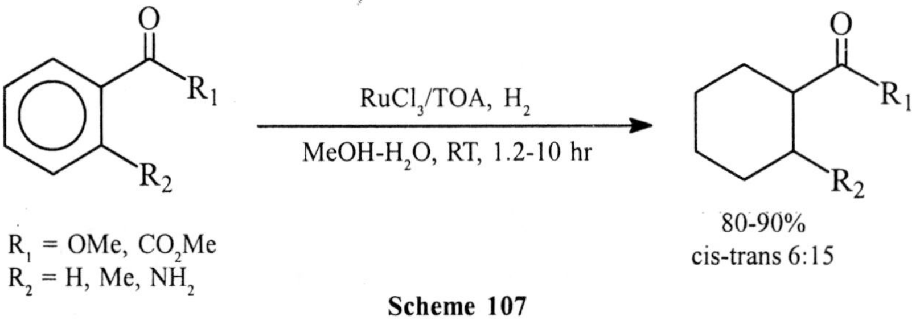

R₁ = OMe, CO₂Me
R₂ = H, Me, NH₂

80-90%
cis-trans 6:15

Scheme 107

12.15.6 Miscellaneous Reductions

(i) Reductive removal of halogen from α-halocarbonyl compounds in aqueous medium can be effected by using sodium dithionite[175], zinc[176], chromous sulfate[177] and sodium iodide.[178]

(ii) 2,3-Epoxyallyl halides can be transformed readily into allylic alcohol (Scheme 108) by zinc-copper couple in H_2O under sonication.

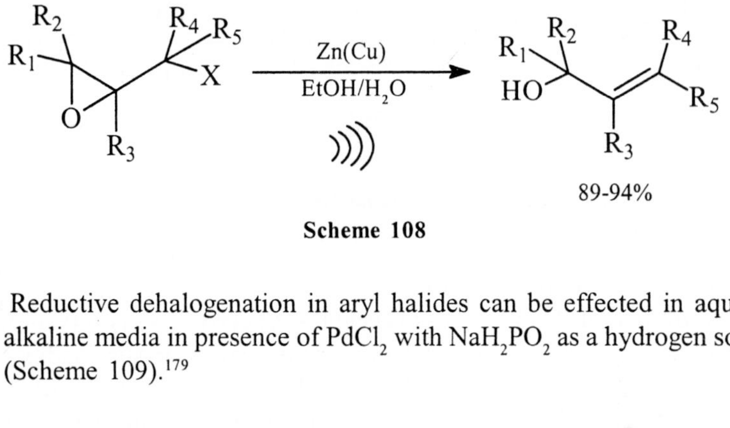

89-94%

Scheme 108

(iiii) Reductive dehalogenation in aryl halides can be effected in aqueous alkaline media in presence of $PdCl_2$ with NaH_2PO_2 as a hydrogen source (Scheme 109).[179]

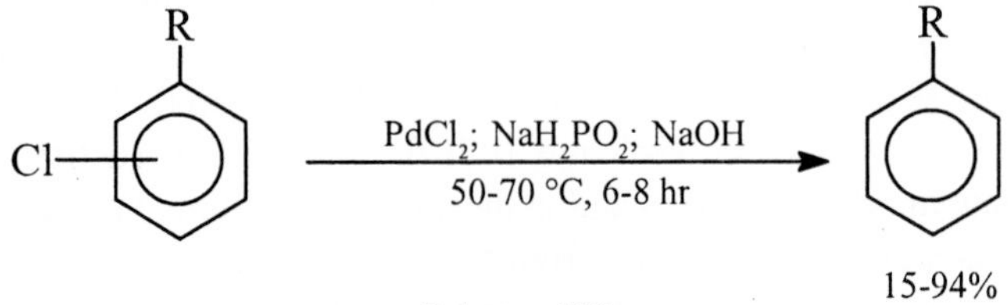

15-94%

Scheme 109

The above method (Scheme 109) does not work in case of nitrogen containing heterocyclic compounds and the yield in case of m-substituted aryl halides is low.

However, m-bromobenzoic acid can be converted into benzoic acid in 90% yield by using water soluble tris[3-(2-methoxyethoxy)propyl] stannane in presence of 4,4'-azobis (4-cyanovaleric acid) (ACVA) or sunlamp as initiator in aqueous NaHCO$_3$.[180] Above debromination can also be effected by using [bis(potassium propanoate)n (hydroxystannate), which in presence of NaBH$_4$ and ACVA affords reductions and free radical cyclisation of aryl and alkenyl bromides (Scheme 110).

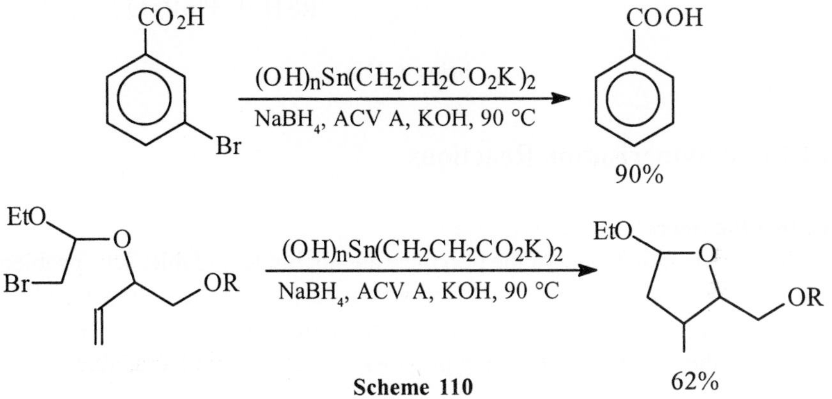

Scheme 110

The hydrogenolysis of halopyridines can be convenient when carried out with 15% aqueous TiCl$_4$ in presence of acetic acid.[181] Aqueous titanium trichloride quantitatively removes cyano group from cyanopyridines. Reductive dehalogenation is also catalysed by SmI$_2$.

(iv) Groups like azide, sulfoxide, disulphide, activated C=C bond and nitroxide can be effectively reduced by using sodium hydrogen telluride (NaTeH) (prepared *in situ* by the reaction of tellurium powder with aqueous ethanolic solution of NaBH$_4$).[182]

(v) Groups like aldehydes, ketones, olefins, nitroxides and azides can be reduced by sodium hypophosphite buffer solution.[183]

(vi) Vinyl sulphones can be stereospecifically reduced to the corresponding olefins with sodium dithionite in aqueous medium (Scheme 111).[184]

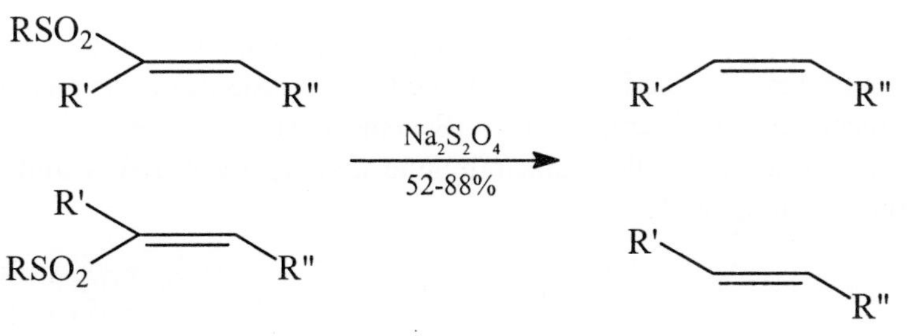

Scheme 111

(vi) Diaryl and dialkyl sulphides can be reduced by triphenylphosphine in aqueous solvents (Scheme 112).[185]

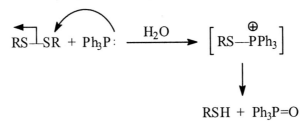

$$RSH + Ph_3P=O$$

Scheme 112

12.16   Polymerisation Reactions

12.16.1 Polymers

It is well known that most of the polymers are not biodegradable. This problem can be approached in two ways. One way is to recycle the polymer and the other way is to convert it again into the monomers and recycle them again. However, the best way is to make polymers which are biodegradable.

12.16.1.1 Recycling of Polymers

Let us now consider recycling, taking the example of the polyester, polyethyleneterephthalate (PET). It is known that PET is obtained by the polymerisation of dimethylterephthalate and ethylene glycol and is used in different products like beverage and food containers, nonfood containers, trays, luggage, boat parts, tapes etc. Fibres of PET are also used in clothing, carpets, blanckets, cord, rope etc. The recycled PET cannot be reused for food containers, as the impurities present are not sanitized at the temperature used in the melt process. Some of the products made of PET are mixtures with other polymers and contain dyes and other materials. These have to be disposed off by only incineration or land-filled. Also, during incineration the products obtained cause atmospheric pollution.

12.16.1.2 Conversion of PET Materials into Monomers

The process for the conversion of scrap PET into monomers is known as the Petrette process.[186] The flow sheet of the Petrette process for the recovery of monomers from PET scrap is given (Scheme 113).

The monomers (DMT and ethylene glycol) are purified and again polymerised to give PET.

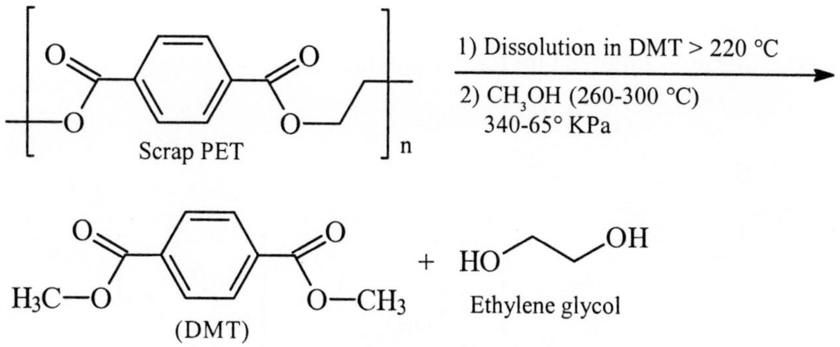

1) Dissolution in DMT > 220 °C

2) CH_3OH (260-300 °C)
   340-65° KPa

+ HO⌒⌒OH

Ethylene glycol

Scheme 113

12.16.1.3 To Make Polymers which are Biodegradable

One of the commonly used biodegradable polymer Thermal Polyaspartate polymer (TPA). Prior to the synthesis of TPA, polyacrylate polymer (PAC) was used as scale inhibitor in water handling processes. If not inhibited, the scale formation lead to loss of energy, non-functioning of pumps, boilers and condensers.

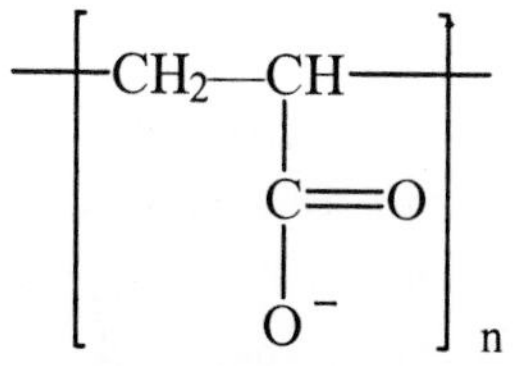

PAC

Though PAC is relatively non-toxic and environmentally benign, it is not biodegradable. This causes a disposal problem in waste water treatment facilities, where the PAC must be removed and land filled. The problem for it's disposal has been solved by Donlar Corporation, which has developed a method for the production of TPA (thermal polyaspartate), a biodegerable alternative to PAC.[187] The Donlar's synthesis[188] consists of heating the aspartic acid followed by hydrolysing the formed succinimide polymer with aqueous base (Scheme 114).

The above synthesis is 'green' because it uses no organic solvent, produces little or no waste and the yield is better than 97%.

12.16.1.4 Manufacture of Polycarbonates

Polycarbonates were earlier prepared by the polymerisation of phenols (e.g., bisphenol-A) using phosgene and methylene chloride. A new process has been developed by Asahi Chemical Industries using solid state polymerisation[189] for producing polycarbonate without phosgene and methylene chloride.[190]

In this process bisphenol-A and diphenyl carbonate directly give low molecular weight prepolymers, which are converted into high molecular weight polymers by crystallisation followed by further polymerisation (Scheme 115).

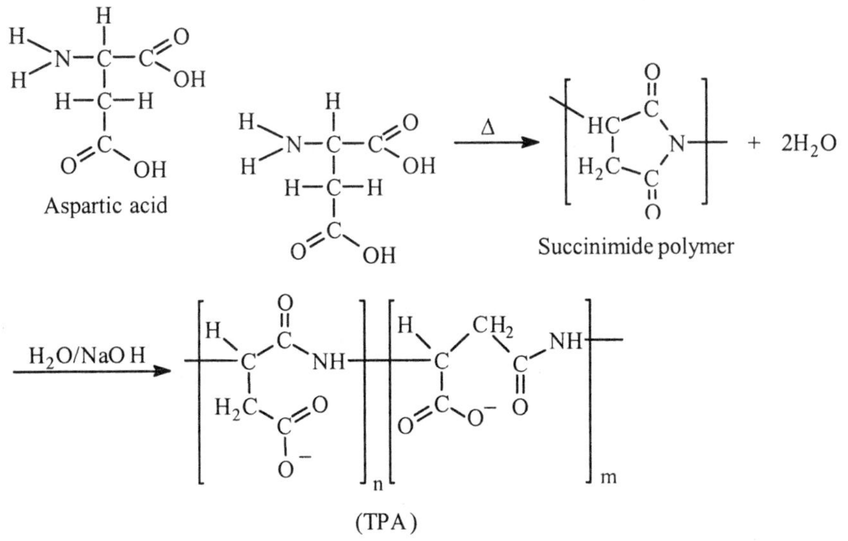

Aspartic acid

Succinimide polymer

(TPA)

Scheme 114

Polycarbonate

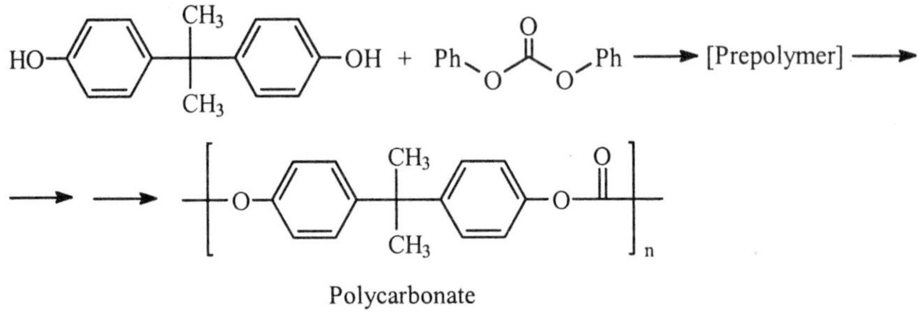

Scheme 115

12.16.1.5 Emulsion Polymerisation
Earlier attempts at emulsion polymerization of norbornenes in aqueous solution using iridium complexes as catalysts gave poor yields (10%).[190] However, it was subsequently found that 7-oxaborane derivatives could be rapidly polymerised in aqueous solution in presence of air using some group VIII coordination complexes as catalysts giving a quantitative yield of a ring-opening metathesis polymerisation product (Scheme 116).[191]

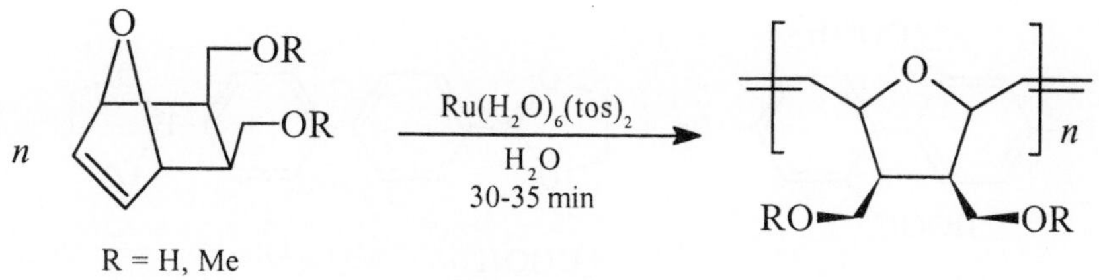

R = H, Me

Scheme 116

Using the above methodology norbornene could also be polymerised.[192,193] The ruthenium catalyst required for the above polymerisation were obtained by the procedure of Berhard-Ludi.[194] The aqueous catalyst solution could be reused; it was found to be more active. Another Ru(IV) catalyst[195] could also be used for emulsion ring-opening polymerisation of norbornene. The polymers obtained by the later catalysts had very high molecular wieght and had high cis-selectivity. Using this method, neogylcopolymers were synthesised (Scheme 117).

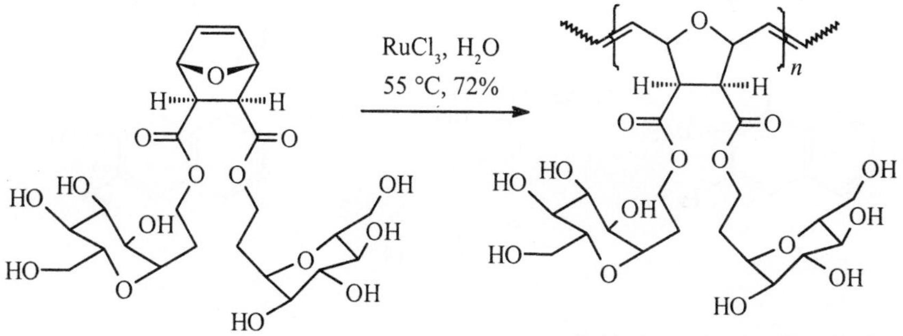

Scheme 117

A non-metallic conducting polymer was synthesised (Scheme 118).[196]

A helical polymer was prepared by using palladium-catalysed coupling between aryl halides with acetylene gas (Scheme 119).[197]

Solid state polymerisation of α-amino acids has been achieved giving rise to high molecular weight polypeptides.

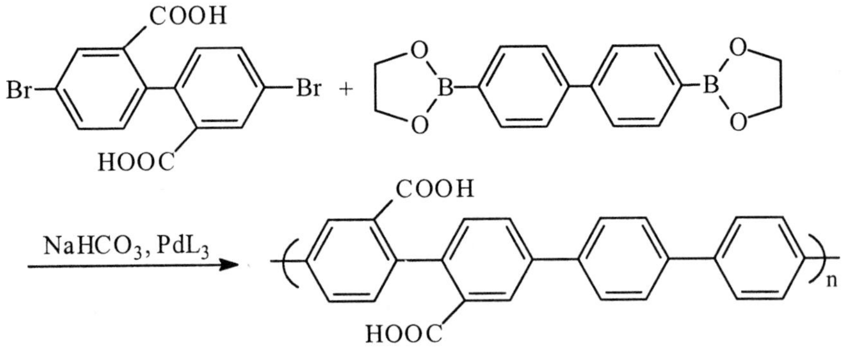

L = P(C$_6$H$_5$)$_2$(m-C$_6$H$_4$SO$_3$Na)

Scheme 118

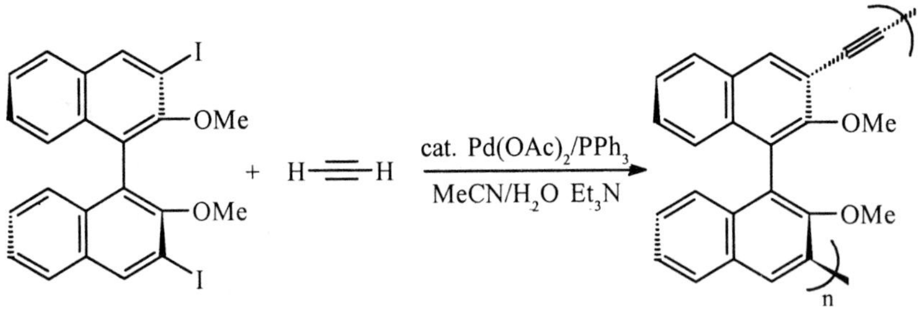

Scheme 119

12.17 Photochemical Reactions

The importance of photochemical reactions can hardly be overemphasized. The earliest known photochemical reaction is the photosynthesis of sugars by plants using sunlight, CO_2 and H_2O in presence of chlorophyll. Some of the common examples of routine photochemical reactions are:

(i) Photochemical cycloaddition of carbonyl compounds and olefins (Paterno-Buchi reaction) to give four membered ether rings (Scheme 120).[198]

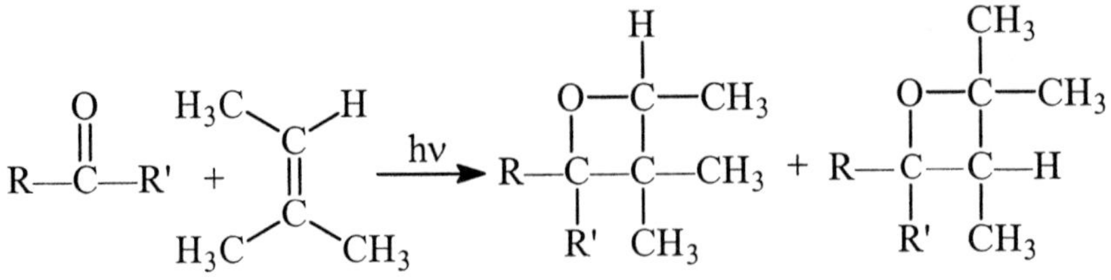

Scheme 120

(ii) Photo fries rearrangements. The phenolic esters in solution on photolysis give a mixture of o- and p-acylphenols.[199]

There are numerous other examples. Most of the photochemical reaction are carried out in solvents like benzene.

In view of the scare of the medium dependence of photochemical reactions, attempts were made for carrying out the reactions in water as a solvent.[200]

It has been reported in 1980's that the photochemical reactions can be conveniently carried out in aqueous phase. Thus, the photodimerisation of thymine, uracil and their derivatives could be carried out in water giving considerably better yields than in other organic solvents (Scheme 121).[201]

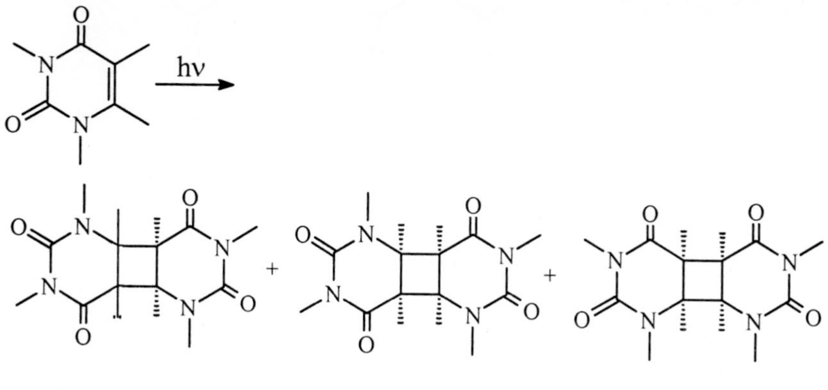

Water	27.8%	63.1%	9.1%	ø = 0.015
Acetonitrile	24.9%	68.2%	6.7%	ø = 0.0047
Methanol	31.4%	68.6%	–	ø = 0.004

Scheme 121. Data taken from reference 201

Organic substrates having poor solubility in water (e.g., stilbenes and alkyl cinnamates) also photodimerize efficiently in water (Scheme 122). The same reaction in organic solvents such as benzene gives mainly cis-trans isomers.[202]

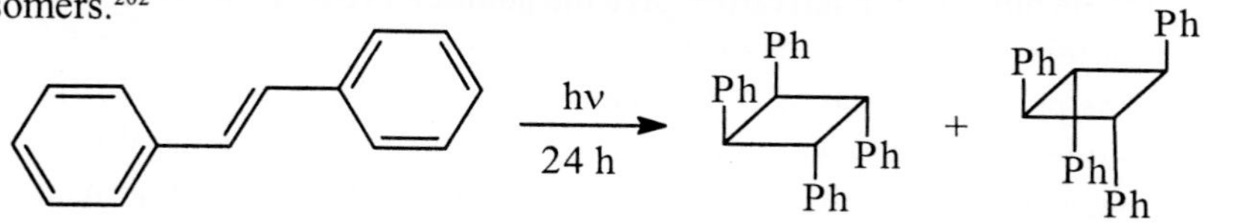

Benzene	0%	0%
Water	12%	10%
Water + LiCl	25%	17%
Water + guanidinium chloride	8%	6%

Scheme 122. Data taken from reference 202

The addition of LiCl (decreasing the hydrophobic effect) increases the yield of dimerisation (Scheme 122), whereas the addition of guanidinum chloride (decreasing the hydrophobic effect) lowers the yield of the product.

Similar results were obtained with alkyl cinnamates.[203]

An interesting example is the photodimerisation of coumarin in water (Scheme 123).

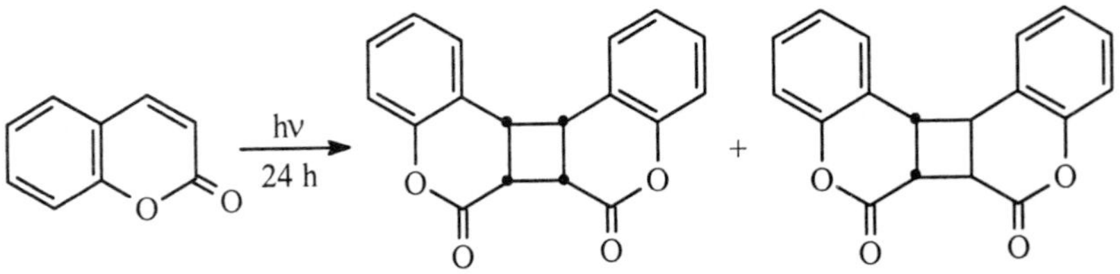

Solvent	Quantum Yield
Benzene	$< 10^{-5}$
Methanol	$< 10^{-5}$
Water	2×10^{-3}

Scheme 123

The yield of the dimerisation of coumarin in water (Scheme 123) is 100 times more than that in benzene or methanol.

It has been found that the micelles (formed by carrying out the reaction in presence of a surfactant/water) have a pronounced effect on regio- and stereo-selectivity of photochemical reactions.

The photodimerisation of anthracene-2-sulphonate in water gives four products A, B, C, D (Scheme 124). However, if the reaction is carried out in presence of β-cyclodextrin, only the isomer A is obtained.[204] Some other examples of photochemical reactions are:

(i) Photoirradiation of dibenzoyldiazomethane in CH_3CN-H_2O in presence of an amino acid derivative gave the addition product (Scheme 125) via the formation of a carbene.[205]

(ii) The photoirridation of o-fluoroanisole in $KCN-H_2O$ gave o-substitution product (catechol monomethyl ether) as the major product along with o-cyanoanisole as the minor product. Similar reaction with p- fluoroanisole gave p-cyanoanisole as the major product along with amount minor of hydroquinol monomethyl ether (Scheme 126).[206]

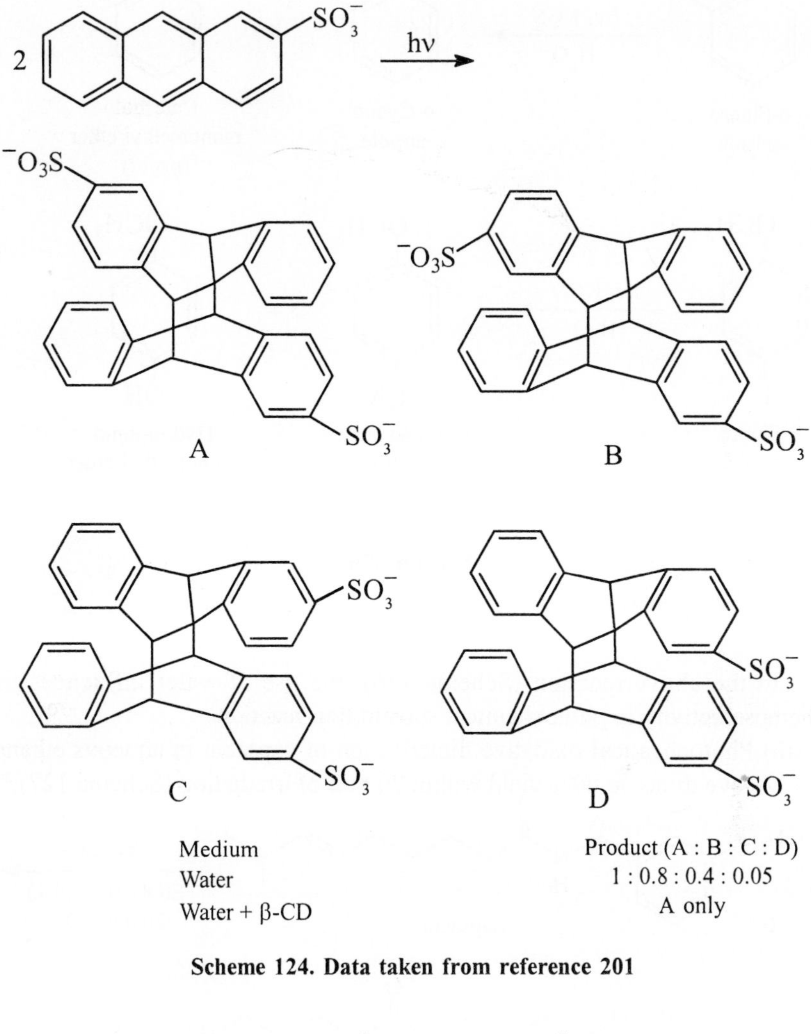

Medium	Product (A : B : C : D)
Water	1 : 0.8 : 0.4 : 0.05
Water + β-CD	A only

Scheme 124. Data taken from reference 201

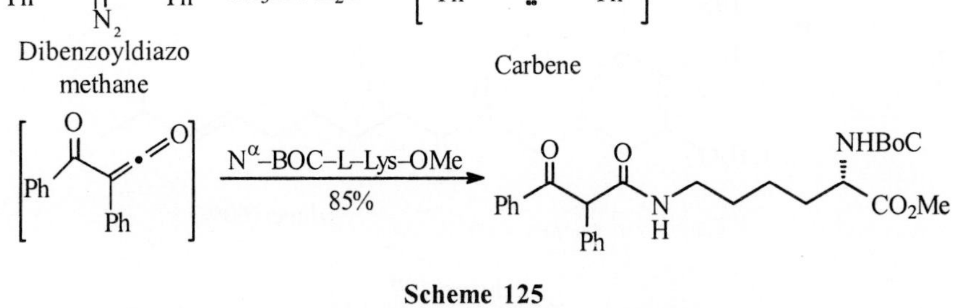

Scheme 125

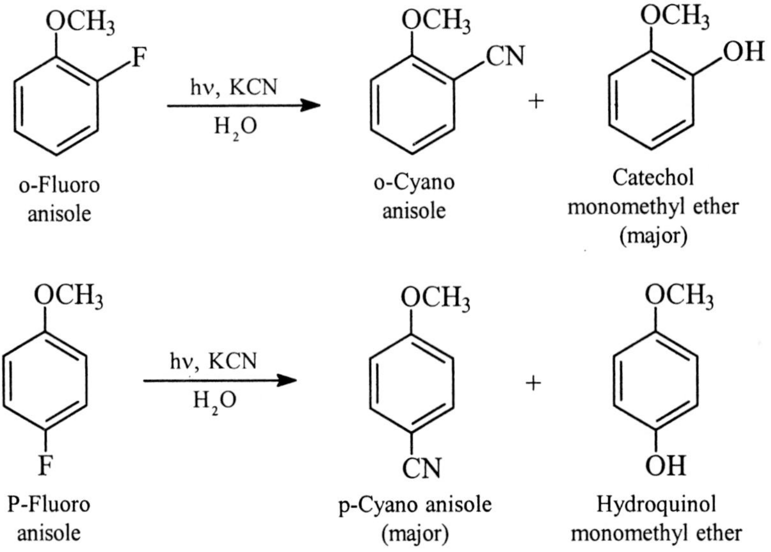

Scheme 126

In the above reaction (Scheme 126), the use of water influences the chemoselectivity in photochemical substitution reactions.

(iii) Photochemical oxidative dimerisation of capsacin in aqueous ethanol gave dimer in 60% yield within 20 min of irradiation (Scheme 127).[207]

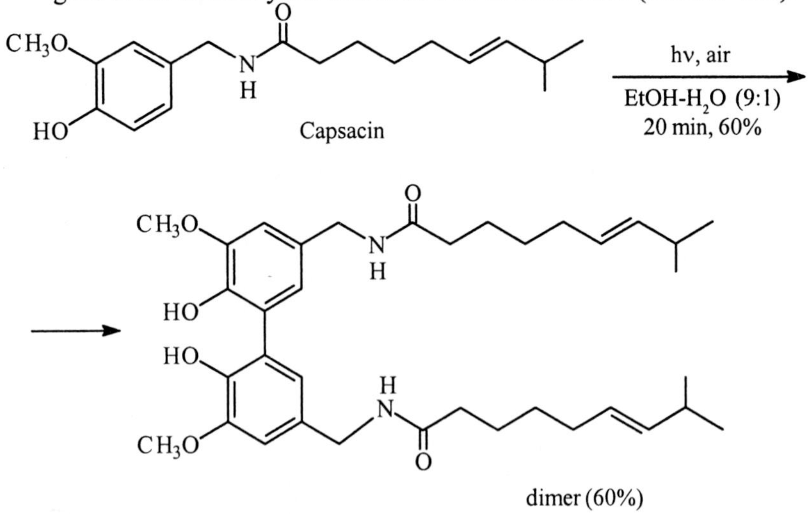

Scheme 127

(iv) Photooxidation of phenol is of interest in environmental chemistry.[208]

Photochemical reactions have also been studied in solid state. Thus the photodimerisation of cinnamic acid to truxillic acid has been achieved in the solid state (Scheme 128).[209]

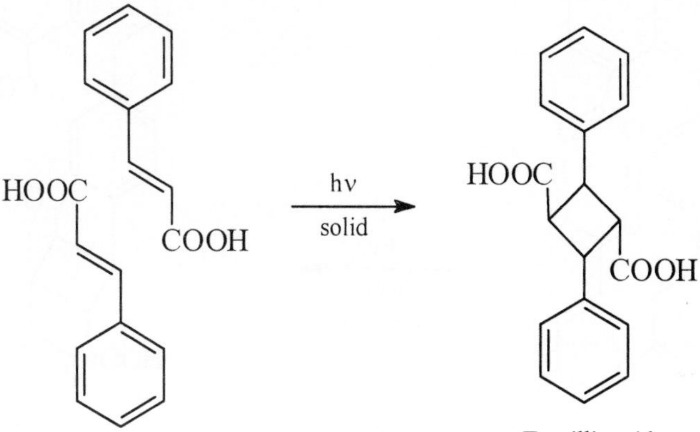

Cinnamic acid
(single cyrstal)

Truxillic acid
(single cyrstal)

Scheme 128

The photodimerisation of cinnamic acid can be controlled by irradiation of its double salts with certain diamines in the solid state. Thus, the double salt crystal of cinnamic acid and o-diaminocyclohexane gave upon irradiation in the solid state, β-truxinic acid as the major product (Scheme 129).[210]

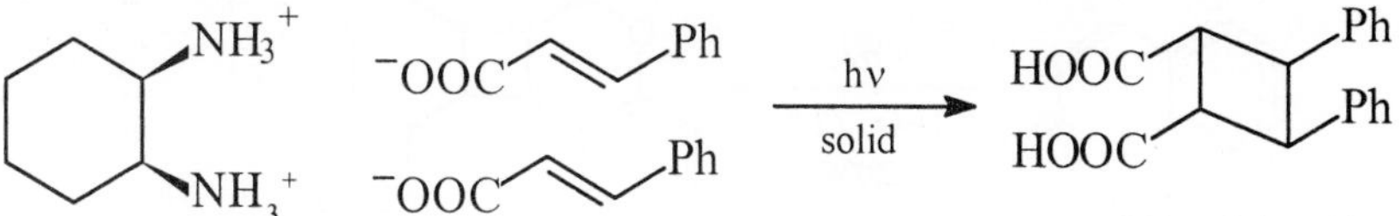

Double salt of cinnamic acid
with o-diamino cyclohexane

β-Truxinic acid

Scheme 129

Similar irradiation of naphthoic acid-derived cinnamic acid (A) in solid state on irradiation for 20-50 hr afforded a single cyclobutane product in 100% yield (Scheme 130).[211]

The photocyclisation of coumarin and its derivatives has been extensively studied.[212] Thus, irradiation of coumarin for 48 hr in the solid state gives a mixture of A, B and C in 20% yields. However, irradiation of an aqueous solution of coumarin for 22 hr affords only the syn-head to head dimer (D) in 20% yield (Scheme 131).

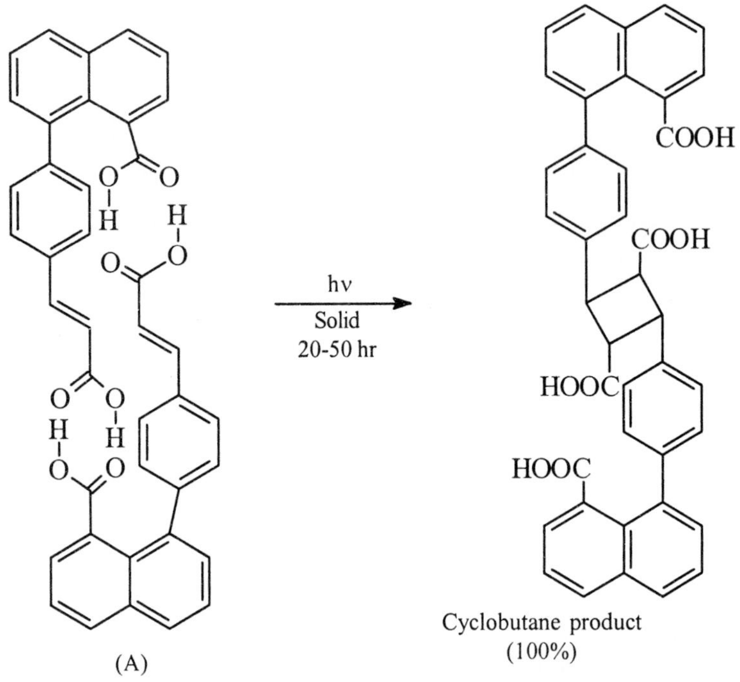

(A)

Cyclobutane product
(100%)

Scheme 130

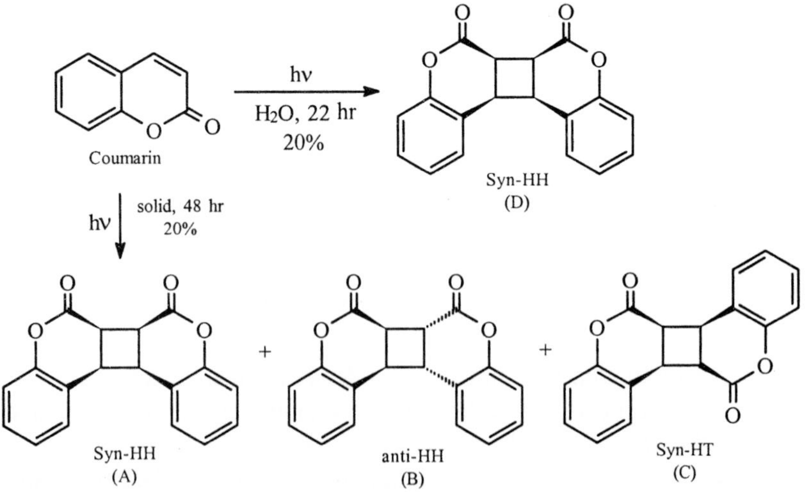

Coumarin

Syn-HH
(D)

hv | solid, 48 hr
       20%

Syn-HH
(A)

+

anti-HH
(B)

+

Syn-HT
(C)

Scheme 131

It is well recorded that benzopinacol can be obtained quantitatively on photoirradiation of 4,4'-dimethyl benzophenone in isopropylalcohol. However, in the solid state photoirradiation gives the dimeric compound (Scheme 132).[213] Besides the representative example of photochemical reactions in solid phase a large number of illustrations are available.[214]

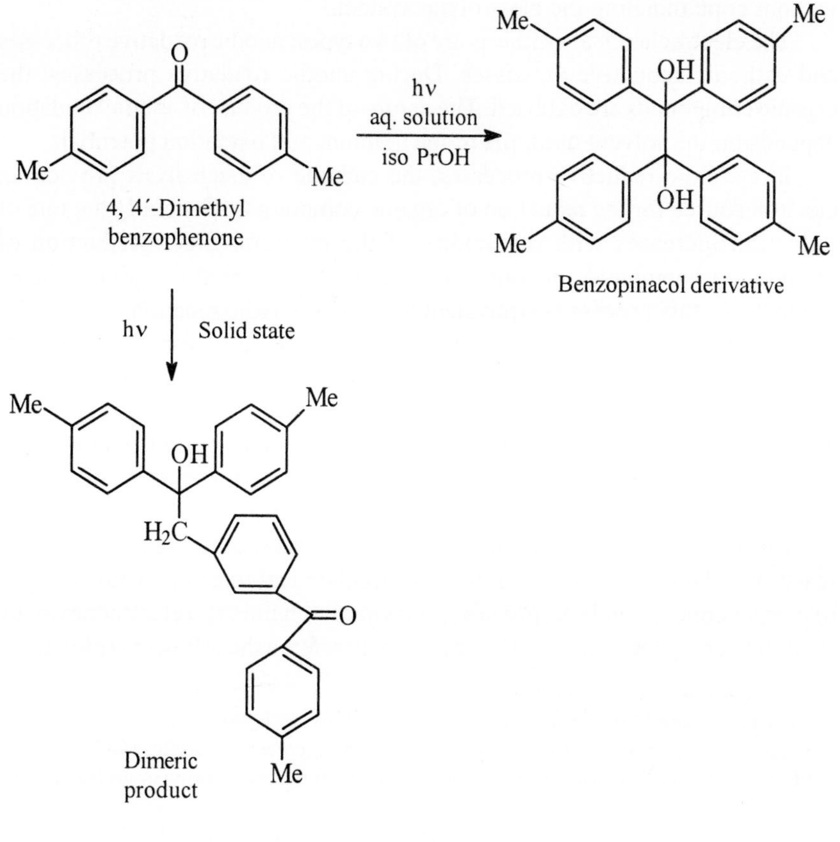

Scheme 132

12.18 Electrochemical Synthesis

12.18.1 Introduction

The earliest electrochemical synthesis is the so-called Kolbe reaction involving the oxidation of carboxylic acids in forming decarboxylated coupling products (alkanes). At present, the electrochemical synthesis has become an independent discipline. A large number of organic reactions (synthesis) have been achieved by this technique. The essential requirement for conducting an electrochemical reaction is the conductivity of the reaction medium. The most commonly used solvent is water, though organic solvents have also been used. However, there is a distinct advantage in using aqueous solutions over organic solvents.[215] In case of organic solvents, during electrolysis, a complex mixture of products get accumulated in the electrolyte, which leads to loss of expensive solvent. On the other hand, electrolysis of water yield O_2/H^+ and H_2/OH^-. In view of this, the electrolysis of water can be performed at a maintained level of pH without contaminating the electrolytic system.

The electrochemical synthesis are of two types: anodic oxidative processes and cathodic reductive processes. During anodic oxidative processes, the organic compounds are oxidised. The nature of the product of anodic oxidation depends on the solvent used, pH of the medium and oxidation potential.

In cathodic reductive processes, the cathode of electrolysis provide an electron source for the reduction of organic compounds. Generally the rate of reduction increases with the acidity of the medium. Electroreduction of unsaturated compounds in water or aqueous-organic mixtures give reduced products — this process is equivalent to catalytic hydrogenation.

An electrochemical process uses a anode made of metal that resists oxidation, such as lead, nickel or most frequently platinum. The anode is usually in the shape of a cylinder made of wire guage. The usual electrolytes are dilute sulphuric acid or sodium methoxide prepared *in situ* from methanol and sodium. The direct current voltage is 3-100 V, the current density is 10-20 A/dm^3, and the temperature of the medium is 20-80 °C.

Electrochemical reactions are practically as diverse as non-electrochemical reactions. Thus, the combination of electrochemical reactions with catalysts (electrochemical catalytic process), enzymatic chemistry (electroenzymatic reactions) are quite common. The readers may refer to the following references:

- A.N. Frumkin, in Advances in Electrochemistry and Electrochemical Engineering, P. Delahay and C.W. Tobias, eds., Interscience, New York, Vol. 3.
- Topics in Current Chemistry, E. Steckhan, ed., Springer-Verlag, Berlin, 1987, Vol. 142.
- E. Steckhan, in Topics in Current Chemistry, E. Steckhan, ed., Springer-Verlag, Berlin, Vol. 170.
- D.K. Kyriacou, Basics of Electroorganic Synthesis, Wiley, New York, 1981.
- H. Lund and M.M. Baizer, eds., Organic electrochemistry, Marcel Dekker, New York, 1991.

Following are given some representative examples of electrochemical synthesis.

12.18.2 Synthesis of Adiponitrile

Adiponitrile is used as an important synthon for hexamethylene diamine and adipic acid, which are used for the manufacture of Nylon-66.

It is obtained commercially by the electroreductive coupling of acrylonitrile. By this process about 90% of adiponitrile is obtained.[216] In this process a concentrated solution of certain quaternary ammonium salts (QASs), such as tetraethylammonium-p-toluene sulfonate is used together with lead or mercury cathode (Scheme 133).

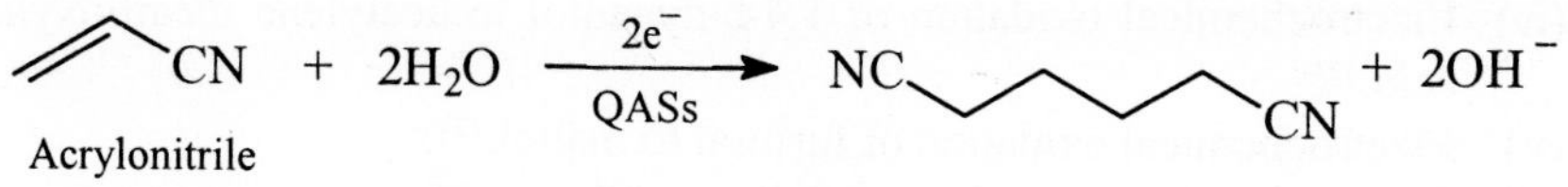

$$CH_2=CH-CN + 2H_2O \xrightarrow[\text{QASs}]{2e^-} NC\text{-}(CH_2)_4\text{-}CN + 2OH^-$$

Acrylonitrile

Scheme 133

It will be appropriate to mention here that selective hydrocyanation of butadiene catalysed by Ni(O)/triarylphosphite complexes give adiponitrile (Scheme 134).[217]

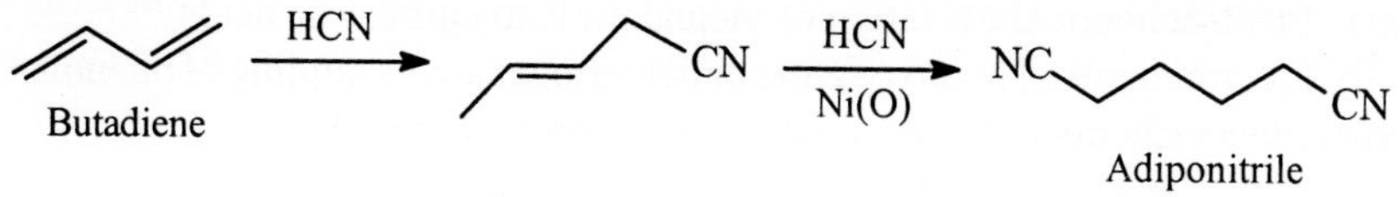

$$\text{Butadiene} \xrightarrow{HCN} CH_2=CH\text{-}CH=CH\text{-}CN \xrightarrow[\text{Ni(O)}]{HCN} NC\text{-}(CH_2)_4\text{-}CN$$

Adiponitrile

Scheme 134

12.18.3 Synthesis of Sebacic Acid

Sebacic acid is an important intermediate in the manufacture of polyamide resins. It is obtained on a large scale by saponification of castor oil.[218] It is now obtained by electrochemical process involving the following three steps (Scheme 135).

$$HO_2C\text{---}(CH_2)_4CO_2H \xrightarrow[\text{esterification}]{CH_3OH} CH_3O_2C(CH_2)_4CO_2H \longrightarrow$$

Adipic acid   Monomethyl ester of
                               adipic acid

$$\xrightarrow[55\,°C]{\text{Electrolysis}} CH_3O_2C(CH_2)_8CO_2CH_3 \xrightarrow{\text{Hydrolysis}} HO_2C(CH_2)_8CO_2H$$

Dimethyl ester of   Sebacic acid
sebacic acid

Scheme 135

In the above process, anodic coupling of the monomethyl ester of adipic acid takes place. The electrolyte is a 20% aqueous solution of monomethyl adipate, neutralised with sodium hydroxide. The anode is platinum-plated with titanium and the cathode is of steel.

12.18.4 Miscellaneous Electrochemical Reactions

(i) Electrochemical reduction of glucose for the manufacture of sorbitol and mannitol.[220]

(ii)   Electrochemical reduction of phthalic acid to the corresponding dihydrophthalic acids.[221]
(iii) Electrochemical coupling of acetone to yield pinacol.[222]
(iv)   Electrochemical oxidation of 1,4-butynediol to acetylene dicarboxylic acid.[223]
(v)   Electrochemical oxidation of furfural to maltol.[224]
(vi)   Electrochemical epoxidation of alkenes.[225]
(vii) Electrochemical conversion of alkenes into ketones.[226]
(viii) Electrochemical oxidation of aromatic rings and side chains to carboxylic acids.[227]
(ix)   Electrochemical oxidation of primary alcohols to carboxylic acids.[228]
(x)   Electrochemical oxidation of secondary alcohols to ketones.[229]
(xi)   Electrochemical oxidation of vicinal diols to carboxylic acids.[230]
(xii) Electrochemical hydroxylation or dehydrogenative coupling of phenols.[231]
(xiii) Electrochemical Kolbe synthesis of hydrocarbons.[232]

12.19 Miscellaneous Reactions in Aqueous Phase

12.19.1 Isomerisation of Alkenes
Alkenes are known to isomerise in presence of transition metal complexes. It has been found that isomerisation of allylic alcohols (or ethers) can be performed in aqueous media in presence of $Ru(II)(H_2O)_6(tos)_2$ (tos = p-toluene sulfonate) (Scheme 136).[233]

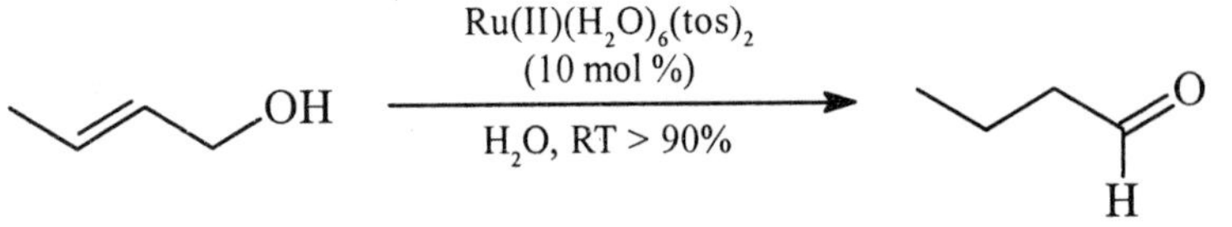

Scheme 136

In the above reaction the initially formed enols and enolethers are unstable and are instantaneously hydrolysed to give the corresponding carbonyl compounds.
    Some other examples are given as follows (Scheme 137).[234]

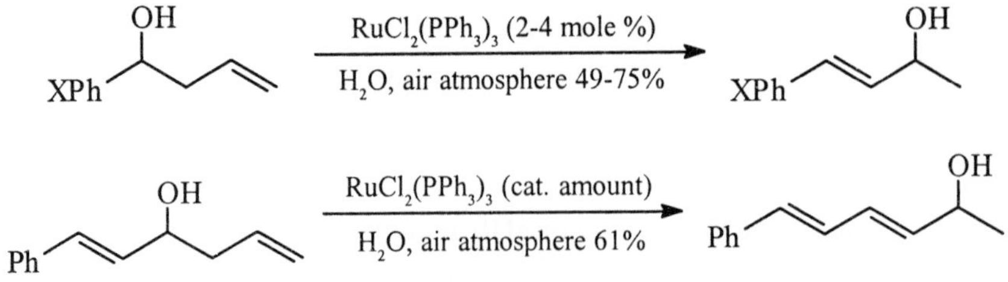

Scheme 137

In both the given examples (Scheme 137) the substrates undergo structural reorganisation involving reshuffling of both the hydroxyl group and the olefin in water. These reactions can be considered as olefin migration followed by an allylic rearrangement.

12.19.2 Carbonylation

Carbonylation is a very important process for the preparation of carboxylic acids (and their derivatives), aldehydes and ketones. It was earlier carried out in presence of transition metal catalysts.[235]

The aryl halides can be converted into the corresponding carboxylic acid by carbonylation in presence of water. The use of PTC is a well established technique for carbonylation of organic halides.[236] Carbonylation of organic halides using various types of phase transfer techniques has been extensively reviewed.[237]

Aryl halides can be carbonylated to give the corresponding carboxylic acids (Scheme 138) under very mild conditions in presence of inorganic bases (like alkaline metal hydroxides, carbonates, acetates etc.) and certain palladium catalysts (like $Pd(OAc)_2$, K_2PdCl, $Pd(NH_3)_4Cl$, $PdCl_2(PPh_3)_2$ etc.). Best results are obtained with simple palladium salt using K_2CO_3 as base (Scheme 138).

$$ArX \quad \xrightarrow[\text{DMF-H}_2\text{O (2:1) 25-50 °C}]{Pd(OAc)_2 \ (1 \text{ mole \%}), \ CO, \ K_2CO_3} \quad ArCOOH$$

$Ar = p\text{-}ZC_6H_4$ ($Z = NO_2$, Cl, CN, Me, NH_2 etc.)

Scheme 138

Water soluble aryl iodides can be carbonylated in H_2O in presence of soluble palladium salt or complexes and K_2CO_3 as base at 25-50 °C (Scheme 139).

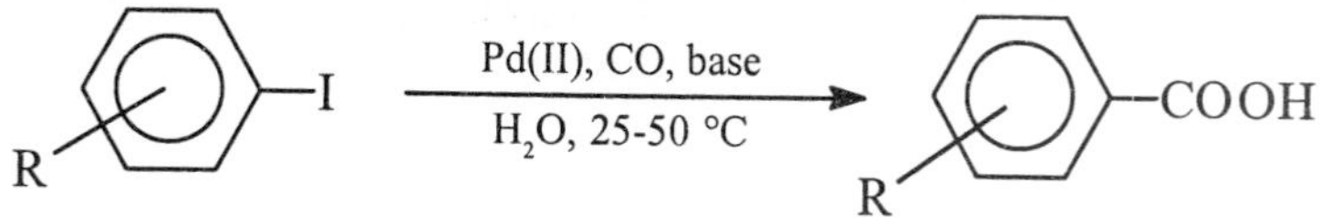

$$\text{(aryl iodide)} \quad \xrightarrow[\text{H}_2\text{O, 25-50 °C}]{Pd(II), \ CO, \ base} \quad \text{(aryl carboxylic acid)}$$

R = m-, p–COOH
o-, m-, p–OH

Scheme 139

Ortho-iodobenzoic acid cannot be carbonylated under the above conditions (Scheme 139). It, can, however be carbonylated in presence of excess iodide ion to give phthalic acid (Scheme 140).

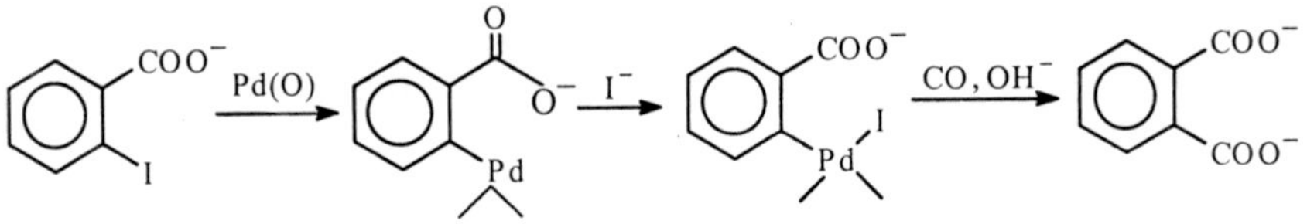

Scheme 140

In case of water insoluble aryl iodides, the iodine atom is first oxidized to the iodyl group by $NaClO_3$, the formed iodyl derivative (having slightly enhanced solubility in water) are readily carbonylated under mild conditions to give carboxylic acids (Scheme 141).[238]

$$ArI \xrightarrow{NaClO_3} ArIO_2 \xrightarrow[\text{CO (1 atm), } H_2O, \text{ 40-50 °C}]{Na_2PdCl_4, \text{ NaOH}} ArCOOH$$

Scheme 141

Carbonylation of allylic and benzylic chlorides was carried out by transition-metals (as catalysis) to give β,γ-unsaturated acids.[239] However, the above method gave low yields. It is found that carbonylation of benzyl bromide and chloride could be carried out by stirring with an aqueous sodium hydroxide and an organic solvent using a PTC and a cobalt catalyst. Even benzylic mercaptan could be carbonylated to give esters under high pressure and temperature (Scheme 142).[240]

$$PhCH_2X \xrightarrow[\text{PTC}]{[Co(CO)_4]^- \text{ aq. NaOH}} PhCH_2CO_2Na$$

$$X = Cl \text{ or } Br$$

$$ArCH_2SH + CO + R'OH \xrightarrow[\substack{850\text{-}900 \text{ psi (lb/in}^2) \\ 190 \text{ °C, 24 hr, 25-83\%}}]{Co_2(CO)_8, H_2O} ArCH_2COOR'$$

Scheme 142

Carbonylation of allyl bromides and chlorides has been achieved in presence of a nickel catalyst in aqueous NaOH at atmospheric pressure.[241] Subsequently it is found that Palladium-catalysed carbonylation of allyl chloride proceeded smoothly in two-phase aqueous NaOH/benzene under atmospheric pressure at room temperature (Scheme 143).[242]

$$CH_2=CHCH_2Cl + CO + ROH \xrightarrow[19-98\%]{Pd} CH_2=CHCH_2COOR$$

R = H, CH_3, C_2H_5 etc.

Scheme 143

In the above method shown in Scheme 143, addition of surfactants (e.g., n-$C_7H_{15}SO_3Na$ or n-$C_7H_{15}CO_2Na$) accelerate the carbonylation.[243]

Some other carbonylation reactions are:
(i) Carbonylation of 1-perfluoroalkyl-substituted 2-iodoalkanes in presence of transition metal catalysts in aqueous media give β-perfluoroalkyl carboxylic acid (Scheme 144).[244]

$$R_f\text{--}CH_2CHR'I + CO + H_2O \xrightarrow[\text{base } 42-89\%]{Pd, Co, or Rh cat.} R_f\text{--}CH_2CHR'CO_2H$$

R_f= perfluoroalkyl group

Scheme 144

(ii) γ-Lactones have been obtained by the carbonylation of terminal alkynes in water in presence of rhodium carbonyl (Scheme 145).[245]

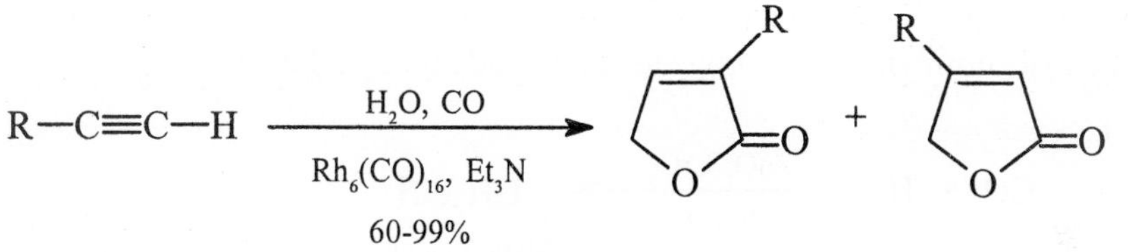

Scheme 145

(iii) The reaction of styrene oxide with carbon monoxide is catalysed by a cobalt complex in presence of methyl iodide to give enol (Scheme 146).[246]

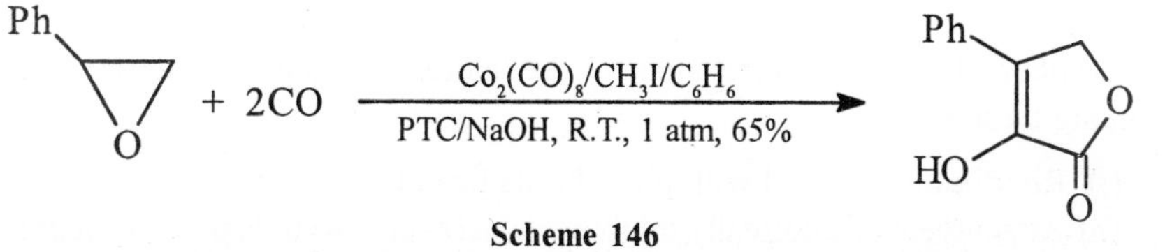

Scheme 146

(iv) Carbonylation of methane under acidic conditions by oxygen and CO in water, catalysed by palladium, platinum or rhodium catalysts gives acetic acid.[247]

(v) 5-Hydroxymethyl furfural can be selectively carbonylated to the corresponding acid by using a water soluble palladium catalyst (Scheme 147).[248]

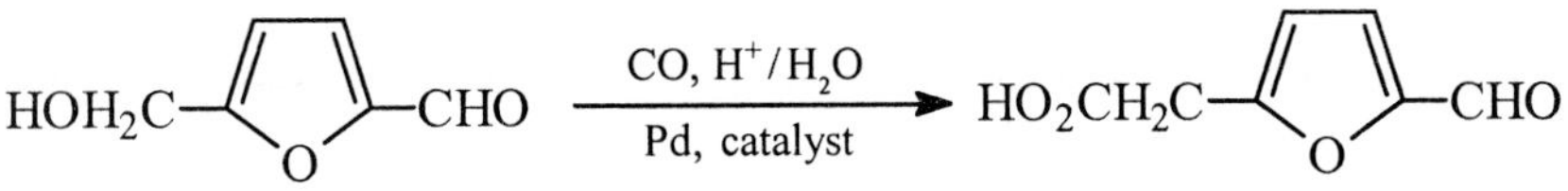

Scheme 147

(vi) Carbonylation of 1-(4-isobutylphenyl)ethanol gives ibuprofen (Scheme 148).[249]

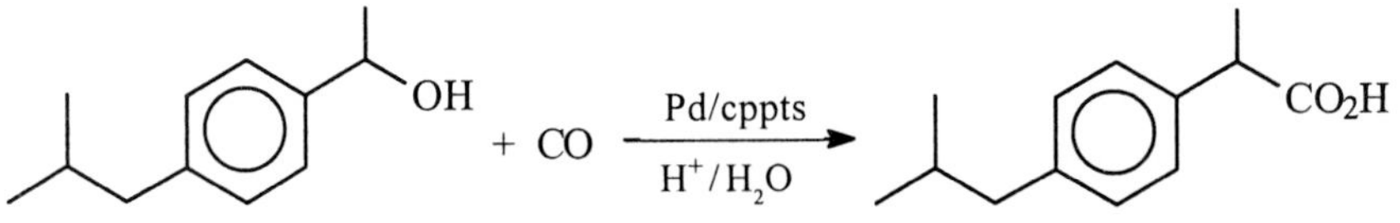

Ibuprofen

Scheme 148

12.19.3 Hydroformylation of Olefins

Hydroformylation of olefins is a chief industrial process for the manufacture of aldehydes and alcohols by reaction with CO and H_2 in presence of a catalyst.[250]

Carbon monoxide and hydrogen have been known to be used for the manufacture of methyl alcohol. Also, the first product to be manufactured by the hydroformylation of propene is butyraldehyde (Scheme 149).[251]

$$CO + H_2 \xrightarrow[300\ °C]{ZnO,\ CrO_3} CH_3OH$$
Methyl alcohol

$$CH_3CH{=}CH_2 + CO + H_2 \xrightarrow[\text{catalyst}]{\text{Rhodium}} CH_3CH_2CH_2CHO$$
Propene   Butyraldehyde

Scheme 149

A number of catalysts were tried for the hydroformylation of olefins. Main among these are:

(i) Rhodium combined with phosphorus ligand
(ii) Attachment of a normally soluble catalyst to an insoluble polymer support.
(iii) Transition-metal complexes with water soluble phosphine ligands and water as immiscible solvent for the hydroformylation. Most of the catalysts had a problem arising out of leaching of the catalyst into the organic phase.

Finally a variety of 1-alkenes were hydroformylated with a highly water-soluble tris-sulphonated ligand, $P(m\text{-}PhSO_3Na)$.[252] Some more effective catalysts involving the use of other sulphonated phosphine ligands have also been developed.[253]

Another unique approach is the concept of support aqueous phase (SAP) catalysis.[254] In this approach, a thin, aqueous film containing a water soluble catalyst adheres to silica gel with a high surface area. The reaction occurs at the liquid-liquid interface. Using this technique, hydroformylation of alkenes like octene, dicyclopentadiene is possible. Through SAP approach, hydroformylation of acrylic acid derivatives is of considerable industrial applications (Scheme 150).[255]

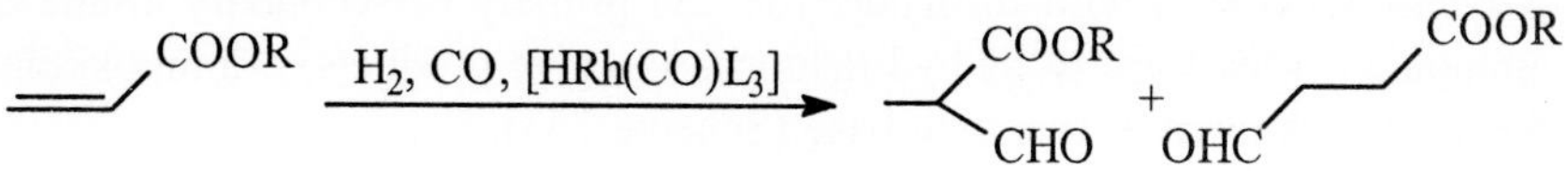

Scheme 150

The major product obtained (Scheme 150) is formylpropionic acids, which are precursors of methacrylate monomers and can be used for a number of important pharmaceuticals. A number of other applications of hydroformylation have been reviewed.[256]

12.19.4 Homologation of 1,3-dihydroxyacetone

In aqueous medium, it is not necessary to protect functional groups. Thus homologation of 1,3-dihydroxyacetone with formaldehyde in presence of base gives the homologated derivatives A and B (Scheme 151).[257]

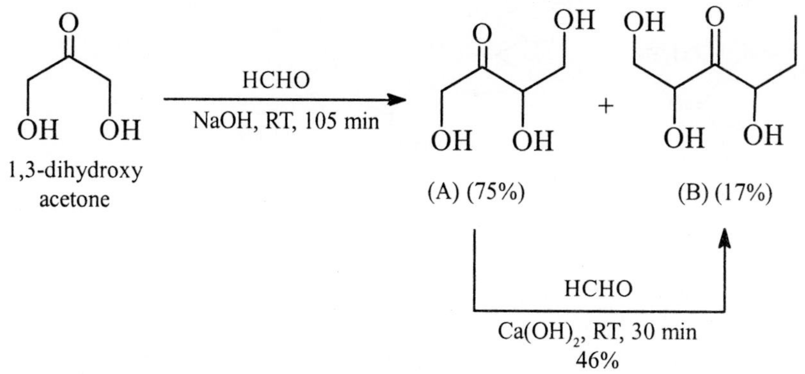

Scheme 151

The above reaction is known as Tollens reaction.

12.19.5 Weiss-Cook Reaction

The reaction of dimethyl 3-oxoglutarate with glyoxal in aqueous acidic solution gives [3.3.0] octane, 3,7-dione-2,4,6,8-tetracarboxylate; and on acid catalysed hydrolysis followed by decarboxylation it gives cis-bicyclo[3.3.0] octane-3,7-dione (Scheme 152).[258] The reaction is believed to involve a double Knoevenagal reaction that gives an α,β-unsaturated γ-hydroxycyclopentenone, which reacts with another molecule of dimethyl 3-oxoglutarate by Michael addition.

12.19.6 Mannich Type Reactions

The original Mannich reaction consisted in the reaction of a compound containing at least one active hydrogen atom (ketones, nitroalkanes, β-ketoesters, β-cyano acids etc.) with formaldehyde and primary or secondary amine or ammonia (in the form of its hydrochloride) to give products, β-aminoketone derivatives, known as mannich base (Scheme 153).[259]

A new approach (or modification) using preformed iminium salts and imines has been developed. The imines react with enolate (especially trimethylsilyl ethers) to give β-amino ketones. In this reaction $TiCl_4$ was used as a promoter.[260] The general scheme for the synthesis of β-aminoketones is given in Scheme 154.

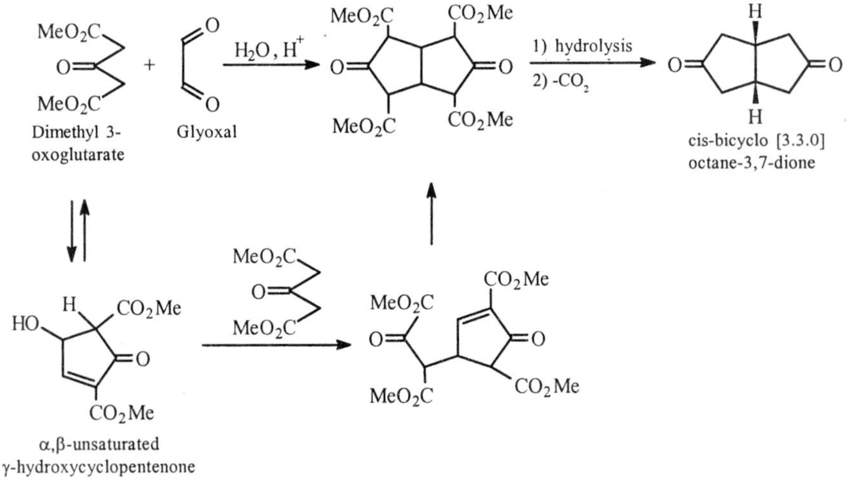

Scheme 152

$$C_6H_5{-}\overset{O}{\overset{\|}{C}}{-}CH_3 + H{-}\overset{O}{\overset{\|}{C}}{-}H + (CH_3)_2\overset{+}{N}H_2Cl^- \longrightarrow C_6H_5{-}\overset{O}{\overset{\|}{C}}{-}CH_2CH_2\overset{+}{N}H(CH_3)_2Cl^- + H_2O$$

Acetophenone   Formaldehyde   Dimethylamine hydrochloride   β-N,N-Dimethylamino ketone (HCl salt)

Scheme 153

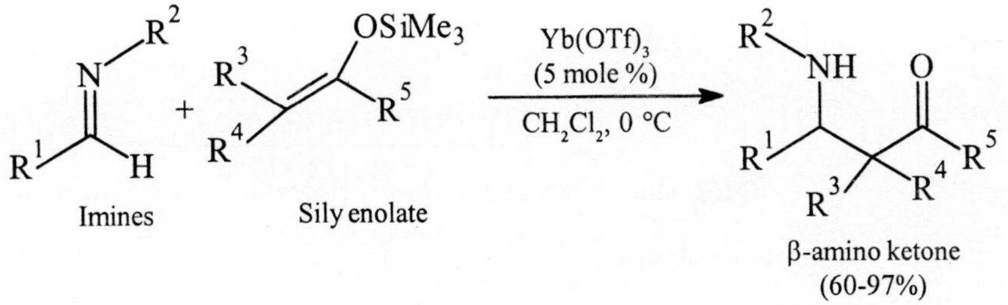

Scheme 154

Vinyl ethers also reacted with imines and amines in presence of catalytic amount of Yb(OTf)$_3$ to give corresponding β-amino ketones (Scheme 155).[261]

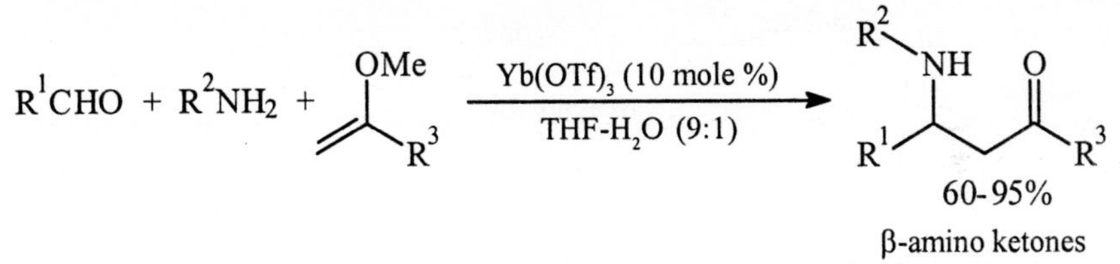

Scheme 155

The Mannich type reaction was also used for the synthesis of β-amino esters from aldehydes using Yb(OTf)$_3$ as catalyst (Scheme 156).[262]

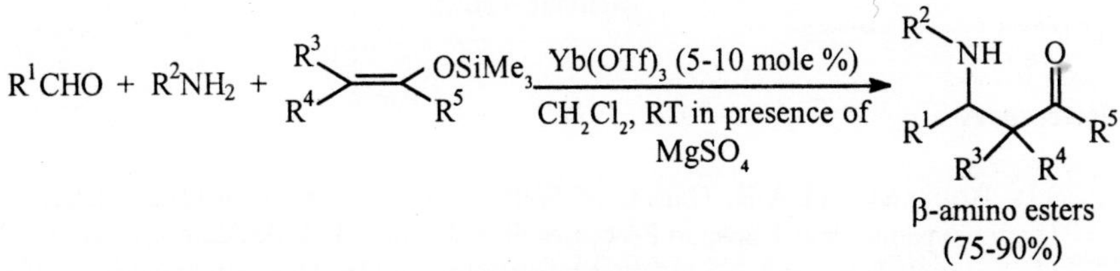

Scheme 156

12.19.7 Conversion of o-nitrochalcones into Quinolines and Indoles

Reduction of o-nitrochalcone under WGSR conditions[263] followed by the cyclisations give rise to the formation of quinolines and indoles (Scheme 157).[264]

12.19.8 Synthesis of Octadienols

The aqueous isomerization of butadiene gives octadienols (Scheme 158).[265]

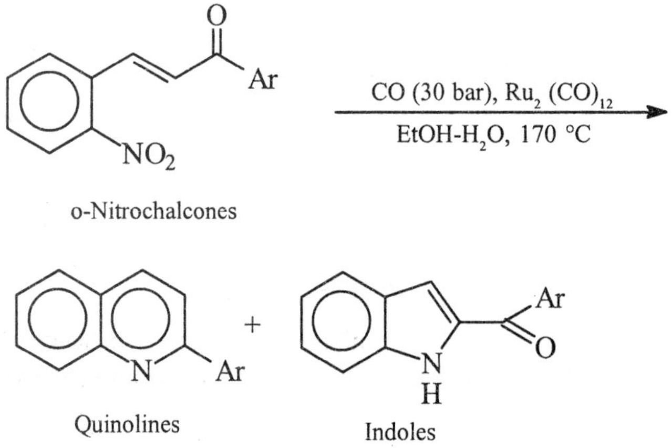

o-Nitrochalcones

Quinolines   Indoles

Scheme 157

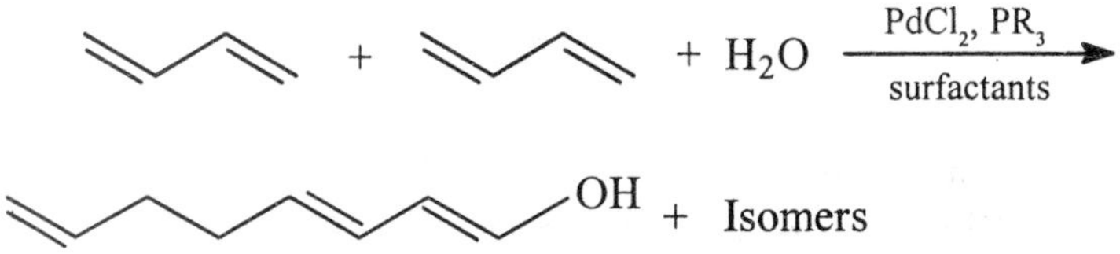

Scheme 158

References

1. S.B. Brummer and A.B. Gancy, in Water and Aqueous Solutions: Structure, Thermodynamics and Transport Processes, R.A. Horne, ed., Wiley-Interscience, 1970.
2. J.L. Kavanau, Water and Solute-Water Interactions, Holden-Day, San Francisco, 1964.
3. R. Brestow (a review), *Acc. Chem. Res.*, 1991, **24**, 159; P.A. Grieco (review), *Aldrichim. Acta*, 1991, **24**, 59.
4. O. Diels and K. Alder, *Liebigs Ann. Chem.*, 1931, **490**, 243; R.B. Woodward and H. Baer, *J. Am. Chem. Soc.*, 1948, **70**, 1161.
5. H. Hopff and C.W. Rautenstrauch, *U.S. Patent 2,262,002*; *Chem. Abstr.*, 1942, **36**, 1046.
6. D.C. Rideout and R. Breslow, *J. Am. Chem. Soc.*, 1980, **102**, 7816.
7. P.A. Grieco, K. Yoshida and P. Gardner, *J. Org. Chem.*, 1983, **26**, 3137.
8. S. Otto and J.B.F.N. Engberts, *Tetrahedron Lett.*, 1995, **36**, 2645.
9. R. Breslow and T. Guo, *J. Am. Chem. Soc.*, 1988, **110**, 5613.
10. T. Dunams, W. Hockstra, M. Pentaleri and D. Liotta, *Tetrahedron Lett.*, 1988, **29**, 3745.
11. C.K. Pai and M.B. Smith, *J. Org. Chem.*, 1995, **60**, 3731.
11a. P.P. Garner, 1999, Diels-Alder reactions in aqueous media (book) 'Organic Synthesis in Water', Paul A. Grieco, Ed., Blackie Academic and Professional, New York, pp. 1-47.
12. A.K. Saksena, V.M. Girijavallabhan, Y.T. Chen, E. Joe, R.E. Pike, J.A. Desai, D. Rane and A.K. Ganguly, *Heterocycles*, 1993, **35**, 129.
13. D.R. Williams, R.D. Gaston and I.B. Horton, *Tetrahedron Lett.*, 1985, **26**, 1391.

14. M.I. Fremery and E.K. Fields, *J. Org. Chem.*, 1963, **28**, 2537. For some several reviews see P.S. Bailey, *Chem. Rev.*, 1958, **58**, 925; R. Giegel, *Rec. Chem. Prog.*, 1957, **18**, 11.

15. D. Roger and S. Weireb, Hetero-Diels-Alder Methodology in Organic Synthesis, Academic Press, 1987.

16. P.A. Grieco and S.D. Larsen, *J. Am. Chem. Soc.*, 1985, **107**, 1768.

17. W. Oppolzer, *Angew. Chem. Int. Ed. Engl.*, 1972, **11**, 1031.

18. D.T. Parker, Hetero Diels-Alder Reactions, in 'Organic Synthesis in Water', Paul A. Grieco, Ed., Blackie Academic and Professional, New York, 1998, pp. 47-80.

19. G.A. Lee, *Synthesis*, 1982, 508.

20. K.V.R. Kishan and G.R. Desiruja, *J. Org. Chem.*, 1987, **52**, 4641; G.R. Desiruja and K.V.K. Kishan, *J. Am. Chem. Soc.*, 1989, **111**, 4838.

21. L. Claisen and O. Eisleb, *Annalen*, 1913, **401**, 21; L. Claisen and E. Tietze, Berichte, 1925, **58**, 275; C.D. Hund and L. Schmerling, *J. Am. Chem. Soc.*, 1937, **59**, 107.

22. R.B. Woodward and R.B. Hoffmann, *Angew. Chem. Int. Ed. Engl.*, 1969, **8**, 781.

23. P. Wipf, Comprehensive Organic Synthesis, B.M. Trost, I. Felming and L.A. Paquette, Eds., Pergman Press, New York, 1991, **Vol. 5**, p. 827; F.E. Zieglar, *Chem. Rev.*, 1988, **88**, 1423; S.T. Rhoads and N.R. Raulins, *Org. React.*, 1975, **22**, 1.

24. W.N. While and E.F. Wolfartt, *J. Org. Chem.*, 1970, **35**, 2196.

25. P.R. Andrews, G.D. Smith and I.G. Young, *Biochemistry*, 1973, **12**, 3492; S.D. Coply and J.J. Knowles, *J. Am. Chem. Soc.*, 1987, **109**, 5008.

26. E. Brandes, P.A. Grieco and J.J. Gajewski, *J. Org. Chem.*, 1989, **54**, 515.

27. J.J. Gajewski, J. Jurayj, D.R. Kimbrough, M.E. Gande, B. Ganem and B.K. Carpenter, *J. Am. Chem. Soc.*, 1987, **109**, 1170.

28. P.A. Grieco, F.B. Brandes, S. McCann and I.D. Clark, *J. Org. Chem.*, 1989, **54**, 5849.

29. J.E. McMurry, A. Andrus, G.M. Ksander, J.H. Musser and M.A. Johnson, *Tetrahedron*, 1981, **37**, 319 (supp) I.

30. P.A. Grieco, E.B. Brandes, S. McCann and J.D. Clark, *J. Org. Chem.*, 1989, **54**, 5849.

31. A. Lubineau, J. Augè, N. Bellanger and S. Caillebourdin, *J. Chem. Soc. Perkin Trans. I*, 1992, **13**, 1631.

32. G. Wittig, U. Schöllkopf, *Ber.*, 1954, **87**, 1318; G. Wittig and W. Haag, *Ber.*, 1955, **88**, 1654; U. Schöllkopf, *Angew. Chem.*, 1959, **71**, 260; S. Trippet, *Quart. Rev.*, 1963, **17**, 406.

33. C. Piechucki, *Synthesis*, 1976, 187; M. Mikolajczyk, S. Grzejszezak, W. Midura and A. Zatorski, *Synthesis*, 1976, 396.

34. M. Mikolajczk, S. Grzejszczyk, W. Midura and A. Zatorki, *Synthesis*, 1976, 396.

35. M. Schimtt, J.J. Bourguignon and C.G. Wermuth, *Tetrahedron Lett.*, 1990, **31**, 2145.

36. A. Michael, *J. Prak. Chem.*, 1887, **35(2)**, 349; E.D. Bergmann, D. Ginsburg and R. Pappo, *Org. Reactions*, 1959, **10**, 179.

37. Z.G. Hajos and D.R. Parrish, *J. Org. Chem.*, 1974, **39**, 1612; U. Elder, G. Sauer and R. Wiechert, *Angew. Chem. Int. Edn. Engl.*, 1971, **10**, 496.

38. N. Harada, T. Sugioka, U. Uda and T. Kuriki, *Synthesis*, 1990, 53.

39. J.F. Lavelle and P. Deslongchamps, *Tetrahedron Lett.*, 1988, **29**, 6033.

40. A. Lubineau and J. Auge, *Tetrahedron Lett.*, 1992, **33**, 8073.

41. R. Ballini, *Synthesis*, 1993, 687.

42. K. Sussang Karn, G. Fodor, I. Karle and C. George, *Tetrahedron*, 1988, **44**, 7047.

43. M. Makosza, *Tetrahedron Lett.*, 1966, 5489.

44. F. Toda, H. Takumi, M. Nagami and K. Tanaka, *Heterocycles*, 1988, **47**, 469.

45. H. Sakuraba, Y. Tanaka and F. Toda, *J. Incl. Phenom*, 1991, **11**, 195.

46. A.T. Nielsen and W.J. Houliban, *Org. Reactions*, 1968, **16**, 1; W. Foerst, Ed., Newer methods of preparative organic chemistry, 1971, **6**, 48.

47. T. Mukaiyama, K. Narasak and K. Banno, *Chem. Lett.*, 1973, 1011; T. Mukaiyama, K. Banno and K. Narasaka, *J. Am. Chem. Soc.*, 1974, **96**, 7503; T. Mukaiyama, *Org. React.*, 1982, **28**, 203.
48. K. Banno and T. Mukaiyama, *Chem. Lett.*, 1976, 279.
49. A. Lubineau, *J. Org. Chem.*, 1986, **51**, 2142; A. Lubineau, E. Meyer, *Tetrahedron*, 1988, **44**, 6065.
50. S. Kobayashi and I. Hachiya, *J. Org. Chem.*, 1994, **59**, 3590; For a review on lanthanides catalysed organic reactions in aqueous media, See S. Kobayashi, *Synlett.*, 1994, 589.
51. P.T. Baurona, K.G. Rosauer and L. Dai, *Tetrahedron Lett.*, 1995, **26**, 4009.
52. T.B. Ayed and H. Amri, *Synth. Commun.*, 1995, **25**, 3813.
53. S. Kobayashi, *Chem. Lett.*, 1991, 2187.
54. S. Kobayashi and I. Hachiya, *Tetrahedron Lett.*, 1992, 1625.
55. S. Kobayashi, Water Stable rare earth Lewis-acid catalysis in aqueous and organic solvents in organic synthesis in water, Paul A. Grieco, Ed., Blackie Academic and Professional, 1998, pp. 262-302.
56. F. Fringuelli, G. Pani, O. Piematti and F. Pizzo, *Tetrahedron*, 1994, **50**, 11499.
57. F. Knoevenagel, *Ber.*, 1898, **31**, 2596.
58. J.R. Johnson, *Org. Reactions*, 1942, **1**, 210.
59. O. Doebner, 1900, **33**, 2140.
60. Y. Nakono, S. Nik, S. Kinouchi, H. Miyamae and M. Igarashi, *Bull. Chem. Soc. Japan*, 1992, **65**, 2934.
61. J. Auge, M. Lubin and A. Lubineau, *Tetrahedron Lett.*, 1994, **35**, 7947.
62. Grinder, *Ann. Chim. Phys.*, 1892, **26**, 369.
63. J.B. Conant and H.B. Cutter, *J. Am. Chem. Soc.*, 1926, **48**, 1016.
64. P. Karrer, Y. Yen and I. Reichstein, *Helv. Chim. Acta.*, 1993, **13**, 1308.
65. A. Clerici and O. Porta, *Tetrahedron Lett.*, 1982, **23**, 3517.
66. A. Clerici and O. Porta, *J. Org. Chem.*, 1982, **47**, 2852; A. Clerici, O. Porta and M. Riva, *Tetrahedron Lett.*, 1981, **22**, 1043.
67. A. Clerici and O. Porta, *J. Org. Chem.*, 1983, **48**, 1690; *Tetrahedron*, 1983, **39**, 1239; A. Clerici, O. Porta and P. Zago, *Tetrahedron*, 1986, **42**, 561; A. Clerici and O. Porta, *J. Org. Chem.*, 1989, **54**, 3872.
68. P. Delair and J.L. Luche, *J. Chem. Soc. Chem. Commun.*, 1989, 398.
69. K. Kalyanam and V.G. Rao, *Tetrahedron Lett.*, 1993, **34**, 1647.
70. A.J. Lapworth, *J. Chem. Soc.*, 1903, **83**, 995; 1904, **85**, 1206; J.S. Buck, *Organic Reactions*, 1948, **IV**, 269.
71. J. Solodar, *Tetrahedron Lett.*, 1971, 287.
72. W. Tagaki and H. Hara, *J. Chem. Soc. Chem. Commun.*, 1973, 891.
73. H. Stetter and G. Dambkes, *Synthesis*, 1977, 403.
74. E.T. Kool and R. Breslow, *J. Am. Chem. Soc.*, 1988, **110**, 1596.
75. L. Claisen and A. Claparede, *Ber.*, 1981, **14**, 2460; J.G. Schmidt, *Ber.*, 1881, **14**, 1459; H.O. House, *Modern Synthetic Reactions*, W.A. Benjamin, California, 2nd ed., 1972, 632-639.
76. T. Mukaiyama, K. Banno and K. Narasaka, *J. Am. Chem. Soc.*, 1974, **96**, 7503; T. Mukaiyama, *Chem. Lett.*, 1982, 353; *J. Am. Chem. Soc.*, 1973, **95**, 967; *Chem. Lett.*, 1986, 187; K. Banno and T. Mukaiyama, *Chem. Lett.*, 1976, 279; Organic Reactions, 1982, **28**, 187.
77. A. Lubineau, *J. Org. Chem.*, 1986, **51**, 2142; A. Lubineau and E. Meyer, *Tetrahedron*, 1988, **44**, 6065.
78. F. Fringueli, G. Pani, O. Piermatti and F. Pizzo, *Life Chem. Rep.*, 1995, **13**, 133.
79. T. Jeffery, *Chem. Commun.*, 1984, 1287.

80. N.A. Bumagin, P.G. More and I.P. Beletskaya, *J. Organometallic Chem.*, 1989, **371**, 397.
81. N.A. Bumagin, N.P. Andryukhova and I.P. Beletskaya, *Dolk. Akad. Nauk. SSSR*, 1990, **313**, 107.
82. N.A. Bumagin, L.I. Sukhomlinova, A.N. Vanchikov, T.P. Tolstaya and I.P. Beletskaya, *Bull. Russ. Acad. Sci. Div. Chem. Sci.*, 1992, **41**, 2130; N.N. Demik, M.M. Rabachnik, M.M. Novikova and I.P. Beletskaya, *Zh. Org. Khim.*, 1995, **31**, 64; T. Jeffery, *Tetrahedron Lett.*, 1994, **35**, 3051; S. Lemaine-Audoire, M. Savignac, C. Dupuis and J.P. Genet, *Tetrahedron Lett.*, 1966, **37**, 2003; J. Diminnie, S. Metts and E.J. Parson, *Organometallics*, 1995, **14**, 4023; P. Reardon, S. Metts, C. Crittendon, P. Daugherity and E.J. Parsons, *Organometallics*, 1995, **14**, 3810.
83. A. Strecker, *Ann.*, 1850, **75**, 27; 1854, **91**, 345; D.T. Moury, *Chem. Rev.*, 1948, **42**, 236.
84. D.T. Moury, *Chem. Rev.*, 1948, **42**, 189.
85. K. Weinges, *Chem. Ber.*, 1971, **104**, 3594.
86. T. Paul Anastas and John C. Warner, *Green Chemistry*, Theory and Practice, Oxford University Press, 1998, pp. 98-99.
87. A. Wurtz, *Ann. Chim. Phys.*, 1855, **44(3)**, 275; *Ann.*, 1855, **96**, 364; R.E. Buntrock, *Chem. Rev.*, 1968, **68**, 209.
88. C.J. Li and T.H. Chan, *Organometallics*, 1991, **10**, 2548.
89. M. Hudlicky, Oxidations in Organic Chemistry, ACS Monograph 186, American Chemical Society, Washington DC, 1990; V.K. Ahluwalia and R.K. Parashar, Organic Reaction Mechanism, Narosa Publishing House, New Delhi, 2002, pp. 137-219.
90. A.I. Vogel, Textbook of practical organic chemistry including qualitative organic analysis, Longmans, London, 1988.
91. F. Fringuelli, R. Germani, F. Pizzo and G. Savelli, *Tetrahedron Lett.*, 1989, **30**, 1427.
92. A.S. Kende, P. Delair, B.E. Blass, *Tetrahedron Lett.*, 1994, **44**, 8123.
93. K.S. Kirshenbaum and K.B. Sharpless, *J. Org. Chem.*, 1985, **50**, 1979.
94. B. Meunier, *Chem. Rev.*, 1992, **92**, 1411.
95. B. Notari, *Stud. Surf. Sci. Catal.*, 1991, **67**, 243; R.A. Sheldon, *J. Mol. Catal.*, 1980, **7**, 107.
96. K. Otsuka, M. Yoshinaka and I. Yamanaka, *J. Chem. Soc. Chem. Commun.*, 1993, 611.
97. G. Berti in Topics in Streochemistry, N.L. Allinger and E.L. Eliel, Eds., **Vol. 7**, p. 97, John Wiley, 1967.
98. E. Fringuelli, R. Germani, F. Pizzo, F. Santinelli and G. Savelli, *J. Org. Chem.*, 1992, **57**, 1198; F. Fringuelli, F. Pizzo and R. Germani, *Synlett.*, 1991, 475.
99. K.B. Sharpless and R.C. Mickaelson, *J. Am. Chem. Soc.*, 1973, **95**, 6136; B.E. Rossiter, T.R. Verhoeven and K.B. Sharpless, *Tet. Lett.*, 1997, 4733.
100. M. Hudlicky, Oxidation in Organic Chemistry, ACS Monograph 186, American Chemical Society, Washington DC, 1990.
101. J. Muzart, *Synthesis*, 1995, 1325; K.L. Reed, J.T. Gupton and T.L. Solarz, *Synth. Commun.*, 1989, **19**, 3579.
102. G. Beri, in Topics in Stereochemistry, N.L. Allinger, E.L. Eliel, Eds., **Vol. 7**, p. 93, John Wiley, 1967.
103. G.B. Payne and P.H. Williams, *J. Org. Chem.*, 1959, **24**, 54.
104. K. Kirshenbaum and K.B. Sharpless, *J. Org. Chem.*, 1985, **50**, 1979.
105. R. Curci, M. Fiarentino, L. Troisi, J.O. Edwards and R.H. Pater, *J. Org. Chem.*, 1980, **45**, 4758; P.F. Corey and F.E. Ward, *J. Org. Chem.*, 1986, **51**, 1925.
106. T.C. Zheng and D.E. Richandson, *Tetrahedron Lett.*, 1991, **56**, 1253.
107. F. Toda, H. Takumi, M. Nagami and K. Tanaka, *Heterocycles*, 1998, **47**, 469.
108. K.A. Hofmann, *Ber.*, 1912, **45**, 3329; P. Grieco, Y. Ohfume, Y. Yokoyama and W.J. Owens, *J. Am. Chem. Soc.*, 1979, **101**, 4749.

109. N.A. Milas and S. Sussman, *J. Am. Chem. Soc.*, 1936, **58**, 1302; R. Daniels and J.L. Fischer, *J. Org. Chem.*, 1963, **28**, 320.
110. M. Mugdan and D.P. Young, *J. Chem. Soc.*, 1949, 2988.
111. K.B. Sharpless and K. Akashi, *J. Am. Chem. Soc.*, 1976, **98**, 1986.
112. M. Minato, K. Yamamoto and J. Tsuji, *J. Org. Chem.*, 1900, **55**, 766.
113. W.B. Weber and J.P. Shephard, *Tetrahedron Lett.*, 1972, 4907.
114. M. Mugdan and D.P. Young, *J. Chem. Soc.*, 1949, 2988.
115. H. Abedel-Rahman, J.P. Adams, A.L. Boyes, M.J. Kelly, D.P. Mansfield, P.A. Procopiou, S.M. Roberts, D.H. Slee, P.J. Sidebottom, V. Silk and N.S. Watson, *J. Chem. Soc., Chem. Commun.*, 1993, 1841.
116. F. Fringuelli, R. Germani, F. Pizzo and G. Savelli, *Synth. Commun.*, 1989, **19**, 1939.
117. R.G. Mathys, A. Schmid and B. Witholt, *Biotechnol. Bioeng.*, 1999, **64**, 459.
118. R.G. Mathys, O.M. Kuts and B. Witholt, *J. Chem. Tech. Biotech.*, 1998, **71**, 315.
119. M.A. Oglioruso and J.F. Wolfe, *Synthesis of Carboxylic acids and their derivatives*, S. Patai and Z. Rappoport, Wiley, New York, 1991.
120. R.H. Dodd and M. LeHyaric, *Synthesis*, 1993, 295.
121. F. Fringuelli, R. Pellegrino, O. Piermatti and F. Pizzo, *Synth. Commun.*, 1994, **24**, 2665.
122. H.D. Dakin, *OS*, 1941, **1**, 149; J.E. Lettler, *Chem. Rev.*, 1949, **45**, 385; H.D. Dakin, *J. Am. Chem. Soc.*, 1909, **42**, 477.
123. G.W. Kabalka, N.K. Reddy and C. Narayana, *Tetrahedron Lett.*, 1992, **33**, 865.
124. A.V. Baeyer and V. Villiger, *Ber.*, 1899, **32**, 3625.
125. F. Fringuelli, R. Germani, F. Pizzo and G. Savelli, *Gazz. Chem. Ital.*, 1989, **119**, 249.
126. A.E. Thomas and F. Ray, *Tetrahedron*, 1992, **48**, 1927.
127. K. Tanaka and F. Toda, *Chem. Rev.*, 2000, **100**, 1028-29.
128. N. Kornblum and W.J. Jones, *Org. Synthesis*, 1963, **43**, 87.
129. K.S. Webb and V. Seneviratrie, *Tetrahedron Lett.*, 1995, **36**, 2377.
130. A. Albini and S. Pietra, Heterocyclic N-oxides; CRC Press, Boca Raton, Fl. 1991; A. Albini, *Synthesis*, 1993, 263.
131. G.J. Robke and E. Behrman, *J. Chem. Res. (S)*, 1993, 412.
132. A. Sakurai and M. Midorikawa, *Bull. Chem. Soc. Japan*, 1968, **41**, 430.
133. R. Balicki and L. Kaczmarek, *Synth. Commun.*, 1993, **23**, 3149.
134. H. Heaney, *Aldrichim. Acta*, 1993, **26**, 35.
135. A. McKillop and D. Kemp, *Tetrahedron*, 1989, **45**, 3299.
136. G.W. Kabalka, S.M. Deshpande, P.P. Wadgankar and N. Chatla, *Synth. Commun.*, 1990, **20**, 1445.
137. K.L. Reed, J.T. Gupton and T.L. Solarz, *Synth. Commun.*, 1990, **20**, 563.
138. B. Baudoin, Y. Ribeill and N. Vicker, *Synth. Commun.*, 1993, **23**, 2833.
139. F. Fringuelli, R. Pellegrino, O. Piermatti and F. Pizzo, *Synth. Commun.*, 1994, **24**, 2665.
140. K.S. Webb, *Tetrahedron Lett.*, 1994, **35**, 3457.
141. T.C. Zheng and P.E. Richardson, *Tetrahedron Lett.*, 1995, **36**, 833.
142. A. McKillop and J.A. Tarbin, *Tetrahedron*, 1987, **43**, 1753.
143. F. Fringuelli, R. Pellegino and F. Pizzo, *Synth. Commun.*, 1993, **23**, 3157.
144. F. Fringuelli, R. Pellegrino, O. Piermatti and F. Pizzo, *Synth. Commun.*, 1994, **24**, 2665.
145. G.A. Lee and H.H. Freedman, *Tetrahedron Lett.*, 1976, 3985.
146. K. Dimroth and W. Tuncher, *Synthesis*, 1977, 339.
147. Michael B. Smith, Organic Synthesis, McGraw Hill, New York, 1994, pp. 343-484; V.K. Ahluwalia and R.K. Parashar, Organic Reaction Mechanisms, Narosa Publishing House, New Delhi, 2002, pp. 221-263 and the references cited therein.
148. K. Kalck and F. Monteil, *Adv. Organomet. Chem.*, 1992, **134**, 219.
149. K. Wan and M.E. Davis, *Tetrahedron Asymmetry*, 1993, **4**, 2461; K. Wan and M.L. Davis, *J. Chem. Soc. Chem. Commun.*, 1993, 1262.

150. C. Petrier and J.L. Luche, *Tetrahedron Lett.*, 1987, **28**, 1234.
151. C. Petrier and J.L. Luche, *Tetrahedron Lett.*, 1987, **28**, 2351.
152. R.E. Harman, S.K. Gupta and D.J. Brown, *Chem. Rev.*, 1973, **73**, 21.
153. T. Okano, M. Kaji, S. Isotani and J. Kiji, *Tetrahedron Lett.*, 1992, **33**, 5547; The application and chemistry of catalyst by soluble transition metal complex, Wiley, London, 1980.
154. G.A. Molander, *Chem. Rev.*, 1992, **92**, 26; J.A. Soderquist, *Aldrichim. Acta.*, 1991, **24**, 15.
155. G. Papadogianakis, Specialist Periodical Reports, Catalysis, J.J. Spivey (Ed.), The Royal Society of Chemistry, 1997, Vol. 13, pp. 114-193.
156. C. Larpent and G. Meignan, *Tetrahedron Lett.*, 1993, **34**, 4331.
157. V.K. Ahluwalia and R.K. Parashar, Organic Reaction Mechanism, Narosa Publishing House, New Delhi, 2002, pp. 221-263.
158. D. Mitchell, C.W. Doccke, L.A. Hay, T.M. Koening and D.D. Wirth, *Tetrahedron Lett.*, 1995, **36**, 5335.
159. G. Lamaty, M.H. Reviere and J.P. Roque, *Bull. Soc. Chim. Fr.*, 1983, 33.
160. B. Boyer, J.F. Betzer, G. Lamaty, A. Leydet and J.P. Roque, *New J. Chem.*, 1995, **19**, 807.
161. C. Denis, B. Laignel, D. Plusquellec, J.Y. Le Marouille and A. Botrel, *Tetrahedron Lett.*, 1996, **37**, 53.
162. F. Toda, R. Kiyoshige and M. Yagi, *Angew. Chem. Int. Ed. Engl.*, 1989, **28**, 320.
163. M.J. Burk, T.G.P. Harper and C.S. Kalberg, *J. Am. Chem. Soc.*, 1995, **117**, 4423.
164. T. Tsukinok, K. Ishimoto, H. Tsuzuki, S. Mataka and M. Tashiro, *Bull. Chem. Soc. Japan*, 1993, **66**, 3419.
165. M. Mordolol, *Tetrahedron Lett.*, 1993, **34**, 1681.
166. A.K. Singh, R.K. Bakshi and E.J. Corey, *J. Am. Chem. Soc.*, 1987, **109**, 6187; E. Hasegawa and D.P. Curran, *J. Org. Chem.*, 1993, **58**, 5008.
167. A.P. Krapcho and D.A. Seidman, *Tetrahedron Lett.*, 1981, **22**, 179.
168. V. Stagopan and S.B. Chandalia, *Synth. Commun.*, 1989, **19**, 1217.
169. C. Petrier and J.L. Luche, *Tetrahedron Lett.*, 1987, **33**, 5417; R.N. Baruah, *Tetrahedron Lett.*, 1992, **33**, 5417.
170. C. Petrier, I.L. Lavaitte and C. Morat, *J. Org. Chem.*, 1989, **54**, 5313.
171. C. Hasegawa and D.P. Curran, *J. Org. Chem.*, 1993, **58**, 5008.
172. F. Fache, S. Lehueds and M. Lemaine, *Tetrahedron Lett.*, 1995, **36**, 885.
173. Y. Kamochi and T. Kudo, *Chem. Pharm. Bull.*, 1995, **43**, 1422.
174. S. Hanessian and C. Girard, *Synlett.*, 1994, 861.
175. S.K. Chung and Q.Y. Hu, *Synth. Commun.*, 1982, **12**, 261.
176. C.J. Rizzo, N.K. Dunlap and A.B. Smith, *J. Org. Chem.*, 1987, **52**, 5280.
177. C.E. Castro and W.C. Kray, Jr., *J. Am. Chem. Soc.*, 1963, **85**, 2768.
178. A. Ono, E. Fujimoto and M. Ueno, *Synthesis*, 1986, 570.
179. D.V. Davydov and I.P. Belotskaya, *Russ. Chem. Bull.*, 1993, **42**, 572.
180. J. Light and R. Breslow, *Tetrahedron Lett.*, 1990, **31**, 2957; J. Light and R. Breslow, *Org. Synth.*, 1993, **72**, 199.
181. A. Clerici and O. Porta, *Tetrahedron*, 1982, **38**, 1293.
182. N. Petragnani and J.V. Comasseto, *Synthesis*, 1986, 1.
183. S.K. Boyer, J. Bach, J. McKenna and E. Jagdmann, Jr., *J. Org. Chem.*, 1985, **50**, 3409.
184. J. Bressner, M. Julia, M. Launay and J.P. Stacino, *Tetrahedron Lett.*, 1982, **23**, 3265.
185. A. Schonberg and M.Z. Barakat, *J. Chem. Soc.*, 1989, 892; L.E. Overman and E.M. O'Connor, *J. Am. Chem. Soc.*, 1976, **98**, 771.
186. R.E. Michel, Recovery of methyl esters of aromic acids and glycols from thermoplastic polyester scrap using methanol vapor, *Eur. Patent*, 484, 963, May 13, 1992;

R.R. Hepner; R.E. Michel, Process for Separation of Glycerls from Dimethyl Terephttalate, *U.S. Patent, 5,391,263*, Feb. 21, 1995; R.E. Michel, Recovery of Dimethyl Terephttalate from Polymer Waste, *U.S. Patent 5,504,122*, April 2, 1996.

187. W. Hater, Environmental compatible scale inhibitor for the mining industry, *Corrosion*, 1998, 213.

188. K.C. Low, A.P. Wheeler and L.P. Koskan, Advances in Chemistry Series 248; *American Chemical Society*, Washington D.C., 1996.

189. K. Komiya, S. Kukuoka, M. Aminaka, K. Hasegawa, H. Hachiya, H. Okamoto, T. Watanabe, H. Yoneda, I. Fukawa and T. Dozone, 'New process for producing polycarbonate without phosphene and methylene chloride', In Green Chemistry: Designing Chemistry for the environment, P.T. Anastas and T.C. Williamson, eds., American Chemical Society Symposium Series, No. 626 (ed.), American Chemical Society, 1996, pp. 1-17.

190. R.F. Rhinehard and H.P. Smith, *Polym. Lett.*, 1965, **3**, 1049.

191. B.M. Novak and R.H. Grubbs, *J. Am. Chem. Soc.*, 1988, **110**, 7542.

192. S.B.T. Nguyen, L.K. Johnson, R.H. Grubbs and J.W.A. Ziller, *J. Am. Chem. Soc.*, 1992, **114**, 3974.

193. S.T. Nguyen and R.H. Grubbs, *J. Am. Chem. Soc.*, 1993, **115**, 9858; C. Fraser and R.H. Grubbs, *Macromolecules*, 1995, **28**, 7248.

194. P. Bernhard, H. Lehmann and A. Ludi, *J. Chem. Soc. Chem. Commun.*, 1981, 1216; P. Bernhard, M. Biner and A. Ludi, *Polyhedron*, 1990, **9**, 1095.

195. S. Wache, *J. Organomet. Chem.*, 1995, **494**, 235.

196. T.I. Wallow and B.M. Novak, *J. Am. Chem. Soc.*, 1991, **113**, 7411.

197. C.J. Li, D. Wang and W.T. Slaven, *Tetrahedron Lett.*, 1996, **37**, 4459.

198. E. Paterno and G. Chieffi, *Gazz. Chem. Ital.*, 1909, **39**, 341; G. Buchi, C.G. Inman and E.S. Lipinsky, *J. Am. Chem. Soc.*, 1954, **76**, 4327; G.C. Lange, *Tetrahedron Lett.*, 1971, **12**, 712.

199. D. Bellus et al., *Chem. Rev.*, 1967, 599; J.C. Anderson and C.B. Reese, *Proc. Chem. Soc.*, 1960, 217.

200. A.G. Griesbeck, K. Kramer and M. Oelgemöller, *Green Chemistry*, 1999, **1**, 205.

201. For a review of organic photochemistry in organic solvents as well as in aqueous solvent see R. Ramamurthy, *Tetrahedron*, 1986, **42**, 5753.

202. M.S. Sayamala and V. Ramamurthy, *J. Org. Chem.*, 1986, **51**, 3712.

203. Y. Ito, T. Karita, K. Kunimoto and T. Matsuura, *J. Org. Chem.*, 1989, **54**, 587.

204. T. Tamaki, *Chem. Lett.*, 1984, 53; T. Tamaki and T. Kokubu, *Inclusion Phenom*, 1984, **2**, 815.

205. K. Nakatani, J. Shiral, R. Tamaki and I. Saito, *Tetrahedron Lett.*, 1995, **36**, 5363.

206. J.H. Liu and R.G. Weiss, *J. Org. Chem.*, 1985, **50**, 3655.

207. H. Tateba and S. Mihara, *Agric. Biol. Chem.*, 1991, **55**, 873.

208. P.G. Tratnyek and J.J. Hoigne, *Photochem. Photobiol. A. Chem.*, 1994, **84**, 153 and the references cited therein.

209. V. Ramamurthy, Ed., Photochemistry of organized and constraint media, VCH, Weinheimn, German, 1991; CRC Handbook of organic photochemistry and photobiology, W.H. Horspool, Ed., CRC Press, Boca Raton, Fl., 1995.

210. Y. Ito, B. Borecka, J. Trotter and J.R. Scheffer, *Tetrahedron Lett.*, 1995, **36**, 6083; Y. Ito, B. Borecka, G. Olovsson, J. Trotter and J.R. Scheffer, *Tetrahedron Lett.*, 1995, **36**, 6087; Y. Ito, B. Olovasson, *J. Chem. Soc.*, Perkin Trans. I, 1997, 127.

211. K.S. Feldmann and R.F. Compbell, *J. Org. Chem.*, 1995, **60**, 1924.

212. R. Gnanaguru, K. Ramasubbu, K. Venkatesan and V. Ramamurthy, *J. Org. Chem.*, 1985, **50**, 2337; J.N. Moorthy, K. Venkatesan and R.G. Weiss, *J. Org. Chem.*, 1992, **57**, 3292; K. Venkatesan, T.N. Guru Row and K.J. Venkatasan, *J. Chem. Soc. Perkin Trans.*

2, 1996, 1475; K. Vishnumurthy, T.N. Guru Row and K. Venkatesan, *J. Chem. Soc. Perkin Trans. 2*, 1997, 615.

213. Y. Ito, T. Matsuura, K. Tabata and M. Ji-Ben, *Tetrahedron Lett.*, 1987, **43**, 1307.

214. K. Tanaka and F. Toda, *Chem. Rev.*, 2000, **100**, 1044-1066.

215. P.M. Bersier, L. Carlsson and J. Bersier, in Topics in Chemistry, J.D. Bunitz, S. Hafner, J.M. Ito, J.M. Lehn, K.N. Raymond, C.W. Rees and J.T.F., eds., Springer-Verlag, 1994, **Vol. 70**.

216. D.E. Danly and C.J.H. King, in Organic Electrochemistry, 3rd ed., H. Lund and M.M. Baizer, eds., Marcel Dekker, New York, 1991; M.M. Baizer and D.E. Danly, *Chemtech.*, 1980, **10(3)**, 161.

217. W.C. Drinkard and R.V. Lindsay, *U.S. Patent 3,496,215* (1970) to DuPont.

218. Topchiev, Pavlov, *Chem. Abstr.*, 1953, **47**, 8002h; Domingue et al., *J. Chem. Ed.*, 1952, **29**, 446.

219. T. Isoya, R. Kakuta and C. Kawamura, *U.S. Patent 3,896,611* (1975) – Asahi Kasu.

220. R.L. Taylor, *Chem. Met. Eng.*, 1973, **44**, 588.

221. H. Nohe, *Chem. Ing. Tech.*, 1974, **46**, 594.

222. T.T. Sugano, B.A. Schenber, J.A. Walburg and N. Shuster, *U.S. Patent 3,992,269* (1976) to Diamond Snamrock Corpn.

223. D. Degner, in Techniques of Electroorganic Synthesis, Part III, J.N. Walburg and B.V. Tilak, eds., Wiley, New York, 1982.

224. M. Taniquchi, in Recent Advances in Electroorganic Synthesis, S. Tori, ed. Elsevier, Amsterdam, 1987.

225. J. Yoshida, J. Hashmoto and N. Kawabata, *J. Org. Chem.*, 1982, **47**, 3575.

226. J. Tsuji and M. Minato, *Tetrahedron Lett.*, 1987, **28**, 3683.

227. M. Kulka, *J. Am. Chem. Soc.*, 1946, **68**, 2472.

228. J. Kaulen and H.J. Schäfer, *Synthesis*, 1979, 513.

229. T. Shono, Y. Matsumura, J. Hayashi and M. Mizoguchi, *Tetrahedron Lett.*, 1979, 165.

230. H. Ruholi and H.J. Schäfer, *Synthesis*, 1988, 54.

231. F. Fichier and F. Ackermann, *Helv. Chim. Acta*, 1919, **2**, 583.

232. F.L.M. Pattison, J.B. Stothers and A.G. Woolford, *J. Am. Chem. Soc.*, 1956, **78**, 2255.

233. D.V. McGrath and R.H. Grubbs, *Organometallics*, 1994, **13**, 224.

234. C.J. Li, D. Wang and D.L. Chem, *J. Am. Chem. Soc.*, 1995, **117**, 12867.

235. For a review, see M.M.T. Khan, Platinum Metals Rev., 1991, **35(2)**, 70.

236. D.J. Colquhoun, H.M. Thomson and M.V. Twigg, Carbonylation: Direct Synthesis of Carbonyl Compounds, Plenum Press, New York, 1991.

237. H. Alper, *J. Organometallic Chem.*, 1986, **300**, 1.

238. V.V. Grushin and H. Alper, *J. Org. Chem.*, 1993, **58**, 4794.

239. R.F. Heck, Palladium Reagents in Organic Synthesis, Academic Press, London, 1985.

240. H. Alper, *J. Organomet. Chem.*, 1986, **300**, 1 and the references cited therein.

241. F. Joo and H. Alper, *Organometallics*, 1985, **4**, 1775.

242. N.A. Bumagin, K.V. Nikitin and I.P. Beletskaya, *J. Organomet. Chem.*, 1988, **358**, 563.

243. T. Okano, T. Hayashi and J. Kiji, *J. Bull. Chem. Soc. Japan*, 1994, **67**, 2339.

244. H. Urata, O. Kosukegawa, Y. Tshil, H. Yugari and T. Fuchikami, *Tetrahedron Lett.*, 1989, **30**, 4403.

245. T. Joh, K. Doyama, K. Onitsuka, T. Shichara and S. Takahashi, *Organometallics*, 1991, **10**, 2493; T. Joh, H. Nagata and S. Takahashi, *Chem. Lett.*, 1992, 1305.

246. H. Alper, H. Arzoumanian, J.F. Petrignani and M. Saldana Maldonado, *J. Chem. Soc. Chem. Commun.*, 1985, 340.

247. M. Lin and A. Sen, *J. Am. Chem. Soc.*, 1992, **114**, 7307; *Nature*, 1994, **368**, 613.

248. H. Alper, J.A. Currie, H.J. des Abbayes, *J. Chem. Soc. Chem. Commun.*, 1978, 311; V. Galamb, M. Gopal and H. Alper, *J. Chem. Soc. Chem. Commun.*, 1983, 1154.

249. G. Papadogianakis, L. Maat and R.A. Sheldon, *J. Chem. Technol. Biotechnol.*, 1997, **70**, 83.

250. B. Cornits, Hydroformylation, Oxo Synthesis, Roelen Reaction: New Synthesis with carbon monoxide, Spinger-Verlag, Berlin, 1980.

251. B. Cornits and E.G. Kuntz, *J. Organometalic Chem.*, 1995, **502**, 177.

252. E. Kuntz, *U.S. Patent 4,248,802, Rhone-Pouline Ind.*, 1981; *Chem. Abstr.*, 1977, **87**, 101944n; J. Jenck, *Fr. Patent 2,478,078* to Rhone, Poulence Industries (03-12-1980); E. Kuntz, *Fr. Patent 2,349,562* to Rhone-Poulence Industries (04-24-1976).

253. W.A. Herrmann, C.W. Kohlpainter, R.B. Manetsberger and H. Bahrmann (Hoechst AG), DE-B 4220,267A, 1992.

254. J.P. Archancet, M.E. Davis, J.S. Merola, B.E. Hansan, *Nature*, 1989, **339**, 454; J. Haggin, *Chem. Eng. News*, 1992, **70(17)**, 40.

255. G. Fremy, Y. Castanet, R. Grzybek, E. Monflier, A. Mortreux, A.M. Trzeciak and J.J. Ziolkowski, *J. Organomet. Chem.*, 1995, **505**, 11.

256. I.P. Beletskaya and C.V. Cheprakov, Aqueous transition-metal catalysis, in organic synthesis in water, Paul A. Grieco, ed., Blackie Academic and Professional, 1998, pp. 196-205.

257. Y. Shigemasa, K. Yokoyama, H. Sashiwa and H. Saimoto, *Tetrahedron Lett.*, 1994, **35**, 1263.

258. U. Weiss and J.M. Edwards, *Tetrahedron Lett.*, 1968, 4885.

259. C. Mannich and W. Krosche, *Arch. Pharm.*, 1912, **250**, 647; F.E. Blike, *Org. Reactions*, 1942, **1**, 303.

260. I. Ojima, S-I. Inaba and K. Yoshida, *Tetrahedron Lett.*, 1977, 3643.

261. S. Kobayashi and H. Ishitani, *J. Chem. Soc. Chem. Commun.*, 1995, 1379.

262. S. Kobayashi, M. Araki and M. Yasuda, *Tetrahedon Lett.*, 1995, **36**, 5773.

263. R.M. Laini and E.J. Crawford, *J. Mol. Catal.*, 1988, **44**, 357.

264. S. Tollari, M. Fedele, E. Bettethine and S. Cenimi, *J. Mol. Catal. A. Chem.*, 1996, **111**, 37.

265. D.M. Roundhill and S. Ganguly, *Organometallics*, 1993, **12**, 4825; E. Monflier, P. Bourdauducq, J.L. Couturier, J. Kervennel and A. Mortreux, *Appl. Catal. A-Gen.*, 1995, **131**, 167.

13. Organic Synthesis in Solid State

13.1 Introduction
The earlier belief that no reaction is possible without the use of a solvent is no more valid. It has been found that a large number of reactions occur in solid state without the solvent. In fact in a number of cases, such reactions occur more efficiently and with more selectivity compared to reactions carried out in solvents. Such reactions are simple to handle, reduce pollution, comparatively cheaper to operate and are especially important in industry. There is some literature available on different aspects of organic synthesis in solid state.[1-9] It is believed that solvent-free organic synthesis and transformations are industrially useful and largely green.

In the present discussion, the organic synthesis will be presented in two parts:
1. Solid phase organic synthesis without using any solvent.
2. Solid supported organic synthesis.

13.2 Solid Phase Organic Synthesis Without Using Any Solvent
The earliest record of an organic reaction in dry state is the Claisen rearrangement[10] of allyl phenylether to o-allylphenol (Scheme 1).

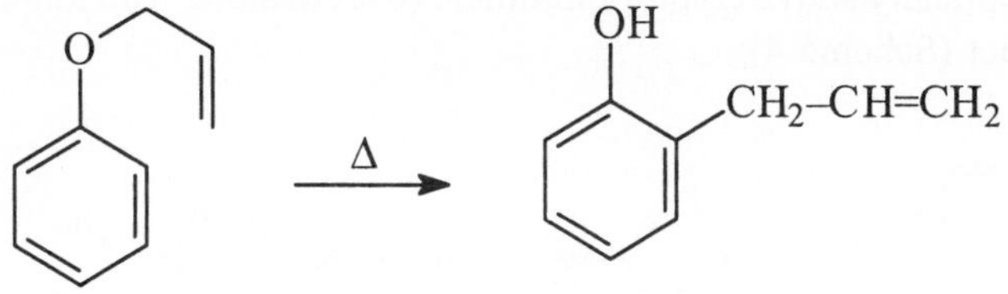

Allyl phenyl ether   o-Allylphenol

Scheme 1

The following gives a brief account of solid phase organic synthesis without the use of any solvent.

13.2.1 Halogenation
Solid phase bromination is known[11] since 1963 but systematic work was done in 1987. It was found[12] that crystalline cinnamic acid on bromination (gas-

solid phase) gives exclusively the erythro isomer, but its chlorination gives the threo and erythro isomers in 88 and 12% yields, respectively (Scheme 2).

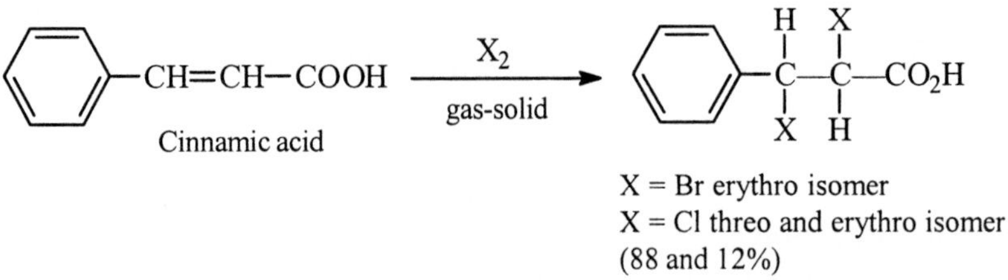

X = Br erythro isomer
X = Cl threo and erythro isomer
(88 and 12%)

Scheme 2

In a similar way bromination of powdered (E)-o-stilbene carboxylic acid with bromine vapour or with powdered pyridine. HBr.Br$_2$ complex in solid state at room temperature gave[12] selectively erytnro-1,2-dibromo-1,2-dihydro-o-stilbene carboxylic acid. However, bromination with bromine in solution gives[12] 4-bromo-3-phenyl-3,4-dihydroisocoumarin as the major product (Scheme 3).

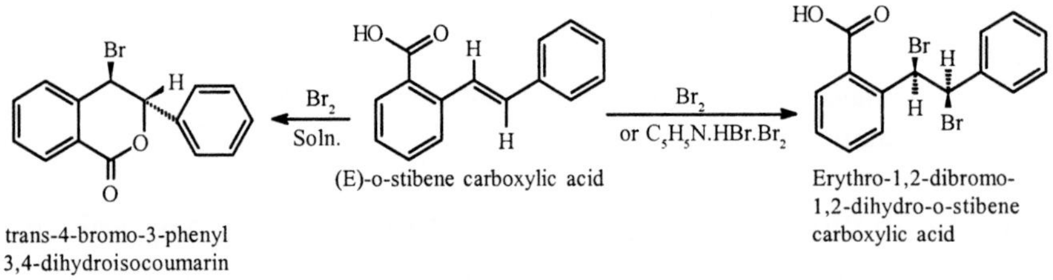

trans-4-bromo-3-phenyl
3,4-dihydroisocoumarin

(E)-o-stibene carboxylic acid

Erythro-1,2-dibromo-
1,2-dihydro-o-stibene
carboxylic acid

Scheme 3

Bromination of 4,4'-dimethylchalcone with bromine (gas-solid reaction) gave[13] optically active erythro-dibromide (6% ee) along with minor amount of a product (Scheme 4).

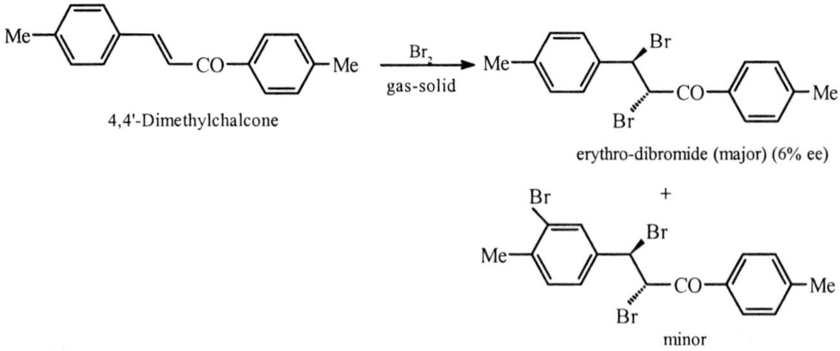

4,4'-Dimethylchalcone

erythro-dibromide (major) (6% ee)

+

minor

Scheme 4

13.2.2 Hydrohalogenation (addition of HBr)

The only example recorded is the gas-solid phase reaction of hydrogen bromide with α- and β-cyclodextrin complexes of ethyl trans-cinnamate to give[14] (R)-(+)-3-bromo-3-phenyl propanoate (46% ee) and (S)-(–)-3-bromo-3-phenyl propanoate (31% ee) (Scheme 5).

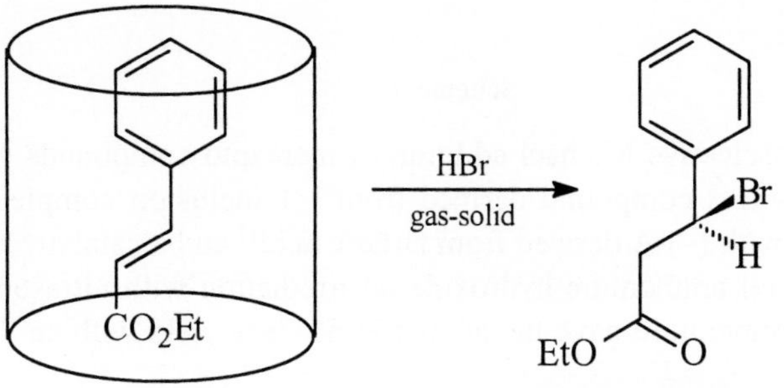

α- and β-cyclodextrin complexes of ethyl trans-cinnamate	Mixture of (R)-(+) and S-(–)-3-bromo-3-phenylpropanoate (ee 46% and 31% respectively)

Scheme 5

13.2.3 Michael Addition

A number of 2'-hydroxy-4',6'-dimethylchalcones undergo a solid state intramolecular Michael type addition to yield[15] the corresponding flavonones (Scheme 6).

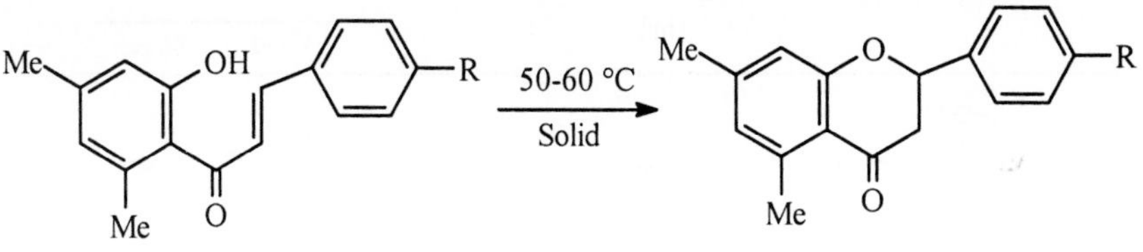

2'-hydroxy-4',6'-dimethylchalcones   5,7-Dimethyl flavones

R=H, Cl or Br

Scheme 6

The Michael addition of chalcone to 2-phenylcyclohexanone under PTC conditions give[16] 2,6-disubstituted cyclohexanone derivative in high distereoselectivity (99% ee) (Scheme 7).

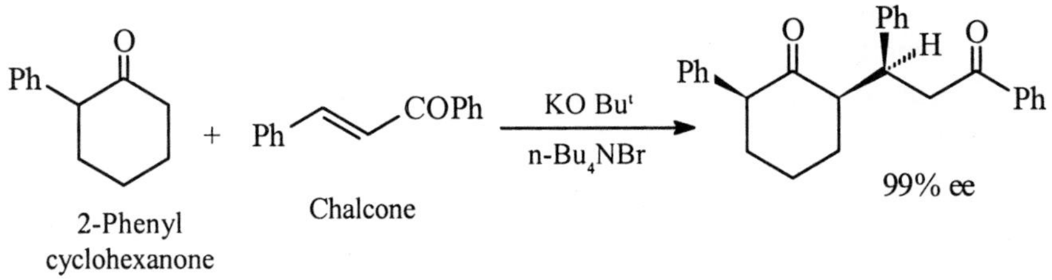

Scheme 7

The enantioselective Michael addition of mercapto compounds with an optically active host compound derived from 1:1 inclusion complex of 2-cyclohexenone with (–)-A derived from tartaric acid[17] and a catalytic amount of benzyltrimethyl ammonium hydroxide on irradiation with ultrasound for 1 hr. at room temperature gave the adduct in 50-78% yield with ee 75-80% (Scheme 8).

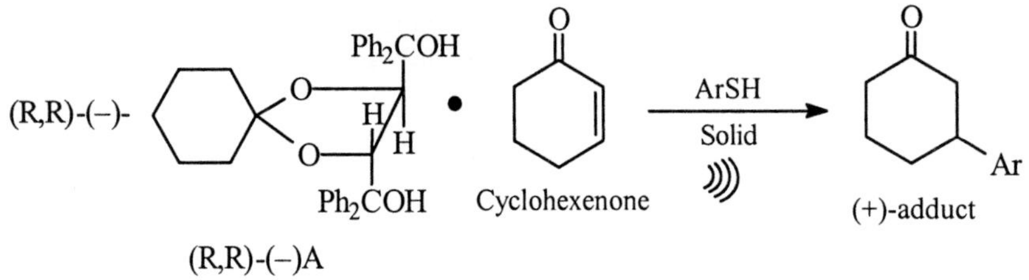

Scheme 8

Using the above procedure (Scheme 8) following adducts were obtained:

Ar	Reaction time (hr)	Adduct	
		Yield	% ee
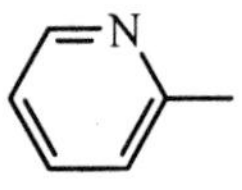 (pyridinyl)	24	51	80
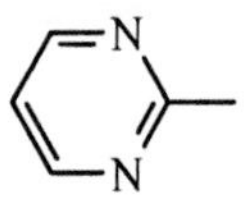 (pyrimidinyl)	36	58	78
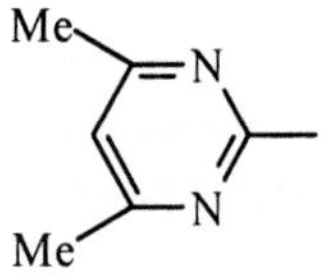 (dimethylpyrimidinyl)	36	77	74

The Michael addition of thiols to 3-methyl-3-buten-2-one in its inclusion crystal with (–)-A also occurred enantioselectively (Scheme 9).

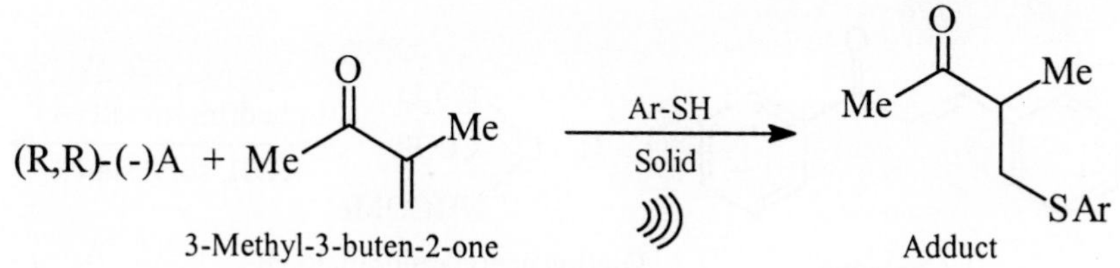

Scheme 9

Using different thiols following adducts were obtained:

Ar	Product	
	Yield	% ee
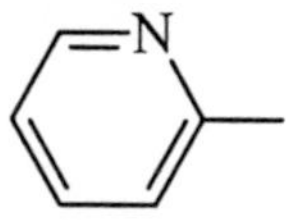	76	49
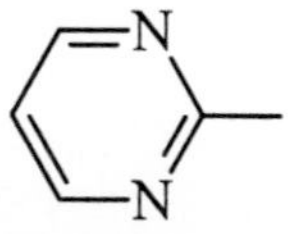	93	9
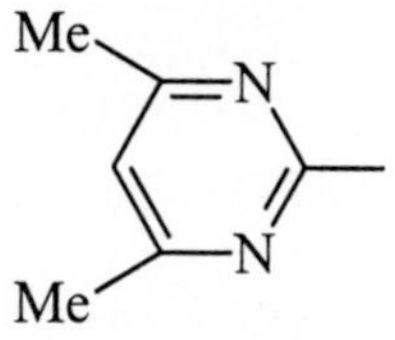	89	4
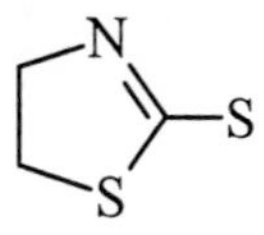	78	53

Michael-addition of diethyl(acetylamido) malonate to chalcone using asymmetric phase transfer catalyst (ephedrinium salts) in presence of KOH in the solid state has been carried out.[17a] The yield is 56% with ee of 60% (Scheme 10).

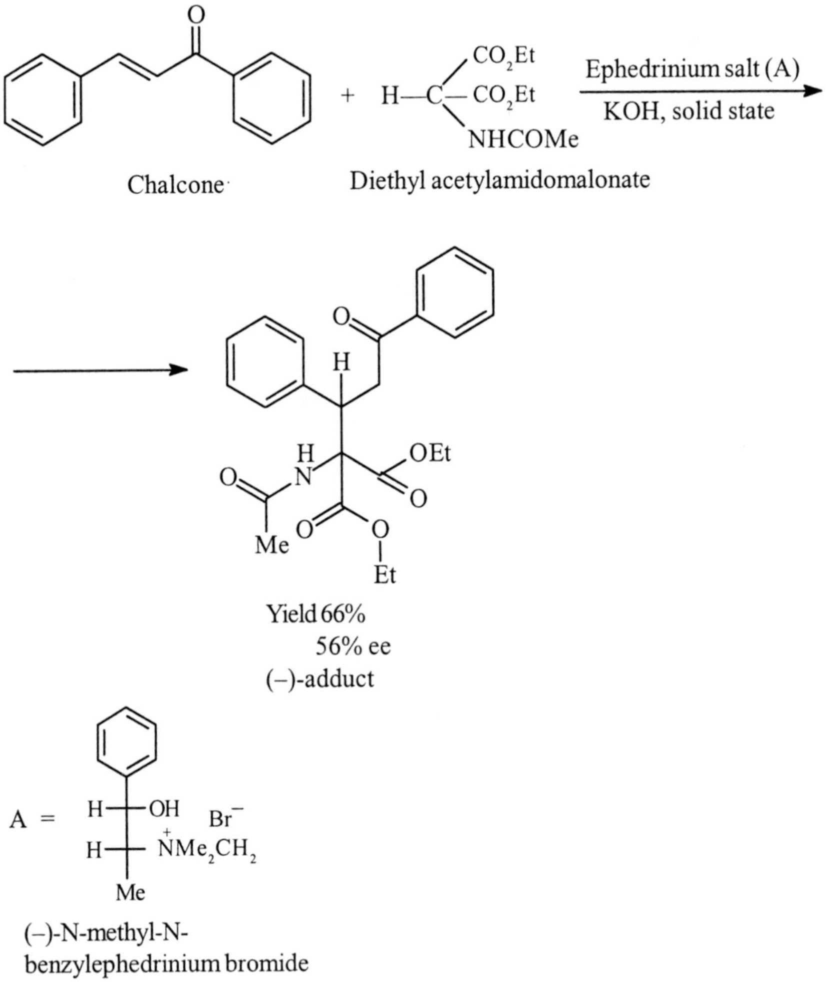

Chalcone   Diethyl acetylamidomalonate

CO$_2$Et
+   H—C—CO$_2$Et   Ephedrinium salt (A)
NHCOMe   KOH, solid state

Yield 66%
56% ee
(−)-adduct

A =

(−)-N-methyl-N-
benzylephedrinium bromide

Scheme 10

13.2.4 Dehydration of Alcohols to Alkenes

The dehydration of alcohols proceed efficiently in the solid state.[18] In this reaction, alcohols of the type PhR^{1}C(OH)CH$_2$R^2 give on dehydration alkenes, PhR^{1}C=CHR^2. The dehydration is carried out by keeping the powdered alcohol in a disiccator filled with HCl gas for 5-6 hr. Using this method[18] the following alcohols were dehydrated (Scheme 11).

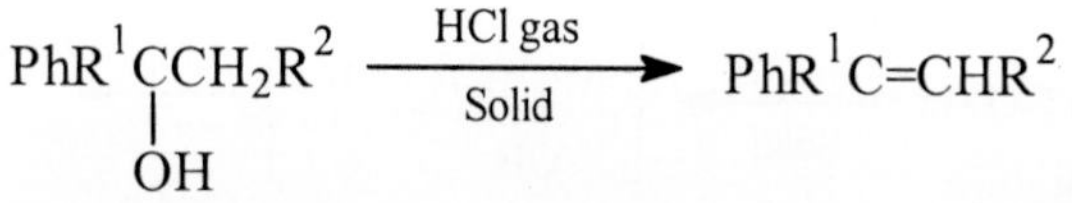

$$PhR^1CCH_2R^2 \xrightarrow[\text{Solid}]{\text{HCl gas}} PhR^1C=CHR^2$$

with OH below the first carbon.

R¹	R²	Yield (%)
Ph	H	99
Ph	Me	99
Ph	Ph	97

Scheme 11

Use of benzene as solvent in the above reaction gave much lower yields (65-75%). The dehydration proceeded much faster by using Cl_3CCO_2H as catalyst.

13.2.5 Aldol Condensation

The aldol condensation of the lithium enolate of methyl 3,3-dimethylbutanoate with aromatic aldehydes gives[19] (the reaction is carried out by mixing freshly ground mixture of the lithium enolate and powdered aldehyde in vacuum for 3 days at room temperature) a 8:92 mixture of the syn and anti products in 70% yield (Scheme 12).

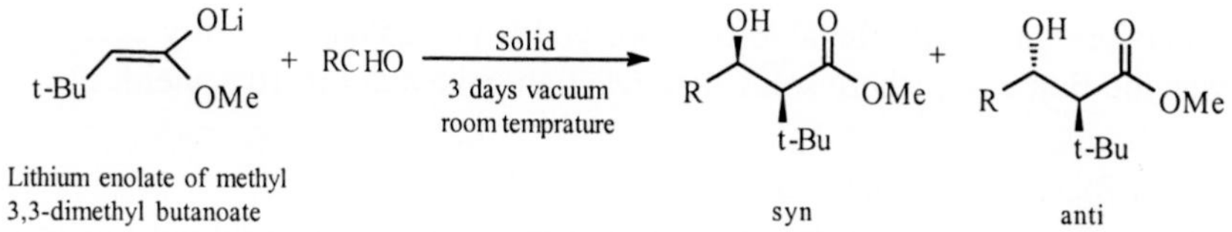

Lithium enolate of methyl
3,3-dimethyl butanoate

syn

anti

$R = 2\text{-OCH}_3C_6H_4\text{-}, 4\text{-Cl-}C_6H_4$
$4\text{-NO}_2C_6H_4\text{-}, 3\text{-NO}_2C_6H_4\text{-}, 2\text{-NO}_2C_6H_4\text{-}$
$4\text{-NO}_2\text{-2-thienyl}$

Scheme 12

In the absence of any solvent, some aldol condensation proceed[20] more efficiently and stereoselectively. In this method, appropriate aldehyde and ketone and NaOH is grounded in a pestle and mortar at room temperature for 5 min. The product obtained is the corresponding chalcone. In this method the initially formed aldol dehydrates more easily to the chalcone in the absence of solvent (Scheme 13).

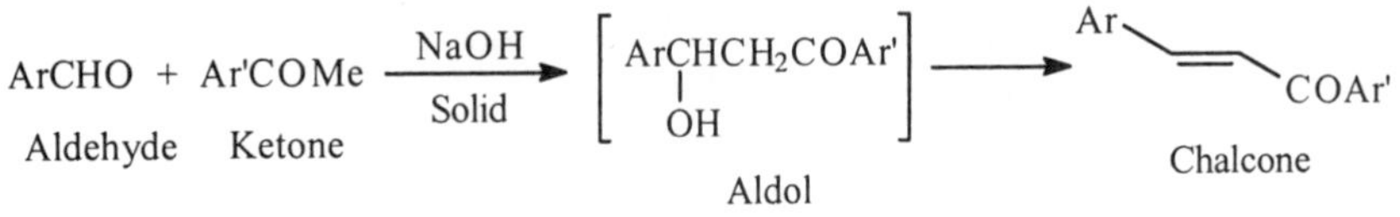

$$\text{ArCHO} + \text{Ar'COMe} \xrightarrow[\text{Solid}]{\text{NaOH}} \left[\begin{array}{c} \text{ArCHCH}_2\text{COAr'} \\ | \\ \text{OH} \end{array} \right] \longrightarrow \text{Ar} \diagdown \!\!=\!\!\diagdown \text{COAr'}$$

Aldehyde   Ketone   Aldol   Chalcone

Scheme 13

Using the above method following chalcones were prepared:

Ar	Ar'	Reaction time (min)	Yield aldol	(%) of chalkone
Ph	Ph	30	10	–
p-MeC$_6$H$_4$	Ph	5	–	97
p-MeC$_6$H$_4$	p-MeC$_6$H$_4$	5	–	99
p-ClC$_6$H$_4$	Ph	5	–	98
p-ClC$_6$H$_4$	p-MeOC$_6$H$_4$	10	–	79
p-ClC$_6$H$_4$	p-BrC$_6$H$_4$	10	–	81
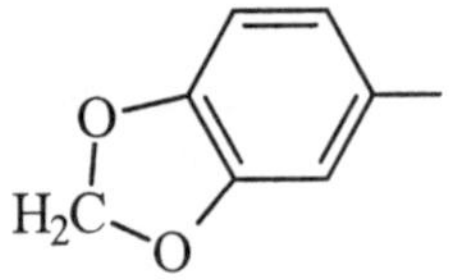	p-BrC$_6$H$_4$	10	–	91

Use of alcohol as a solvent in the above method using conventional procedure gave only the aldol in poor yields (10-25%). The only exception was the first case, where solid state reaction gave aldol in 10% yield.

13.2.6 Dieckmann Condensation

Dieckman condensation reactions of diesters have been carried out[21] in solid state in presence of a base (like Na or NaOEt) using high-dilution conditions in order to avoid intermolecular reaction. It has been found[22] that the Dieckman condensation of diethyl adipate and pimelate proceed very well in absence of the solvent; the reaction products were obtained by direct distillation of the reaction mixture. In this method the diester and powdered Bu^{t}OK were mixed using a pestle and mortar for 10 min. The solidified reaction mixture was neutralised with p-TsOH.H$_2$O and distilled to give cyclic compounds (Scheme 14).

Any of the bases like Bu^{t}OK, Bu^{t}ONa, EtOK or EtONa could be used in the above reaction.

13.2.7 Grignard Reaction

The results obtained by carrying out the usual Grignard reaction are different than that obtained in the solid state.[23] Thus the reaction of ketone (e.g. benzophenone) with Grignard reagent (the reaction carried out by mixing ketone and powdered grignard reagent, obtained by evaporating the solution of the Grignard reagent (prepared as usual) in vacuo) in solid state gives more of the reduced product of the ketone than the adduct (Scheme 15).

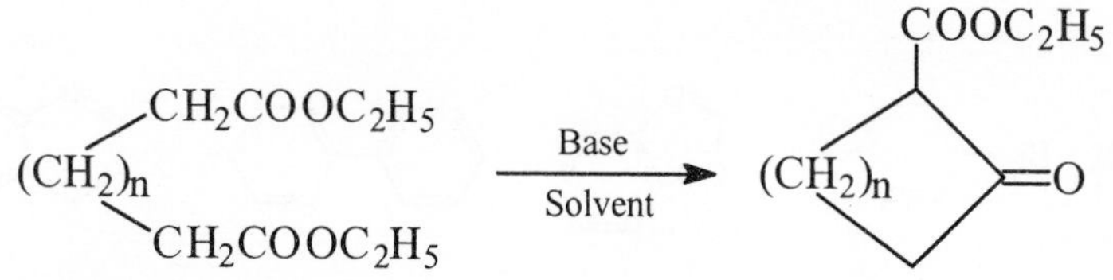

$$
\underset{\substack{\text{Diethyl adipate (n = 2)}\\\text{Diethyl pimelate (n = 3)}}}{(CH_2)_n \begin{array}{l} CH_2COOC_2H_5 \\ CH_2COOC_2H_5 \end{array}} \xrightarrow[\text{Solvent}]{\text{Base}} \underset{60\text{-}70\%}{(CH_2)_n \begin{array}{c} COOC_2H_5 \\ \diamondsuit {=}O \end{array}}
$$

Scheme 14

$$
Ph_2CO + RMgX \xrightarrow[\text{Solid}]{0.5 \text{ hr}} \underset{\substack{\text{Adduct}\\(A)}}{Ph_2RCOH} + \underset{\substack{\text{Reduced product}\\\text{of ketone (B)}}}{Ph_2CHOH}
$$

Scheme 15

Following results are obtained:

Grignard reagent RMgX		% products obtained in solid state	
R	X	(A)	(B)
Me	I	No reaction	
Et	Br	30	31
i-Pr	Br	2	20
Ph	Br	59	–

13.2.8 Reformatsky Reaction
Treatment of aromatic aldehydes with ethyl bromoacetate and Zn-NH$_4$Cl in the solid state give the corresponding Reformatsky reaction products[24] (Scheme 16).

13.2.9 Wittig Reaction
The well known Wittig reaction has been reported[25] to occur in solid phase. In this procedure a 1:1 mixture of the finely powdered inclusion compound of cyclohexanone or 4-methyl cyclohexanone and (–)-B (derived from tartaric acid[17] and a catalytic amount of benzyltrimethyl ammonium hydroxide) was heated at 70 °C with Wittig reagent carbethoxymethylene triphenylphosphorane to give optically active 1-(carbethoxymethylene) cyclohexane (Scheme 17) or

the corresponding 4-methyl compound.

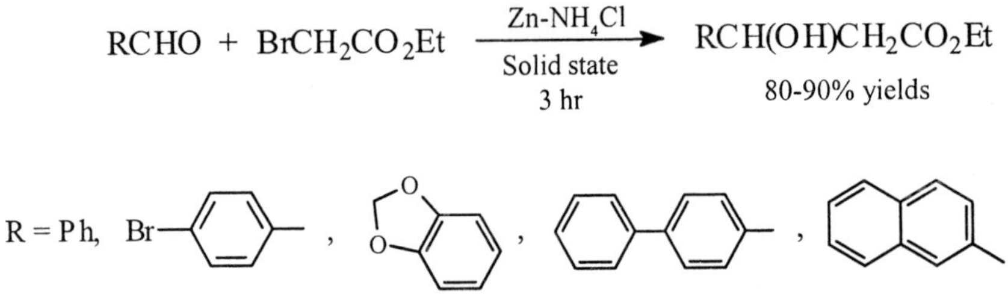

$$RCHO + BrCH_2CO_2Et \xrightarrow[\substack{\text{Solid state} \\ \text{3 hr}}]{\text{Zn-NH}_4\text{Cl}} RCH(OH)CH_2CO_2Et$$

$$80\text{-}90\% \text{ yields}$$

R = Ph, Br— ⬡ — , ⬡ , ⬡—⬡— , ⬡⬡—

Scheme 16

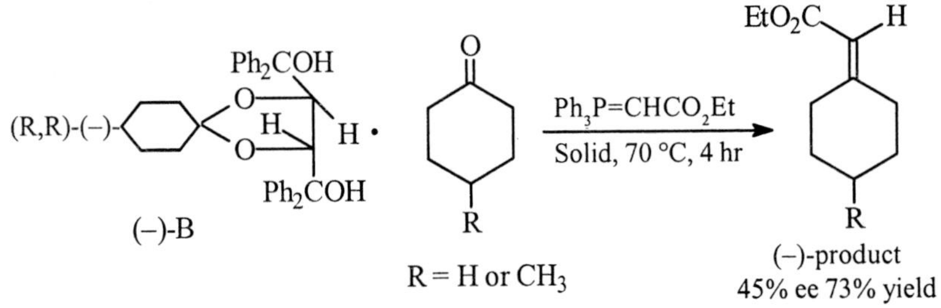

Scheme 17

In a similar way inclusion compound of 3,5-dimethylcyclohexanone and (–)-A (derived from tartaric acid[17] and a catalytic amount of benzyltrimethyl ammonium hydroxide) on reaction with the Wittig reagent gave optically active 3,5-dimethyl-1-(carbethoxymethylene) cyclohexane (Scheme 18).

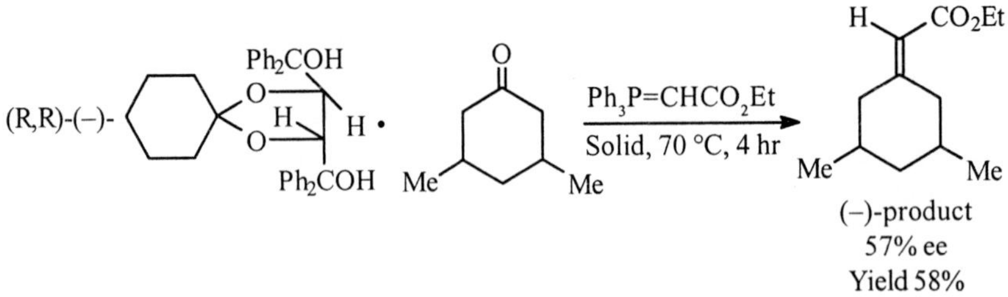

Scheme 18

13.2.10 Armoatic Substitution Reactions

13.2.10.1 *Nuclear Bromination*

Nuclear bromination of phenols with N-bromosuccinimide (NBS) is accomplished[18] in the solid state. Thus, the reaction of 3,5-dimethylphenol with 3 mol equivs of NBS in the solid state for 1 min gave the tribromo derivative in 45% yield. However, if the reaction is conducted in solution a mixture of mono and dibromo derivatives is obtained (Scheme 19).

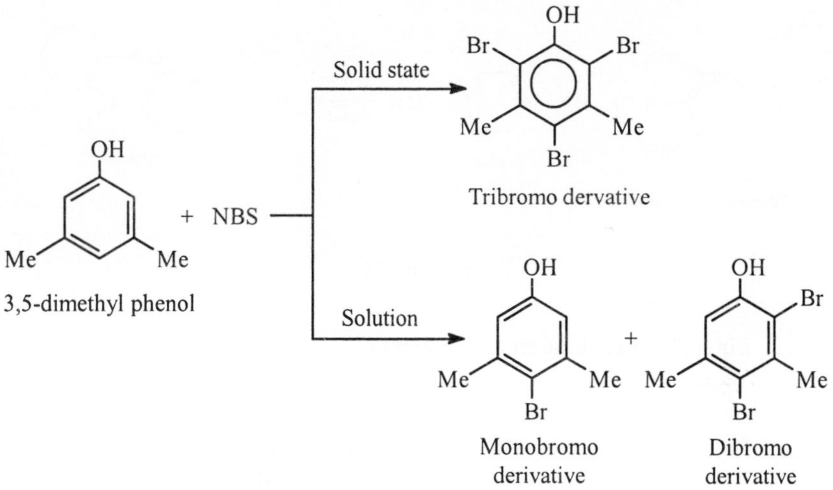

Scheme 19

13.2.10.2 Nitration

The nitration of aromatic compounds with stoichiometric quantity of nitric acid and acetic anhydride (in absence of solvent) at 0-20° C in presence of zeolite beta catalyst gave[26] the nitration product as given (Scheme 20).

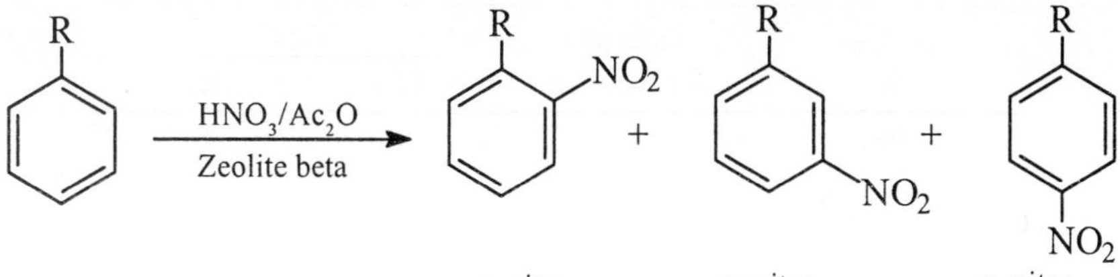

R	Yield		Ratio of	
	(%)	o–	m–	p–
Me	>99	18	3	79
t-Bu	92	8	trace	92
Cl	>99	7	0	93
NO$_2$	13	1	92	7

Scheme 20

13.2.10.3 Synthesis of Amines

The halogen in an aromatic compound (e.g., 4-chloro-3,5-dinitrobenzoic acid) can be made to react with amino group (e.g., p-aminobenzoic acid) in solid state at 180° C to give[27] the corresponding amine (Scheme 21).

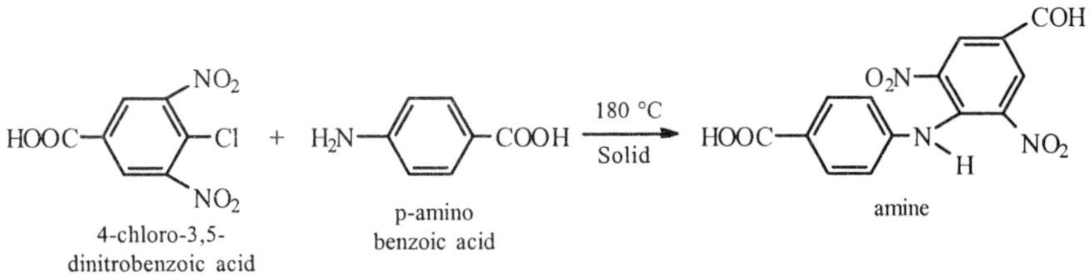

Scheme 21

13.2.11 Pinacol-Pinacolone Rearrangement

Pinacol-pinacolone rearrangements proceed[28] faster and more selectively in solid state. A 1:3 molar ratio of the pinacol and p-toluene sulphonic acid (p-TsOH) (powered mixture) on keeping at 60° C give the products (A) and (B). It is found that the hydride migrates more easily than the phenyl anion and the yield of A is higher than that of B in all the reactions (Scheme 22).

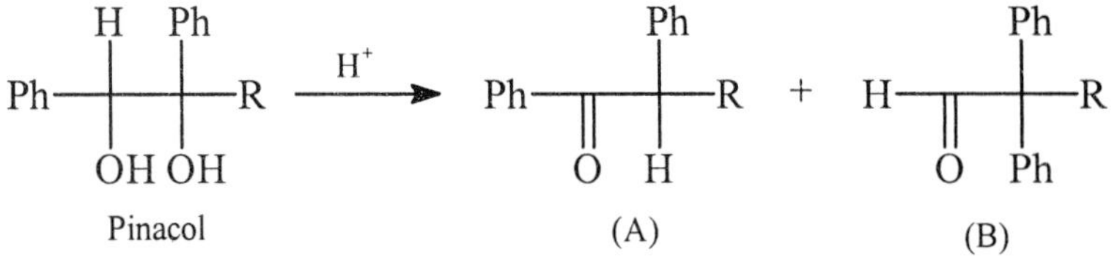

Piancol R	Time (hr) for reaction	Yield	
		(A)	(B)
Ph	2.5	89	8
o-MeC$_6$H$_4$	0.5	45	29
m-MeC$_6$H$_4$	0.3	70	30
p-MeC$_6$H$_4$	0.7	39	19
p-MeOC$_6$H$_4$	0.7	89	0
p-ClC$_6$H$_4$	1.0	54	41

Scheme 22

However, pinacol-pinacolone rearrangement in presence of CCl$_3$CO$_2$H (in place of TsOH) gives major amount of the isomeric product (B). The reaction is considerably enhanced if the water formed during the reaction is continuously removed under reduced pressure.

13.2.12 Benzil-Benzilic Acid Rearrangement

Normally the above rearrangement has been carried out by heating benzil and alkali metal hydroxides in aqueous organic solvent. It is found that the rearrangement proceeds more efficiently and faster in the solid state[29] (Scheme 23). The reaction takes 0.1 to 6 hr and the yields are 70-93%.

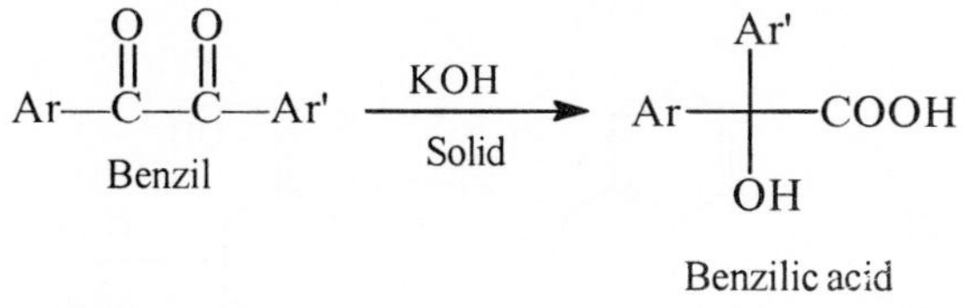

$$Ar-\overset{O}{\underset{}{\overset{||}{C}}}-\overset{O}{\underset{}{\overset{||}{C}}}-Ar' \quad \xrightarrow[\text{Solid}]{KOH} \quad Ar-\overset{Ar'}{\underset{OH}{\overset{|}{C}}}-COOH$$

Benzil   Benzilic acid

Ar = Ar' = Ph
Ar = Ar' = p-ClC$_6$H$_4$
Ar = Ar' = p-NO$_2$C$_6$H$_4$
Ar = Ph; Ar' = p-ClC$_6$H$_4$
Ar = Ph; Ar' = p-NO$_2$C$_6$H$_4$
Ar = Ph; Ar' = p-MeOC$_6$H$_4$

Scheme 23

The effect of the reaction efficiency of the above rearrangement (Scheme 23) in the solid state increases in the following order:

$$KOH > Ba(OH)_2 > NaOH > CsOH$$

13.2.13 Beckmann Rearrangement

Usually, Beckmann rearrangement of oximes of ketones are converted into anilides by heating with acidic reagents like PCl$_5$, HCOOH, SOCl$_2$ etc. However, solid-state Beckmann rearrangement has been reported.[30] In this method oxime of a ketone is mixed with montmorillonite and irradiated for 7 min in a microwave oven to give corresponding anilide in 91% yield (Scheme 24).

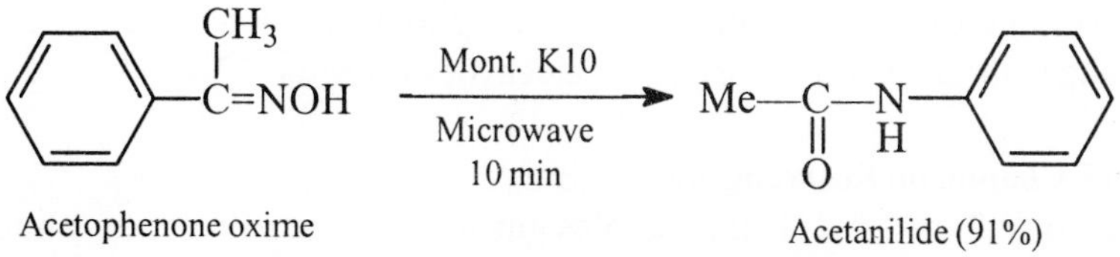

Acetophenone oxime   Acetanilide (91%)

Scheme 24

However, conventional heating gives only 17% yield.

Another interesting example is the preparation[31] of (−)-5-methyl-ε-caprolactam of 88% ee by heating 1:1 inclusion compound of (−)-1, 6-bis(o-chlorophenyl)-1,6-diphenylhexa-2,4-diyne (A) and oxime of p-methylcyclohexanone in ether-petroleum ether with conc. sulphuric acid gave (−)-5-methyl-ε-caprolactam. In place of oxime of p-methylcyclohexanone, use of the oxime of 3,5-dimethyl cyclohexanone gave the corresponding (+)-4,6-dimethyl-ε-caprolactam (Scheme ·26).

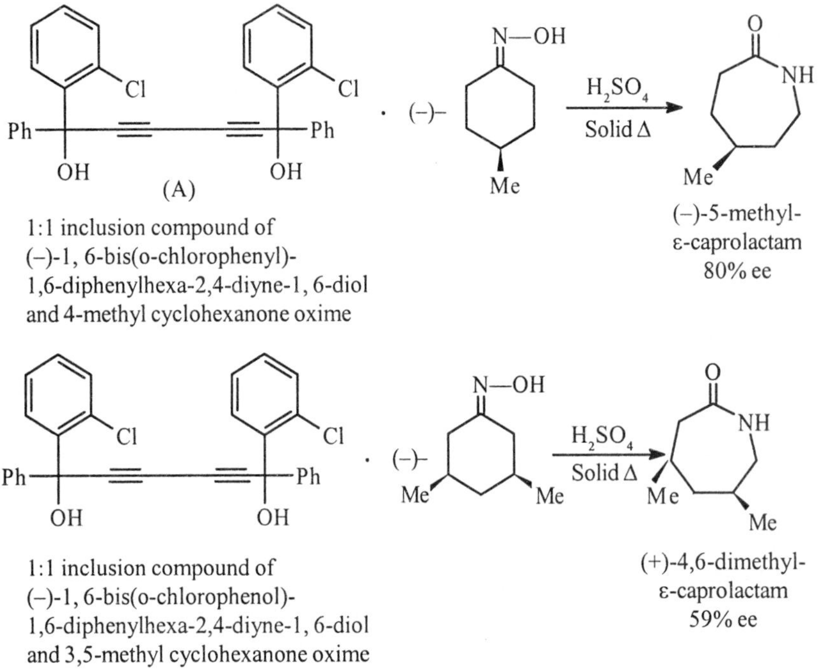

1:1 inclusion compound of
(−)-1, 6-bis(o-chlorophenyl)-
1,6-diphenylhexa-2,4-diyne-1, 6-diol
and 4-methyl cyclohexanone oxime

(−)-5-methyl-
ε-caprolactam
80% ee

1:1 inclusion compound of
(−)-1, 6-bis(o-chlorophenol)-
1,6-diphenylhexa-2,4-diyne-1, 6-diol
and 3,5-methyl cyclohexanone oxime

(+)-4,6-dimethyl-
ε-caprolactam
59% ee

Scheme 26

13.2.14 Meyer-Schuster Rearrangement
Propargyl alcohols on keeping (2-3 hr) with p-toluene sulphonic acid (TsOH) rearranges to give[18] α,β-unsaturated aldehydes (Scheme 27).

13.2.15 Chapmann Rearrangement
5-Methoxy-2-aryl-1,3,4-oxadiazoles on heating in solid state undergo rearrangement to give[32] the corresponding oxadiazole-5-ones (Scheme 28).

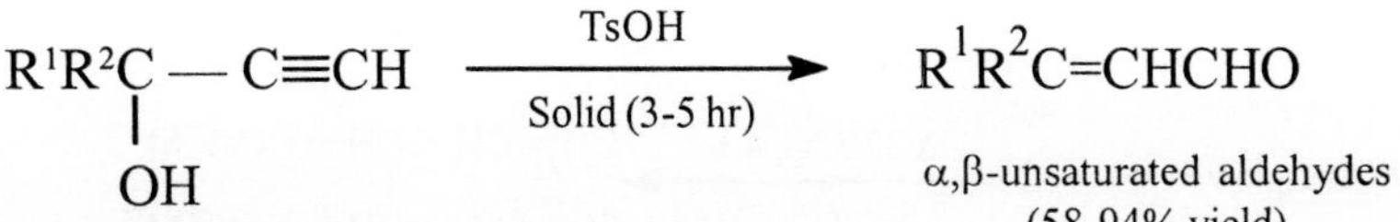

$$R^1R^2C \overset{|}{\underset{OH}{}} - C\equiv CH \xrightarrow[\text{Solid (3-5 hr)}]{\text{TsOH}} R^1R^2C=CHCHO$$

Propargly alcohols

α,β-unsaturated aldehydes
(58-94% yield)

$R^1 = R^2 = Ph$
$R^1 = Ph;\ R^2 = o\text{-ClC}_6H_4$
$R^1 = R^2 = 2,4\text{-Me}_2C_6H_3$

Scheme 27

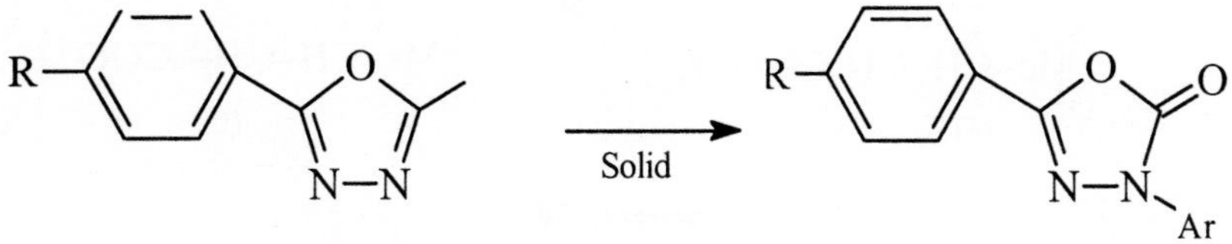

Solid

R = H, Cl, OMe, NO₂

Scheme 28

The same type of rearrangement also occur with methyl cyanurates and thiocyanurates in the solid state[33] (Scheme 29).

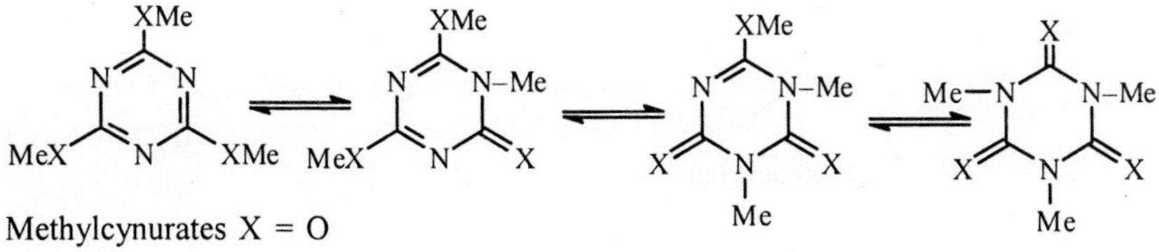

Methylcynurates X = O
Methylthiocynurates X = S

Scheme 29

13.2.16 Miscellaneous Reactions

13.2.16.1 Coupling Reactions

(i) Coupling of salts of crotonic acids
Sodium crotonate on heating in solid state gave hex-1-ene-3,4-dicarboxylate (A). On the other hand, potassium crotonate on heating at 320 °C for 4 hr gives (A) along with the isomeric dimers (B) and (C) (Scheme 30).[34]

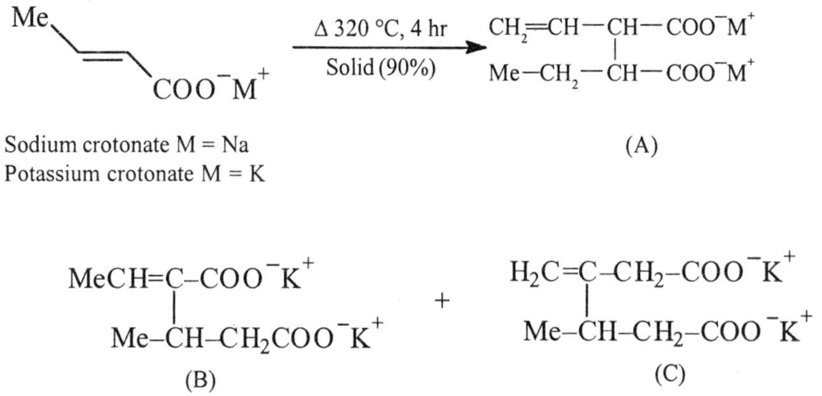

Sodium crotonate M = Na
Potassium crotonate M = K

(A)

(B)   +   (C)

Scheme 30

In a similar way, cross thermal coupling of binary salts of but-3-enoic acid and methacrylic acid (the salts were obtained by mixing equimolar amounts of the two acids with an equimolar amount of metal or alkaline metal earth salts) on heating gave[35] the coupled product, which was isolated as the methyl ester, i.e. (E)-methyl-hex-1-ene-1,5-dicarboxylate (Scheme 31).

but-3-enoate   Methacrylate

(E)-Methylhex-1-ene-
1,5-dicarboxylate

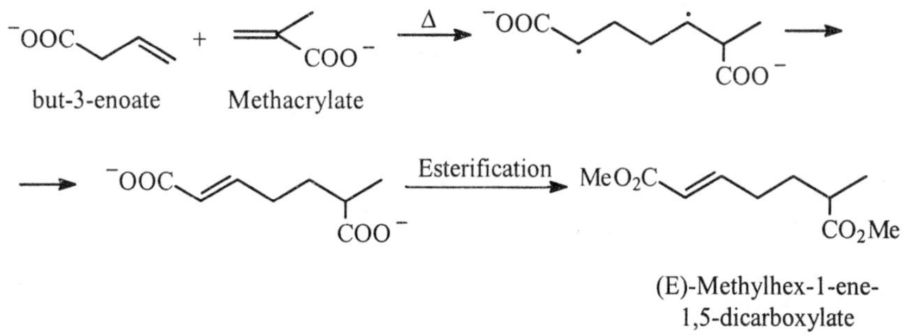

Scheme 31

(ii) Pinacol coupling

Aromatic aldehydes and ketones undergo pinacol coupling in solid state by Zn-ZnCl$_2$ reagent to give[36] a mixture of the reduced product (0-2%) along with major amount (45-80%) of the coupled product (glycol) (Scheme 32).

However, coupling of aromatic ketones under the above solid state conditions (Scheme 32) give only the coupled product (Scheme 33).

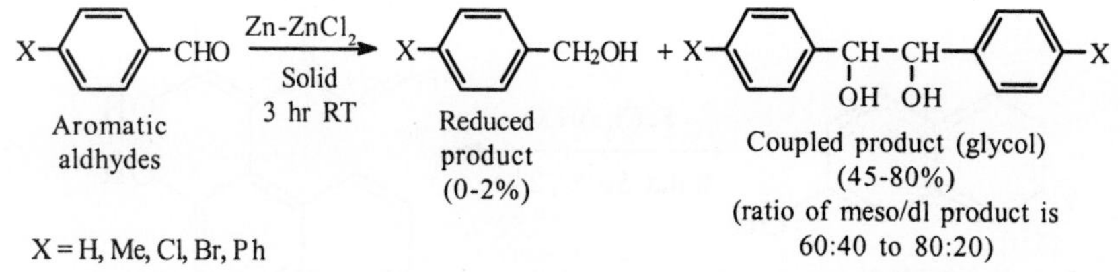

$$X-\boxed{}-CHO \xrightarrow[\substack{\text{Solid} \\ \text{3 hr RT}}]{\text{Zn-ZnCl}_2,} X-\boxed{}-CH_2OH + X-\boxed{}-\underset{\underset{\text{OH OH}}{|\quad|}}{CH-CH}-\boxed{}-X$$

Aromatic aldehydes

X = H, Me, Cl, Br, Ph

Reduced product (0-2%)

Coupled product (glycol) (45-80%)
(ratio of meso/dl product is 60:40 to 80:20)

Scheme 32

$$ArCOAr' \xrightarrow[\substack{\text{Solid} \\ \text{3 hr RT}}]{\text{Zn-ZnCl}_2} ArC\underset{\underset{\text{OH OH}}{|\quad|}}{}C-Ar'$$

30-90% Yield

Ar	Ar'
Ph	Ph
p-MeC$_6$H$_4$	p-ClC$_6$H$_4$
Ph	p-ClC$_6$H$_4$

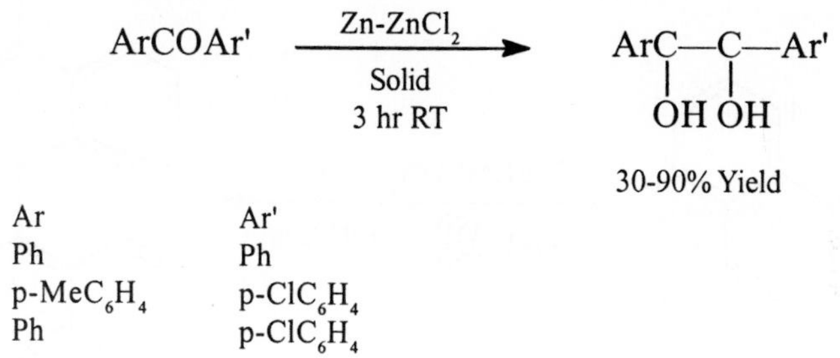

Scheme 33

(iii) Oxidative coupling of phenols

Oxidative coupling of phenols in presence of FeCl$_3$.6H$_2$O proceed much faster in the solid state[37] than in solution. The reaction is carried out by mixing the phenol and FeCl$_3$.6H$_2$O in powdered state and leaving the mixture for 2 hr at 50 °C. Some of the coupling products of phenols are given in the following (Scheme 34).

(iv) Oxidative coupling of acetylenic compounds

Oxidative coupling of acetylenic compounds proceeds more efficiently in the solid state[38] than in water. In this procedure, powedered cuprous aryl acetylide and CuCl$_2$.2H$_2$O was kept for 3 hr to give the coupled product in 40-75% yield (Scheme 35), compared to 10-30% in water.

In a similar way propargyl alcohols could be coupled[38] in solid state (Scheme 36).

(v) Dimerisation of [60] fullerene

Due to extremely very low solubility of fullerens in organic solvents, its chemical transformations is very difficult. The dimerisation reaction of [60] fullerene in presence of KCN proceeded in solid state (Scheme 37) to give[39] [2+2] adduct in 18% yield (Scheme 37).

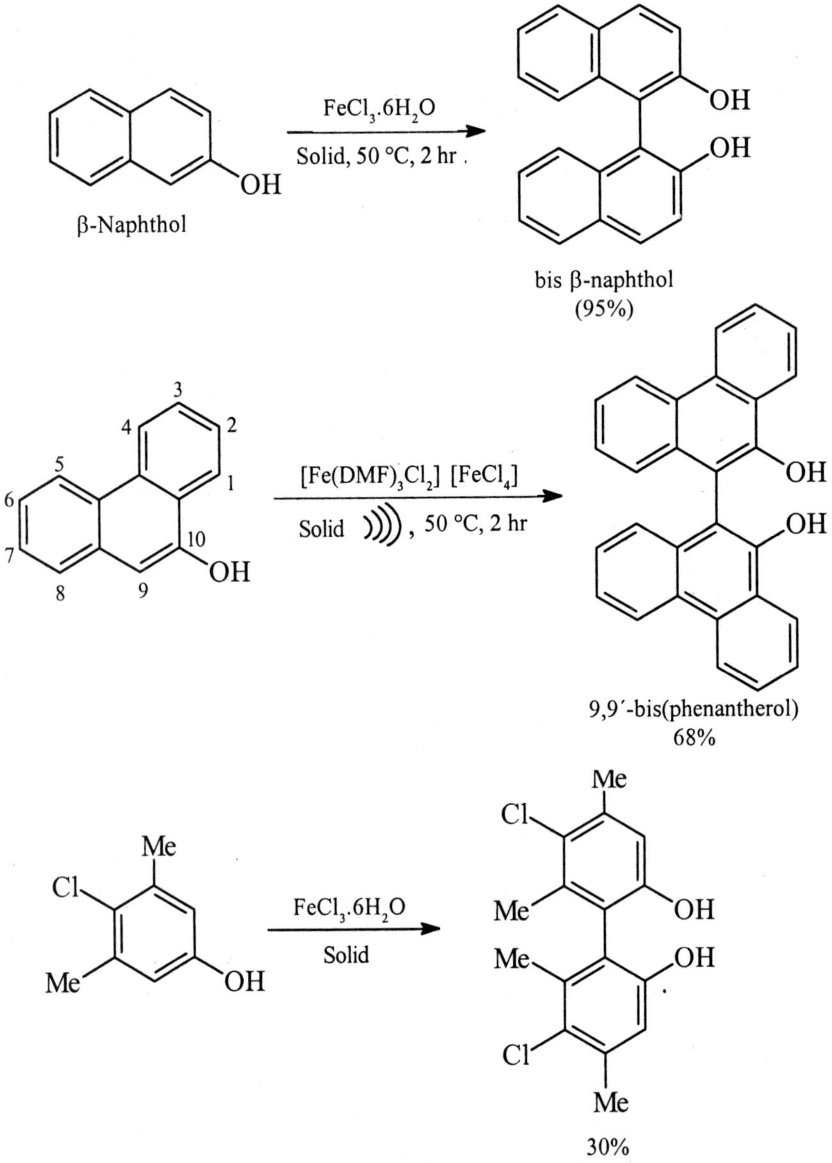

β-Naphthol

bis β-naphthol
(95%)

FeCl$_3$.6H$_2$O

Solid, 50 °C, 2 hr .

[Fe(DMF)$_3$Cl$_2$] [FeCl$_4$]

Solid)))) , 50 °C, 2 hr

9,9′-bis(phenantherol)
68%

FeCl$_3$.6H$_2$O

Solid

30%

Scheme 34

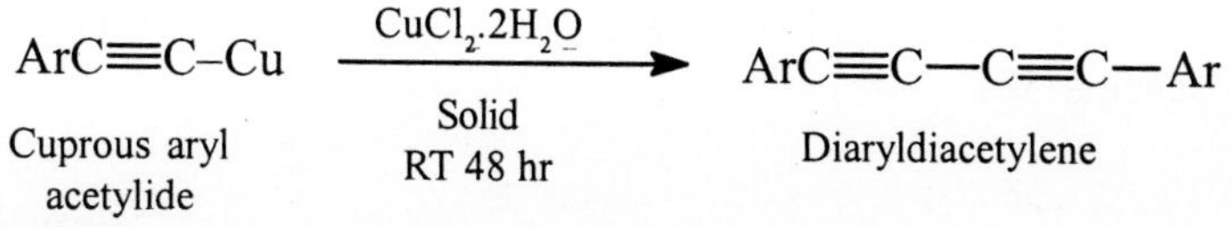

$$ArC{\equiv}C\text{--}Cu \xrightarrow[\substack{\text{Solid}\\ \text{RT 48 hr}}]{CuCl_2.2H_2O} ArC{\equiv}C\text{--}C{\equiv}C\text{--}Ar$$

Cuprous aryl   Diaryldiacetylene
acetylide

Ar = Ph, p-MeC$_6$H$_4$, p-PhC$_6$H$_4$, 2,3,5,6-(Me)$_4$C$_6$H$_2$ and PhOCH$_2$

Scheme 35

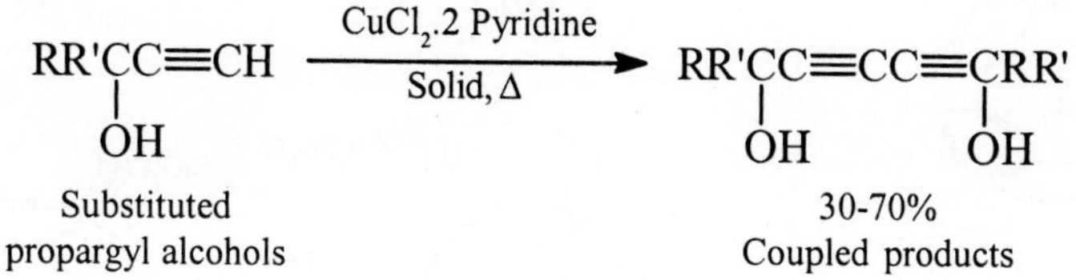

$$\underset{\underset{OH}{|}}{RR'CC}{\equiv}CH \xrightarrow[\text{Solid, }\Delta]{CuCl_2.2\text{ Pyridine}} \underset{\underset{OH}{|}}{RR'CC}{\equiv}CC{\equiv}\underset{\underset{OH}{|}}{CRR'}$$

Substituted   30-70%
propargyl alcohols   Coupled products

R = R$'$ = Ph

R = Ph; R$'$ = o-ClC$_6$H$_4$

R = R$'$ = p-MeC$_6$H$_4$

R = Ph; R$'$ = 2,4-(Me)$_2$-C$_6$H$_3$

R = 2,4-(Me)$_2$-C$_6$H$_3$; R$'$ = o-ClC$_6$H$_4$

R = Ph; R$'$ = Me

Scheme 36

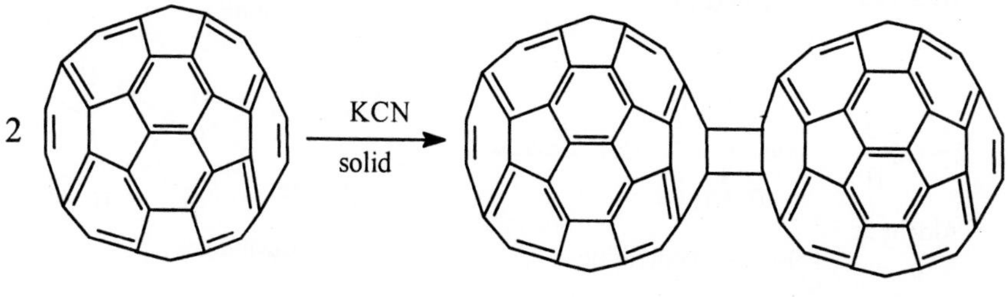

Scheme 37

It is appropriate to mention here that [60] fullerene reacted with ethyl bromoacetate in presence of Zn dust (the reaction was carried out by vigorously agitating the mixture for 20 min at room temperature) and the adduct (A) was obtained[40] in 17.2% yield along with some by products (designated B, C and D) (Scheme 38).

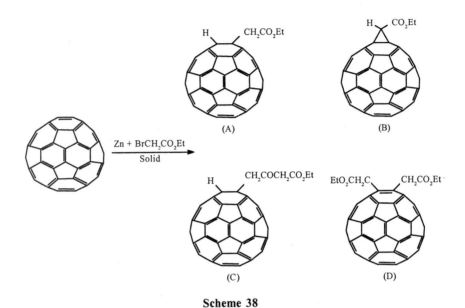

Scheme 38

13.2.16.2 Solid Phase Synthesis of Oxazolidines

The reaction of aldehydes with (–)-ephedrine or (+)-pseudoephedrine by keeping powered reaction mixture at room temperature give[41] quantitatively the corresponding oxazolidines (Scheme 39).

R = PhCH=CH

R = H (Paraformaldehyde)

R = 4-OH-3-OMeC$_6$H$_3$

R = C$_5$H$_5$FeC$_5$H$_4$

R = 4-NO$_2$C$_6$H$_4$ **Scheme 39**

13.2.16.3 Synthesis of Azomethines

Solid state reaction of anilines and aromatic aldehyde (by grinding the reactants) give[42] quantitative yield of various azomethines (Scheme 40).

$$ArNH_2 \quad + \quad Ar'CHO \xrightarrow[\text{Solid}]{\text{Mix}} ArN=CH-Ar'$$

Aromatic amine   Aromatic aldehyde   Azomethines

Ar = 4-MeC$_6$H$_4$ Ar = 4-ClC$_6$H$_4$
4-MeOC$_6$H$_4$ 4-BrC$_6$H$_4$
4-NO$_2$C$_6$H$_4$ 4-NO$_2$C$_6$H$_4$
4-ClC$_6$H$_4$ 4-OHC$_6$H$_4$
4-BrC$_6$H$_4$ 4-OH, 3-OMeC$_6$H$_3$
4-HOC$_6$H$_4$
1-Naphthyl

Scheme 40

13.2.16.4 Synthesis of Homoallylic Alcohols

Treatment of aldehydes with 3-bromopropene and Zn-NH$_4$Cl in absence of any solvent gave[43] the corresponding homoallylic alcohols in 85-99% yields (Scheme 41).

$$RCHO + BrCH_2CH=CH_2 \xrightarrow{Zn-NH_4Cl} RCH(OH)CH_2CH=CH_2$$

Aldehyde   3-Bromopropene   Homoallylic alcohol

R = Ph, β-naphthyl, n-pentyl or trans CH$_3$CH=CH$_2$

Scheme 41

13.2.16.5 Synthesis of Cyclopropane Derivatives

The reaction of chalcones and trimethylsulphonium iodide and KOH in the solid state gave[44] the corresponding cyclopropanes (Scheme 42).

The same cyclopropane derivative was also obtained[44] by the reaction of chalcone with (+)-S-methyl-S-phenyl-N-(p-tolyl)sulfoximide and tert. BuOK at room temperature in 94% yield (Scheme 43).

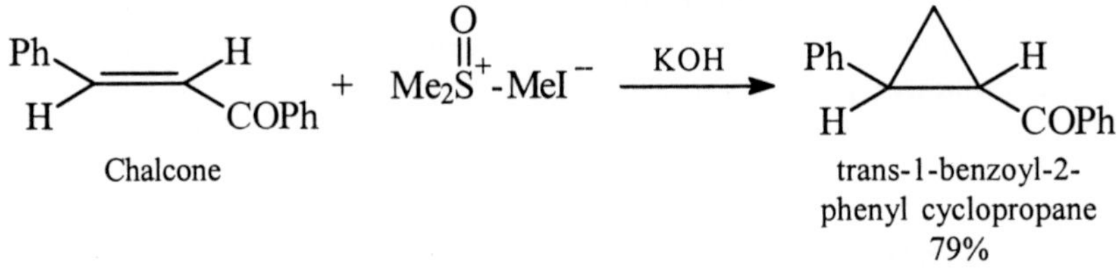

Chalcone

trans-1-benzoyl-2-
phenyl cyclopropane
79%

Scheme 42

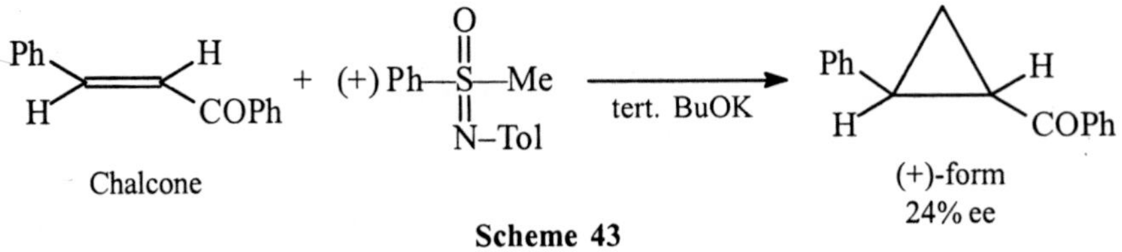

Chalcone

(+)-form
24% ee

Scheme 43

The γ-CD complex of phenylmethyldiazirine (obtained by treatment of γ-CD solution with the phenylmethyldiazirine) on heating in inert atmosphere gave[45] trans-1,2-diphenyl-1-methylcyclopropane (Scheme 44).

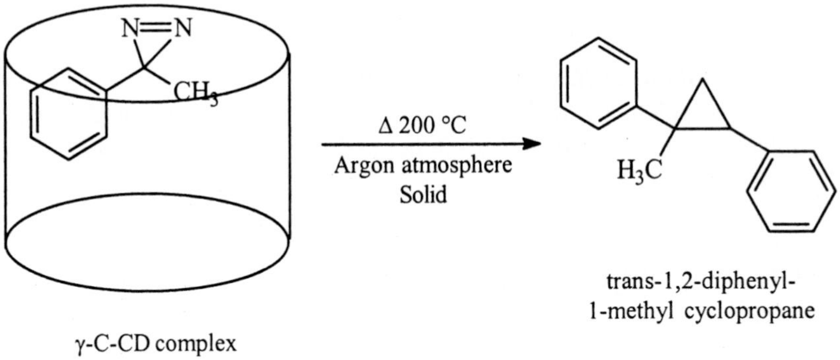

Δ 200 °C

Argon atmosphere
Solid

trans-1,2-diphenyl-
1-methyl cyclopropane

γ-C-CD complex

Scheme 44

13.2.16.6 Synthesis of Oxirazies
The reaction of cyclohexanones with trimethylsulfonium iodide and KOH in the solid phase gave[44] gave the oxirazines (Scheme 45).

13.2.16.7 Synthesis of Aziridines
The reaction of imines with trimethylsulfonium iodide in solid state in presence of KOH gives[44] the corresponding aziridines (Scheme 46).

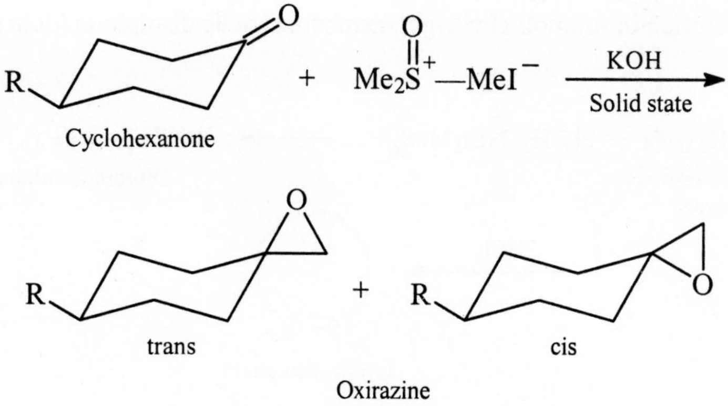

O
R
Cyclohexanone
O
Me2S+—MeI−
KOH
Solid state
O
R
trans
R
O
cis
Oxirazine

R	Yield (%) (trans + cis)	Ratio trans : cis
t-Bu	83	97 : 1
Me	33	91 : 9
Et	75	92 : 8
Ph	82	97 : 3

Scheme 45

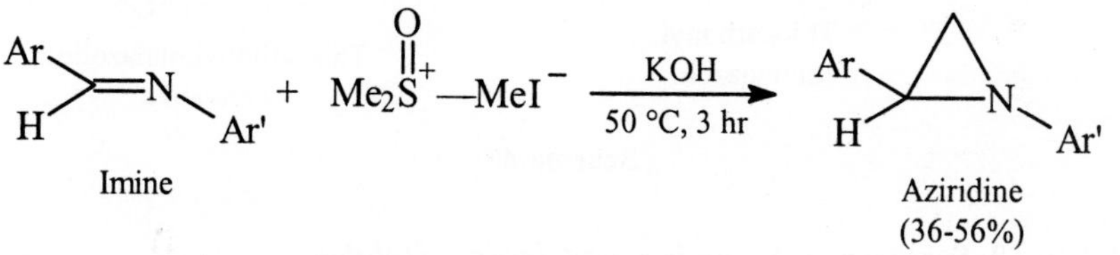

Ar
H
N
Ar'
Imine
O
Me2S+—MeI−
KOH
50 °C, 3 hr
Ar
H
N
Ar'
Aziridine
(36-56%)

Ar = Ar' = Ph
Ar = Ph; Ar' = p-$MeOC_6H_4$
Ar = p-ClC_6H_4; Ar' = p-$MeOC_6H_4$
Ar = p-MeC_6H_4; Ar' = p-$MeOC_6H_4$

Scheme 46

13.2.16.8 Synthesis of Disulphides

Disulphides have been obtained[46] by the solid state reaction of benzyl halides, alkyl halides and acyl halides at room temperature in the presence of benzyltriethylammonium tetrathiomolybdate. For example, butyl iodide, benzyl bromide and benzoyl chloride give the corresponding disulphides in 70-74%

yield. The 1-bromo-6-iodohexane reacted with benzyltriethylammonium tetrathiomolybdate to give exclusively the dibromodisulphide. However, the same reaction in solution give eight membered cyclic disulphide (Scheme 47).

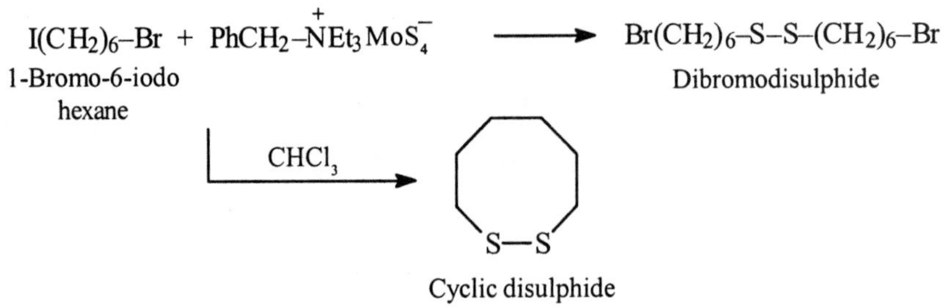

Scheme 47

13.2.16.9 Synthesis of Thiocarbonylimidazolide Derivatives

The title compounds can be prepared[47] by grinding together thiocarbonyldiimidazole and alcohols (Scheme 48).

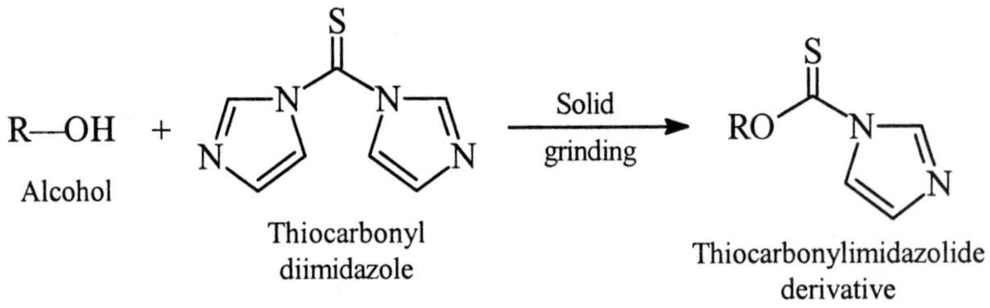

Scheme 48

13.2.16.10 Synthesis of Secondary and Tertiary Halides

These can be prepared[48] from the corresponding secondary and tertiary alcohols on exposing the powdered alcohol to HCl gas in a desiccator (Scheme 49).

13.2.16.11 Formation of Ethers from Alcohols

Treatment of alcohols with p-toluene sulphonic acid (TsOH) in solid state by keeping the powdered mixture (1:1) at room temperature for 10 min give[48] the corresponding ethers in 94-98% yields (Scheme 50).

The formation of ethers in benzene solution gave 40-73% yields while in methanoic solution gave 1-50% yields.

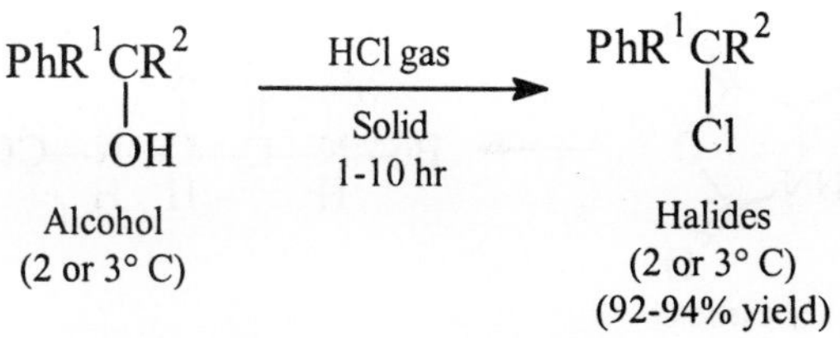

$$\underset{\underset{\text{(2 or 3° C)}}{\text{Alcohol}}}{\overset{\text{PhR}^1\text{CR}^2}{\underset{\text{OH}}{|}}} \xrightarrow[\substack{\text{Solid} \\ \text{1-10 hr}}]{\text{HCl gas}} \underset{\underset{\substack{\text{(2 or 3° C)} \\ \text{(92-94\% yield)}}}{\text{Halides}}}{\overset{\text{PhR}^1\text{CR}^2}{\underset{\text{Cl}}{|}}}$$

$R^1 = Ph;\ R^2 = H$
$R^1 = Ph;\ R^2 = H$
$R^1 = o\text{-ClC}_6\text{H}_4;\ R^2 = H$

Scheme 49

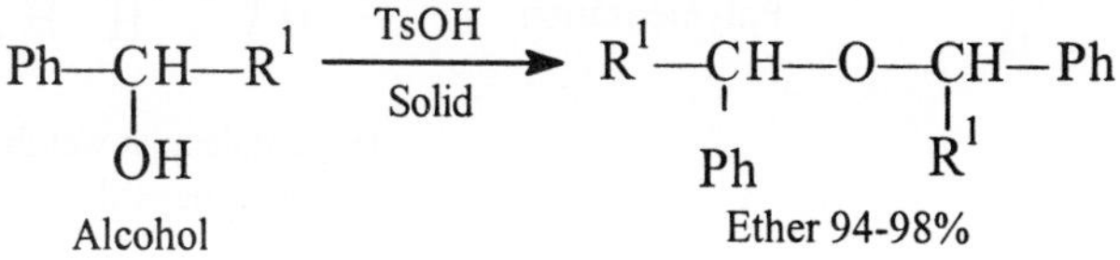

$$\underset{\text{Alcohol}}{\overset{\text{Ph—CH—R}^1}{\underset{\text{OH}}{|}}} \xrightarrow[\text{Solid}]{\text{TsOH}} \underset{\text{Ether 94-98\%}}{R^1\text{—CH—O—CH—Ph}}$$

$R^1 = Ph,\ o\text{-ClC}_6\text{H}_4,\ p\text{-BrC}_6\text{H}_4,\ p\text{-NO}_2\text{C}_6\text{H}_4,\ p\text{-MeC}_6\text{H}_4$

Scheme 50

13.2.16.12 *Synthesis of High Molecular Weight Polypeptides*
N-carboxy anhydrides of α-amino acids, viz. L-leucine, L-alanine, γ-benzyl-L-glutamine and glycine undergo solid state polymerisation[49] by using butylamine as an initiator in a hexane suspension at 20-50 °C (Scheme 51).

13.3 Solid Supported Organic Synthesis

13.3.1 Introduction
In these reactions, the reactants are stirred in a suitable solvent (for example, water, alcohol, methylene chloride etc.). The solution is stirred thoroughly with a suitable adsorbent or solid support like silica gel, alumina, phyllosilicate (M^{n+}-montomorillonite etc.). After stirring, the solvent is removed in vacuo and the dried solid support on which the reactants have been adsorbed are used for carrying the reaction under microwave irradiation.

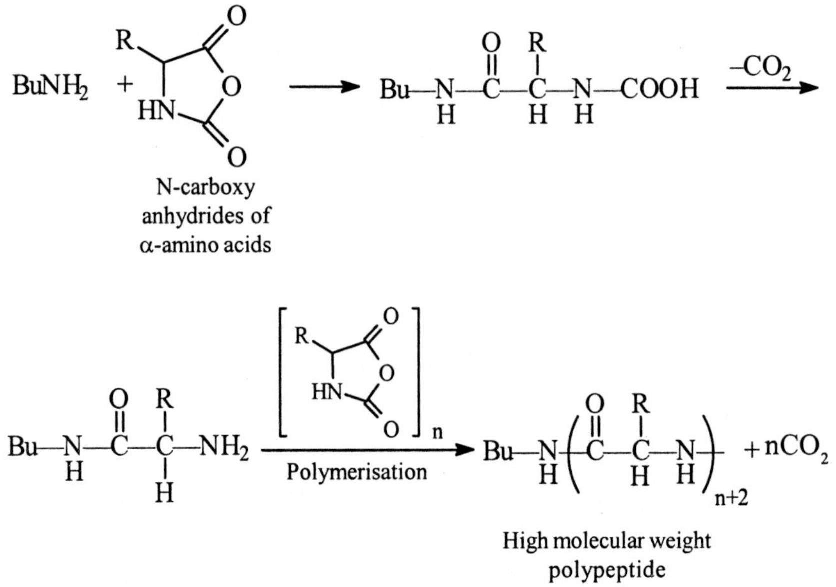

N-carboxy
anhydrides of
α-amino acids

High molecular weight
polypeptide

Scheme 51

Some of the important applications of solid support synthesis are given as follows.

13.3.2 Synthesis of Aziridines

Dry media synthesis under focussed microwave irradiation (MWI) by Michael addition has resulted in various substituted aziridines, though elimination predominates over the Michael addition under MWI when compared to classical heating under same condition (Scheme 52).[50]

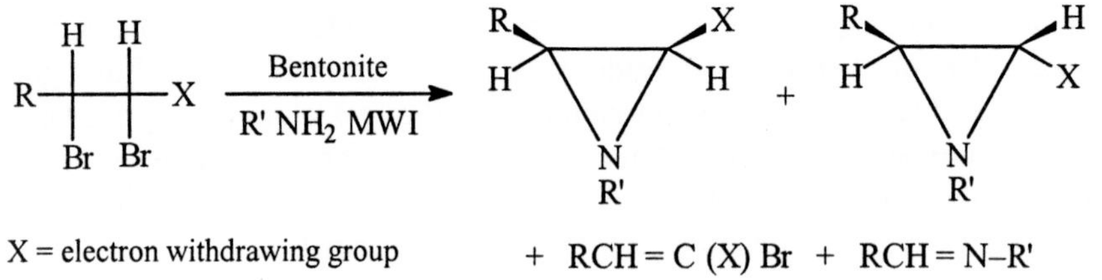

X = electron withdrawing group   + RCH = C (X) Br + RCH = N–R'

Scheme 52

13.3.3 Synthesis of β-lactams

In a preliminary report, Bose et al.[50a] had described an efficient and rapid synthesis of number of β-lactams under microwave irradiation (MWI). Further in closed Teflon vessels using KF and phase transfer catalyst (PTC), β-lactams have been synthesized in few minutes from ketene silyl acetal and aldimines (Scheme 53).[51]

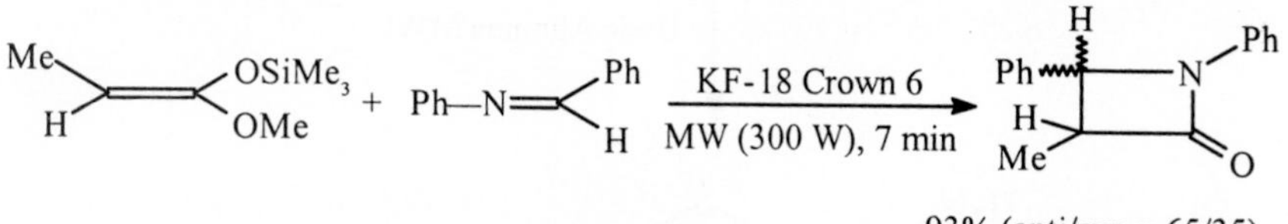

Scheme 53

93% (anti/syn = 65/35)

N(4-hydroxycyclohexyl)-3-mercapto/cyano-4-aryl-azetidine-2-one has been synthesized from N-(4-hydroxycyclohexyl)-arylaldimine by reacting with ethyl α-mercapto/α-cyanoacetate on basic alumina under microwaves (Scheme 54), wherein not only the reaction time were brought down from hours to minutes in comparison to conventional heating but also yields were improved.[52]

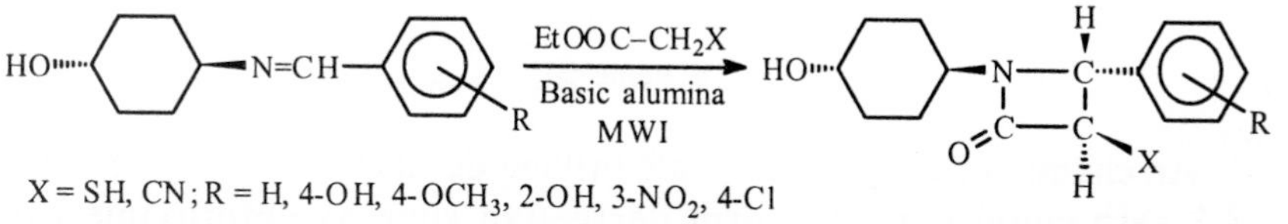

X = SH, CN; R = H, 4-OH, 4-OCH$_3$, 2-OH, 3-NO$_2$, 4-Cl

Scheme 54

Deacylation of cephalosporins, a growing class of β-lactam antibiotics, has been investigated using enzymatic and microwave activated solid phase techniques. The deacylation was achieved in less time with better yields (Scheme 55).[53]

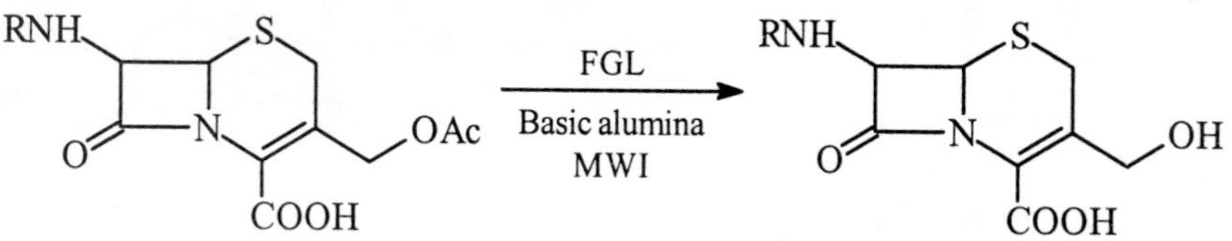

Scheme 55

Reaction of 7-amino-3-[5'-methyl-1',3',4'-thiadiazol-2'-ylthiomethyl] cephalosporanic acid with heterocyclic amines using basic alumina under microwave irradiation (MWI) afforded new cephalosporin analogs in shorter reaction time with improved yield as compared to conventional heating (Scheme 56).[54]

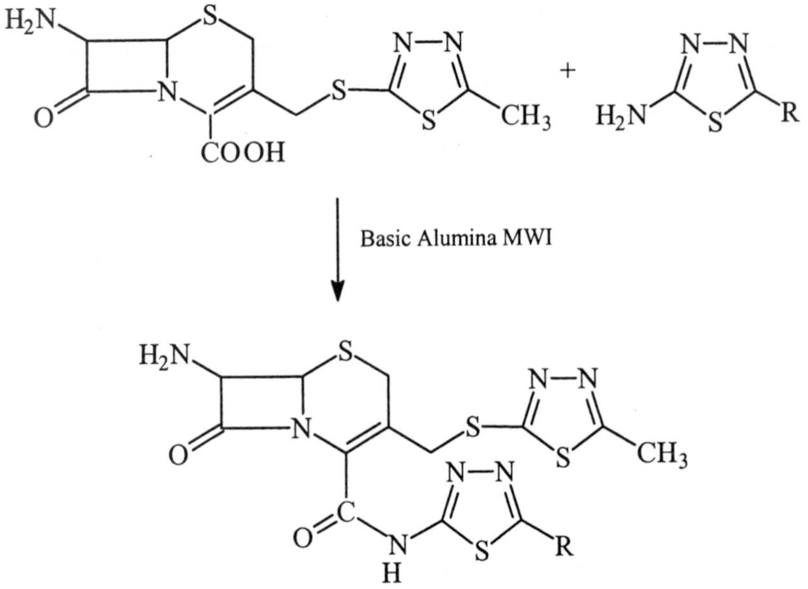

R = CH$_3$, C$_6$H$_5$, C$_9$H$_{19}$, C$_{11}$H$_{23}$, 3-pyridinyl, 4-ClC$_6$H$_4$

Scheme 56

An environmentally friendly safe method developed for the preparation of 3-carbamoyl cephalosporin derivative such as cefuroxime uses o-transcarbomylase, an enzyme of microbial origin for the conversion of 3-hydroxy function to the desired 3-carbomyl group. This new synthesis replaces the conventional chemical route, which employs hazardous isocyanates such as dichlorophosphenyl isocynate or chlorosulfinyl isocyanate to achieve the same conversion (Scheme 57).[55]

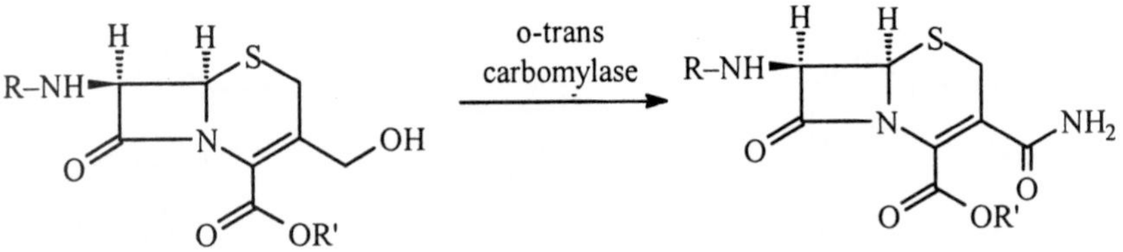

R = H, acyl group
R' = H, or carboxyl protecting group

Scheme 57

13.3.4 Synthesis of Pyrroles

A simple and fast synthesis of a tetrapyrrolic macrocycle under dry media conditions with microwave activation was performed. Pyrrole and benzaldehyde adsorbed on silica gel afforded tetrahydroporphyrin within 10 min, whereas conventional method needed 24 hr (Scheme 58).[56]

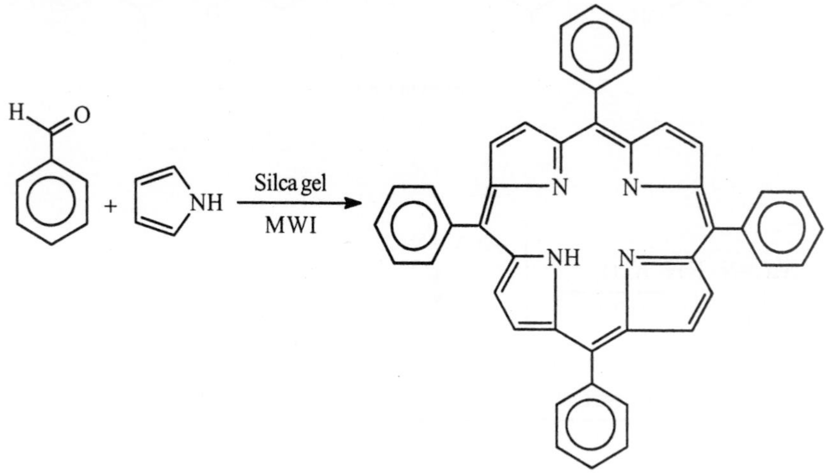

Scheme 58

Synthesis of substituted pyrrole over silica gel under microwave irradiation has been reported (Scheme 59).[57]

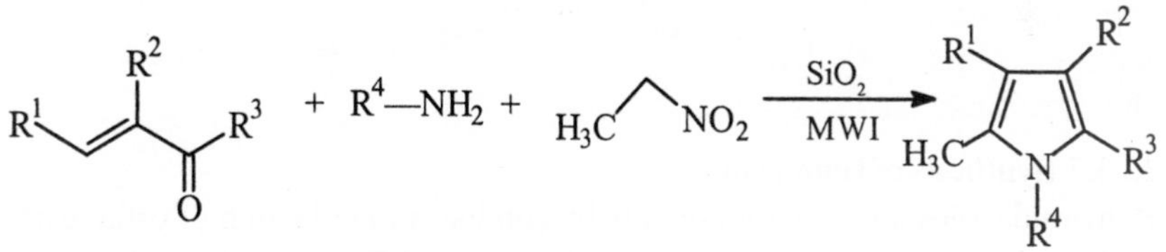

Scheme 59

13.3.5 Synthesis of Furans

Naturally occurring, pharmacologically important 2-aroyl-benzofurans are easily obtainable in the solid state from α-tosyloxyketones and salicylaldehydes in the presence of a base such as KF doped alumina using microwave irradiation (Scheme 60).[58]

13.3.6 Synthesis of Pyrazoles

Diarylnitrilimine on cyclocondensation with alkynes and disubstitued alkenes gave the corresponding pyrazoles when irradiated with microwaves. The nitrilimine was generated *in situ* and thus was a one pot reaction of hydrazonyl chloride over alumina (Scheme 61).[59]

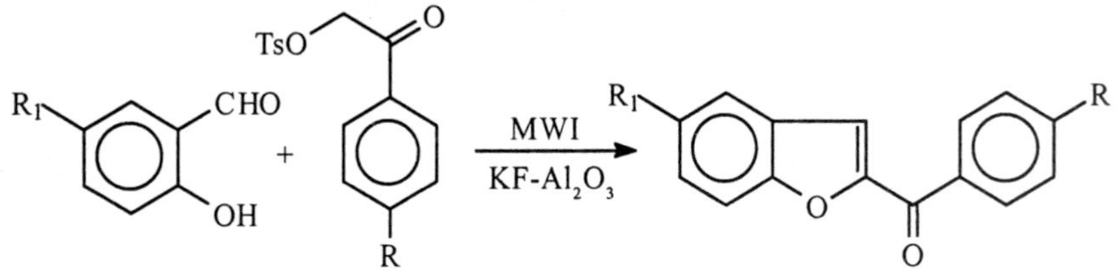

Scheme 60

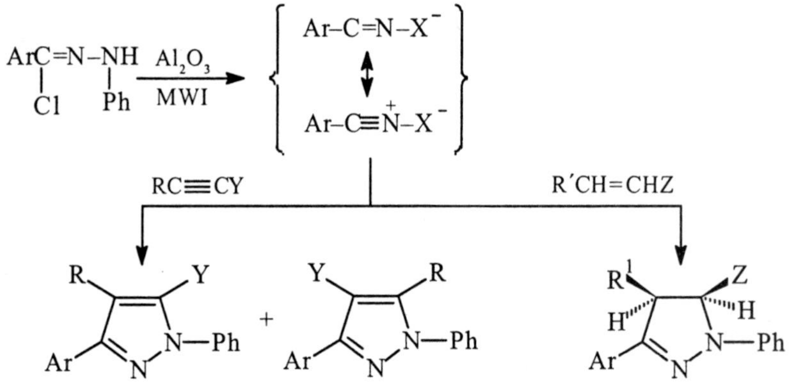

Scheme 61

13.3.7 Synthesis of Imidazoles

Benzimidazoles are prepared readily by condensation reaction of ortho-esters with o-phenylenediamines in the presence of KSF clay under solvent free conditions using focused microwave irradiation (Scheme 62).[60]

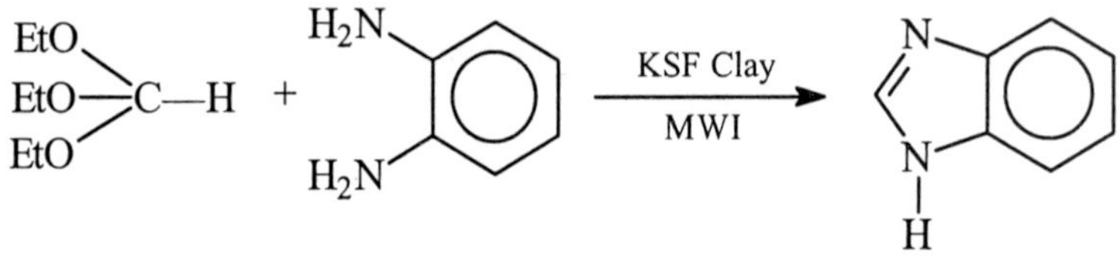

Scheme 62

4-Alkylidene-1H-imidazol-5(4H)-ones are obtained in good to excellent yields by 1,3-dipolar cycloaddition, in solvent-free condition under focussed microwaves from an activated imidate and aldehydes in the presence of catalytic amounts of anhydrous acetic acid (Scheme 63).[61]

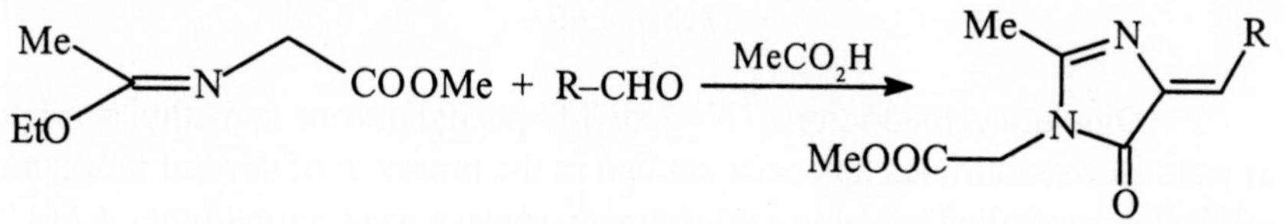

Scheme 63

An expeditious solvent-free synthesis of pyrazolino/iminopyrimidino/ thioxopyrimidino imidazoline derivatives from oxazolones on solid support using microwaves has been described. The reaction time was brought down from hours to minutes with improved yield as compared to conventional heating (Scheme 64).[62]

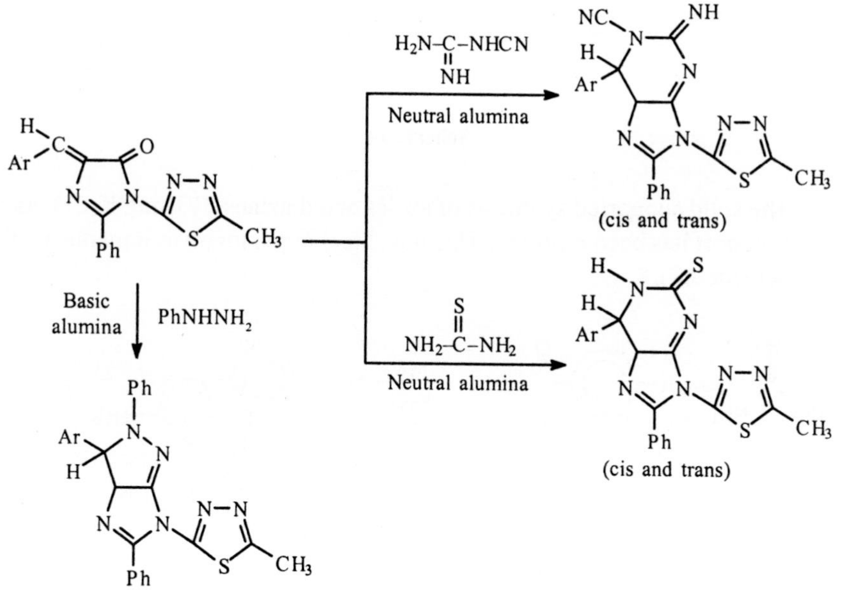

Scheme 64

13.3.8 Synthesis of Azoles

Oxazolines are readily prepared from the carboxylic acids and α,α,α-tris(hydroxymethyl) methylamine under MWI (Scheme 65).[63]

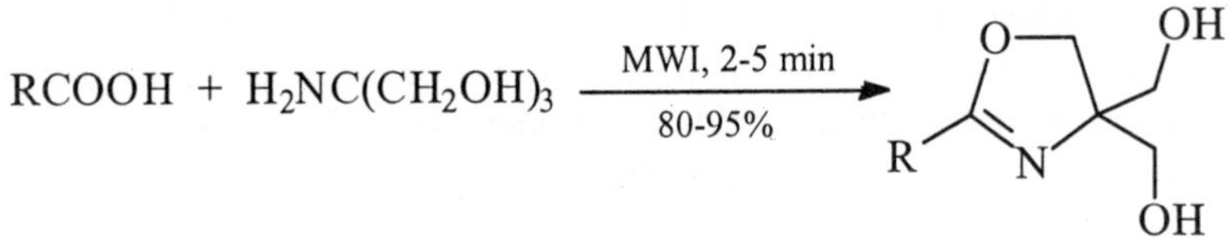

$$RCOOH + H_2NC(CH_2OH)_3 \xrightarrow[\text{80-95\%}]{\text{MWI, 2-5 min}}$$

Scheme 65

1,3-Dipolar cycloaddition of N-methyl-C-phenyl nitrone to methyl acrylate to yield isooxazolidines has been studied in the presence of several inorganic solid supports (silica gel, alumina) using microwave irradiation (Scheme 66).[64]

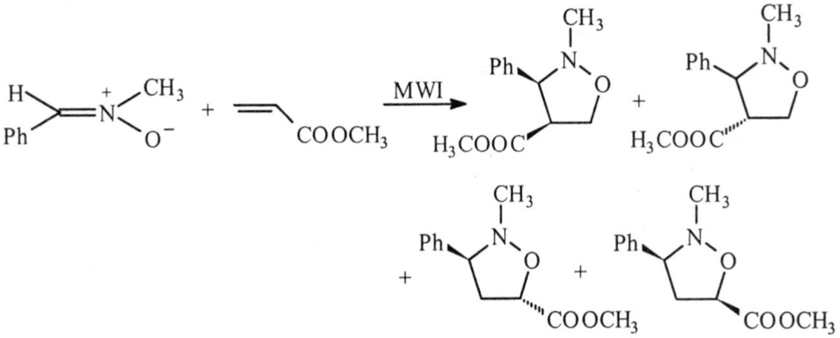

Scheme 66

The solid supported synthesis of azoles and diazines by using K_2CO_3 as a solid support has been reported. This novel technique involves aqueous work up (Scheme 67).[65]

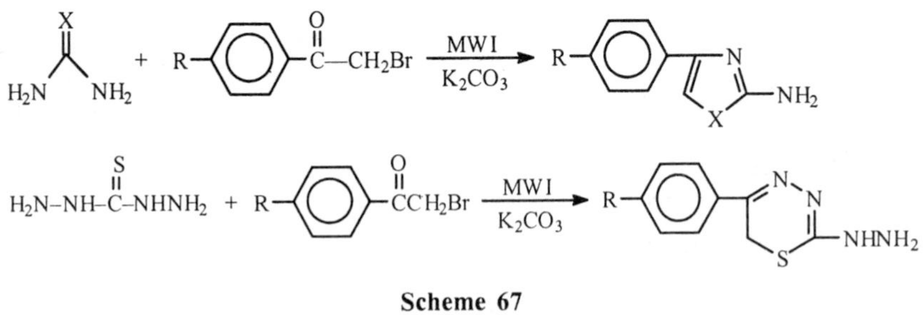

Scheme 67

Thiazole and its derivatives are obtained by reaction of α-tosyloxyketones, which are generated *in situ* from aryl-methyl ketones and [hydroxy (tosyloxy) iodo] benzene with thioamides in the presence of K10 clay using MWI (Scheme 68).[58]

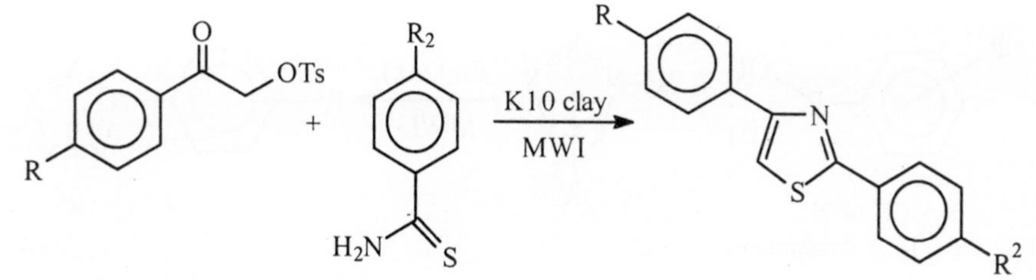

Scheme 68

Amidoximes on reaction with isopropenyl acetate in presence of KSF clay under MWI gave 1,2,4-oxadiazoles in good yield (Scheme 69). The 1,2,4-oxadizaoles were also obtained by MWI from O-acylamidoximes adsorbed on alumina (Scheme 70).[66]

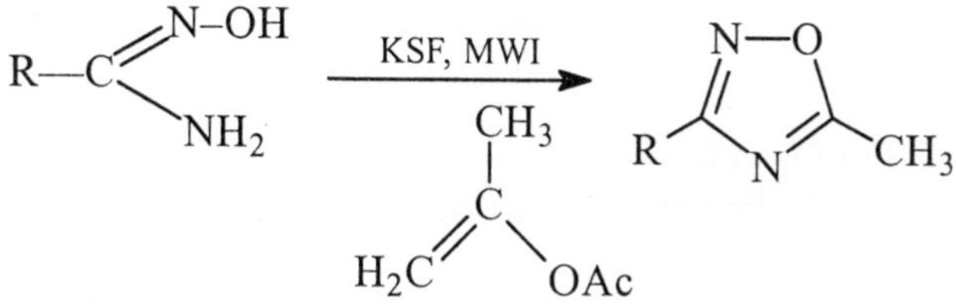

Scheme 69

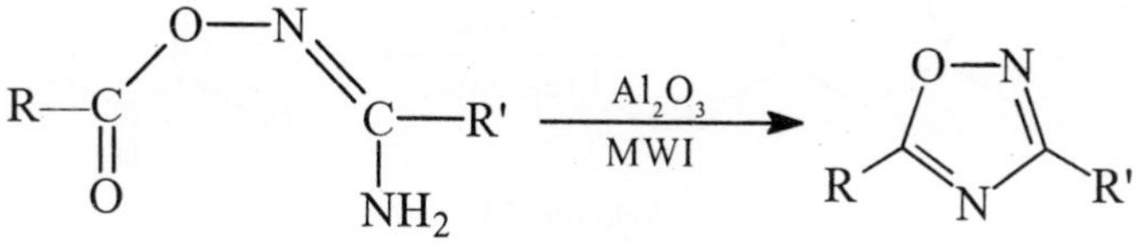

Scheme 70

A general method for solid phase synthesis of N-arylated benzimidazoles, imidazoles, triazoles and pyrazoles has been demonstrated utilizing copper (II) mediated coupling of aryl boronic acids under MWI (Scheme 71).[67]

Microwave accelerated solid state synthesis of spiroindole derivatives has also been described. The products were obtained in few seconds in good yield.[68]

Reaction of aromatic aldehydes with phenyl nitromethane under microwave irradiation (MWI) on basic alumina afforded excellent yields (90-96%) of isoxazoles within 2-3 minutes (Scheme 72).[69]

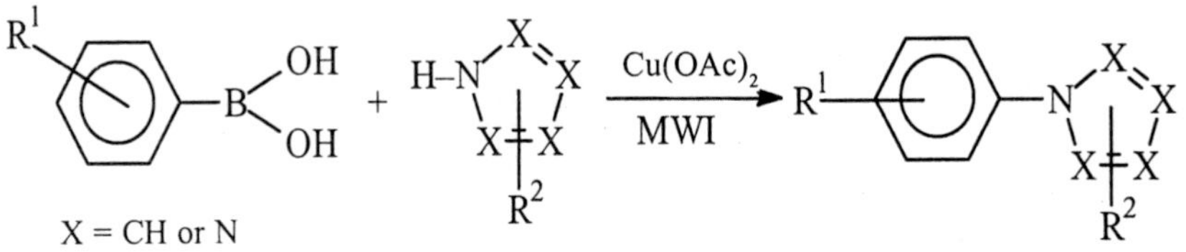

X = CH or N
R¹, R² = substituents

Scheme 71

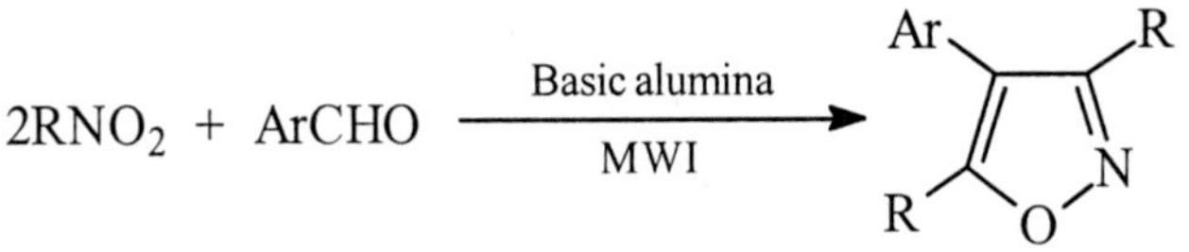

$$2RNO_2 + ArCHO \xrightarrow[MWI]{Basic\ alumina}$$

Scheme 72

13.3.9 Synthesis of Pyridines

A rapid and high yielding protocol for the synthesis of 1,4-dihydropyridine in the presence of silica gel under microwave irradiation (MWI) has been described (Scheme 73).[70]

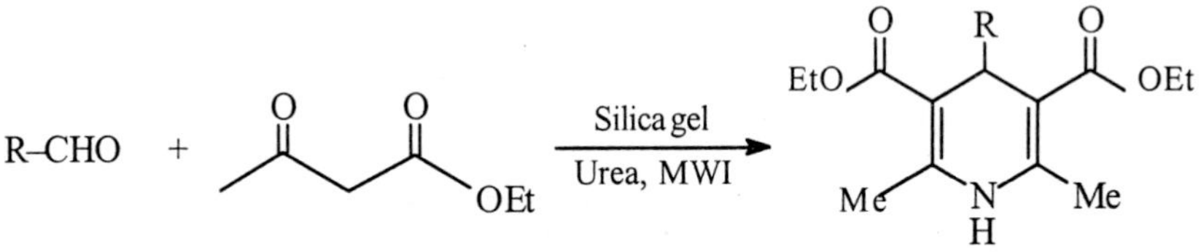

Scheme 73

13.3.10 Synthesis of Chromenes and Flavones

Isoflav-3-enes possessing chromene nucleus are well known estrogens. There is a great demand for the development of ecofriendly synthetic methods for these derivatives. A facile and general method for the one pot synthesis of isoflav-3-enes have been described (Scheme 74).[71]

A solvent-free synthesis of flavones which involves the MWI of o-hydroxydibenzoyl methanes absorbed on montmorillonite K10 clay for 1 to 1.5 min has been achieved. Rapid and exclusive formation of cyclized flavones occurs in good yields (Scheme 75).[72]

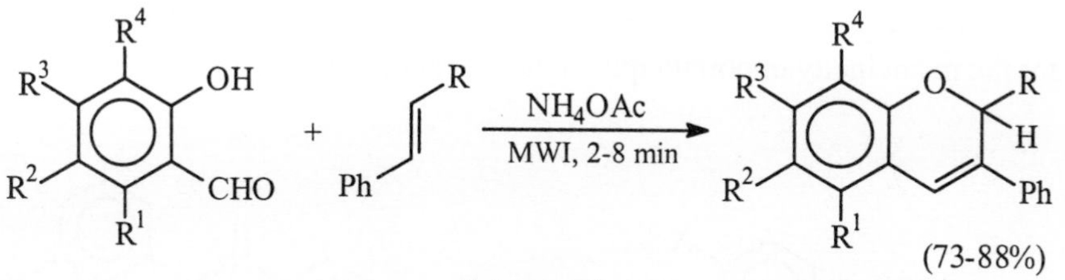

Scheme 74

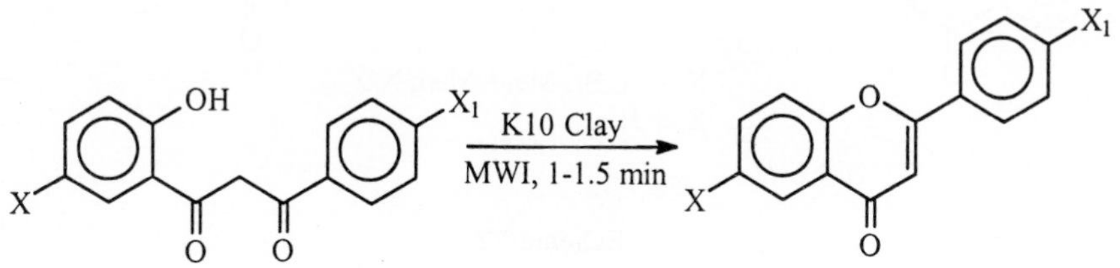

Scheme 75

X = H, OCH$_3$; X$_1$ = H, Me, OMe, NO$_2$

13.3.11 Synthesis of Quinoline

KSF clay catalyzed Friedlander condensation of 2-aminoarylaldehyde or ketones with carbonyl compounds containing α-methylene group has been achieved in solvent free condition under MWI to give polycyclic quinoline derivatives (Scheme 76).[73]

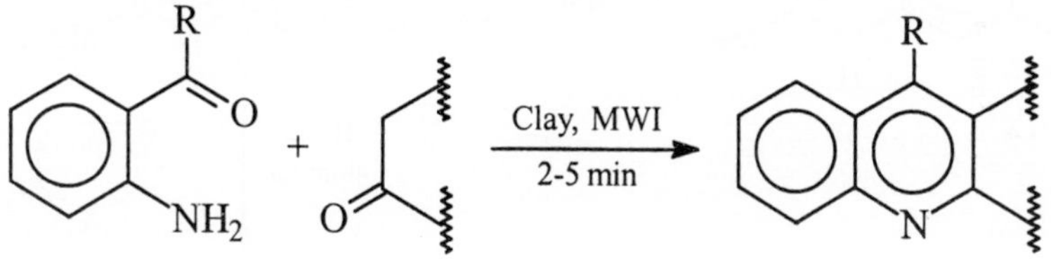

Scheme 76

In yet another solventless cyclisation reaction using montmorillonite K10 clay under MWI condition, readily available 2'-aminochalcones provide easy access to 2-aryl-1,2,3,4-tetrahydro-4-quinolones which are valuable precursor for the medicinally important quinalones (Scheme 77).[74]

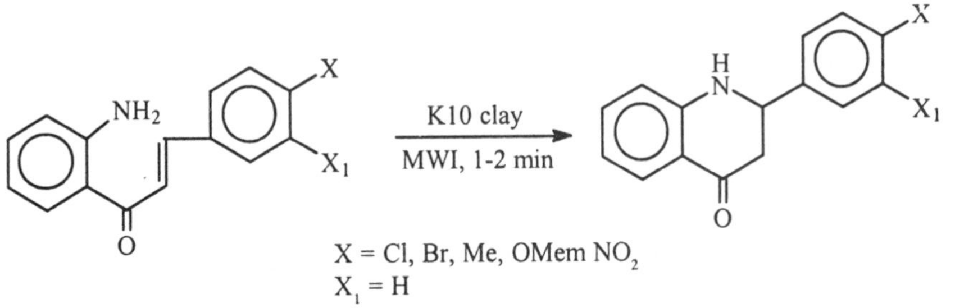

X = Cl, Br, Me, OMem NO$_2$
X$_1$ = H

Scheme 77

In an alumina-supported synthesis of antibacterial quinolines using microwaves wherein the reaction times were brought down from hours to seconds with improved yield as compared to the conventional heating (Scheme 78).[75]

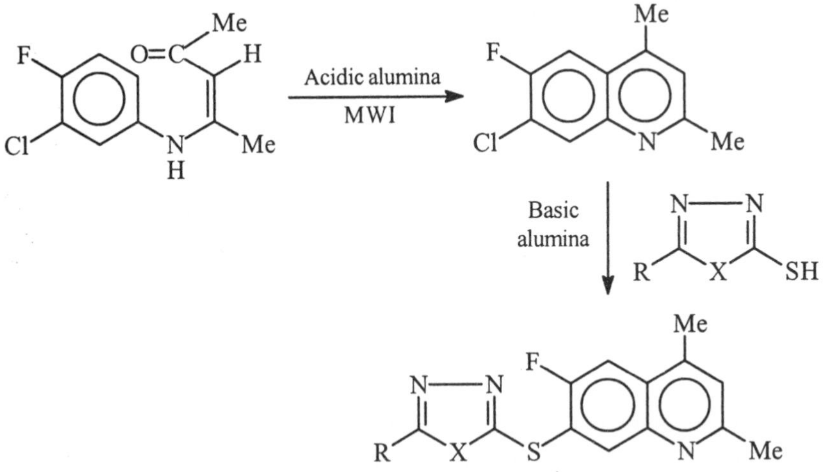

Scheme 78

13.3.12 Synthesis of Pyrimidines

A one pot synthesis of pyrano [2,3-d] pyrimidines from thiobarbituric acids under microwave irradiation using basic alumina has been reported. A significant reduction in reaction time and enhancement in the yield was observed (Scheme 79).[76]

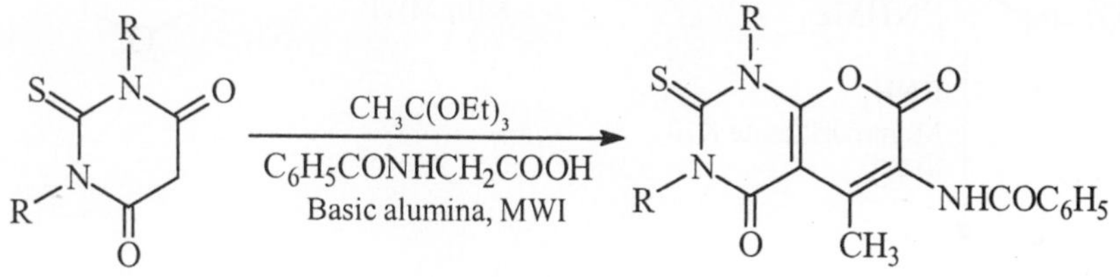

Scheme 79

Pyrimidino [1,6-a] benzimidazoles (Scheme 80) and 2,3-dihydroimidazo [1,2-c] pyrimidines (Scheme 81) under focused microwave irradiation have been synthesised.[77]

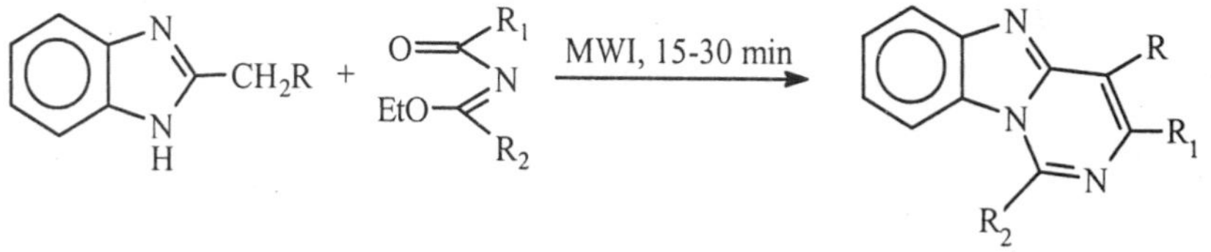

Scheme 80

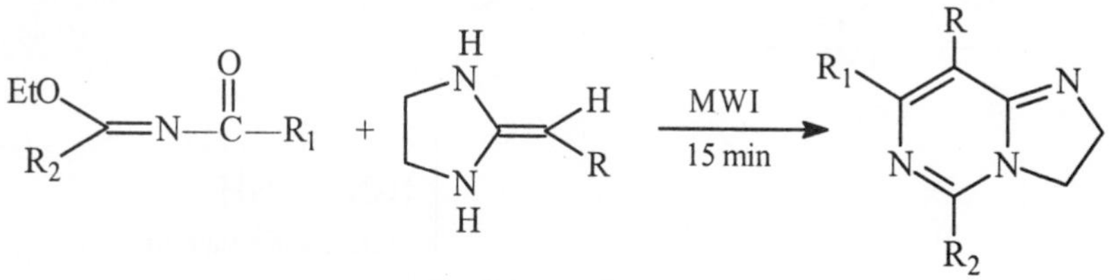

Scheme 81

13.3.13 Synthesis of Oxadiazines

Three component condensation of N,N′-dimethyl urea, paraformaldehyde and primary amines using montmorillonite K10 clay in dry media under microwave irradiation (MWI) lead to the formation of triazones whereas condensation of dimethyl urea and paraformaldehyde supported on montmorillonite K10 using MWI gave 4-oxo-oxadiazinone (Scheme 82).[78]

An efficient synthesis of benzopyranopyrimidines using three different solid supports viz., acidic alumina, montmorillonite, silica gel has been carried out. The products were obtained with improved yields as compared to conventional heating (Scheme 83).[79]

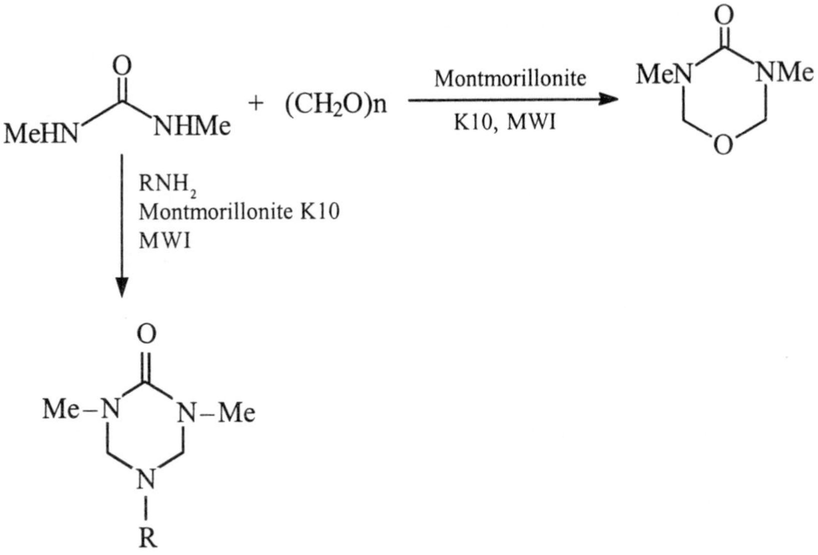

Scheme 82

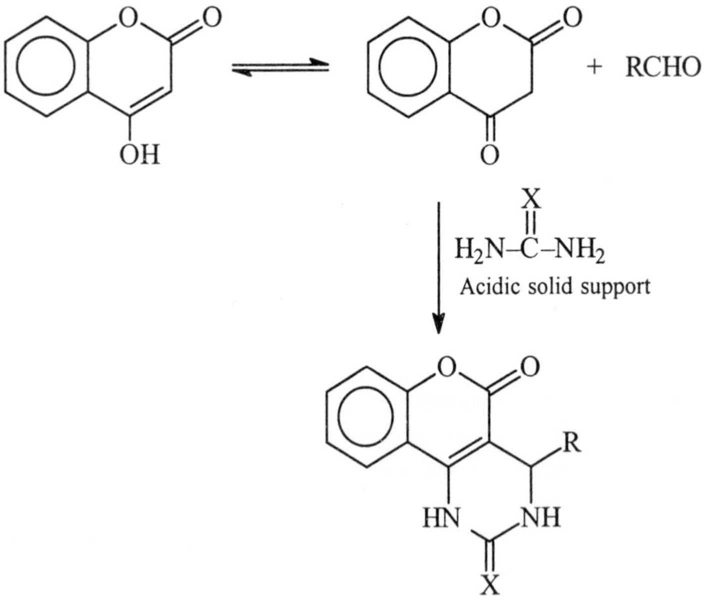

Scheme 83

13.3.14 Synthesis of Thiadiazepines

An environmentally benign synthesis of the 1,2,4-triazolo[3,4-b]-1,3,4-thiadiazepines from substituted triazoles and chalcones on basic alumina under MWI is reported (Scheme 84).[80]

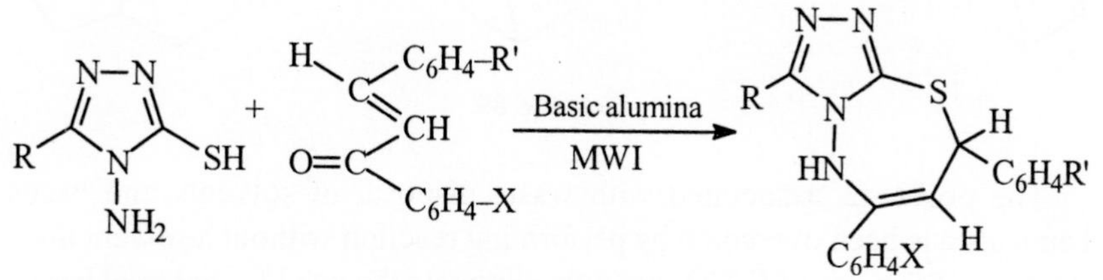

Scheme 84

13.3.15 Miscellaneous Reactions

Scandium tri-fluoromethane sulfonate microencapulated (Lewis acid) have replaced traditional monomeric Lewis acid reactions like Michael, Fridel Craft, Mannich and immino Aldol reactions (Schemes 85-88).[81-84]

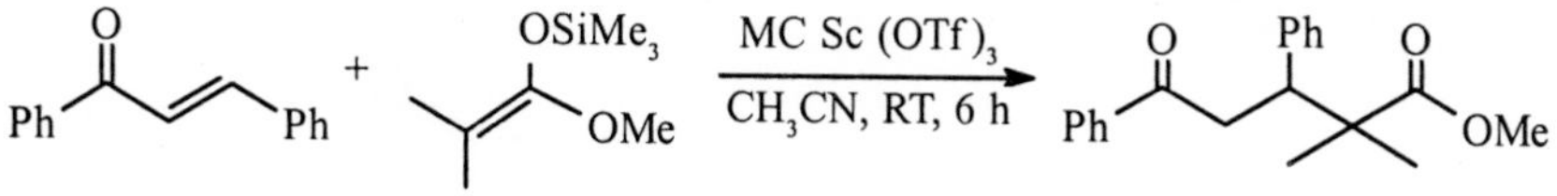

Scheme 85

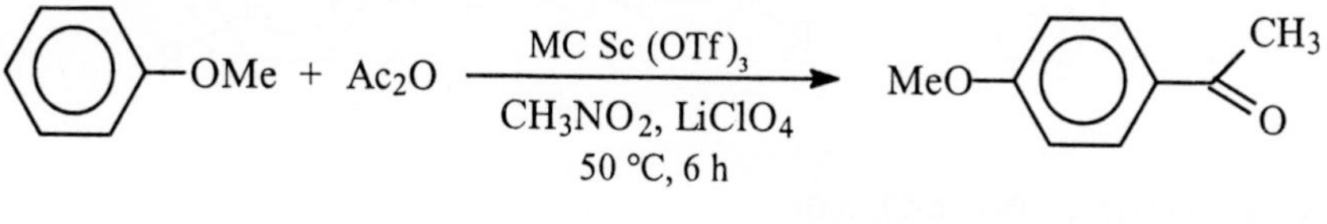

Scheme 86

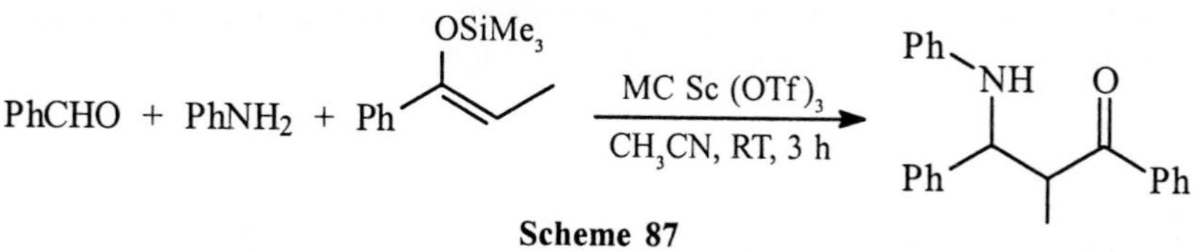

Scheme 87

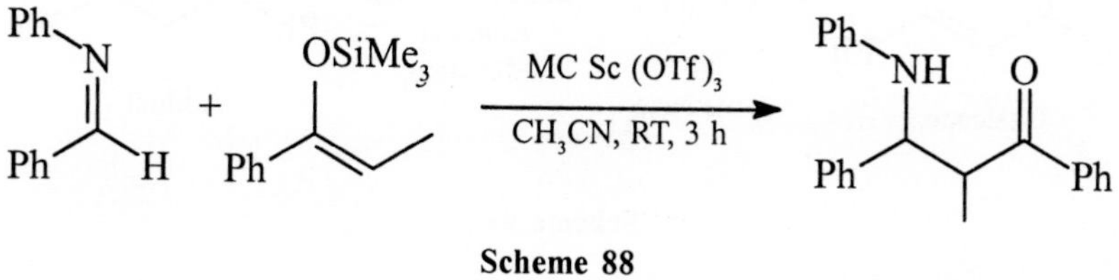

Scheme 88

Industrial applications of basic catalysts are in the allylation of phenol, side chain allylation and isomerization reaction (Scheme 89).[85]

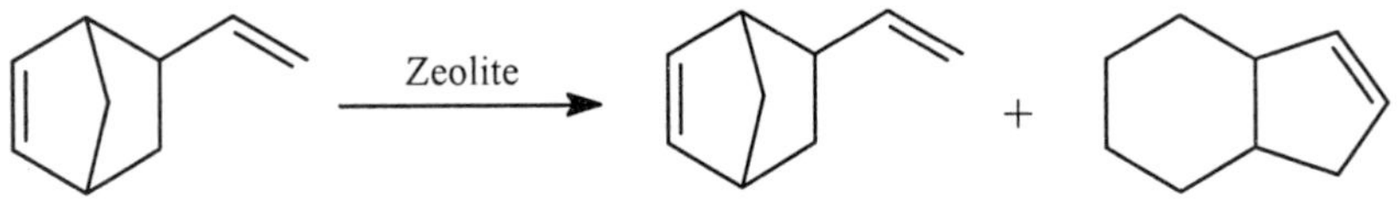

Scheme 89

The problems associated with waste disposal of solvents and excess chemicals has been overcome by performing reaction without a solvent under microwave. The use of K_2CO_3 not only eliminate the need for external base to netralize HCl evolved but also enables aqueous workup (Scheme 90).[86]

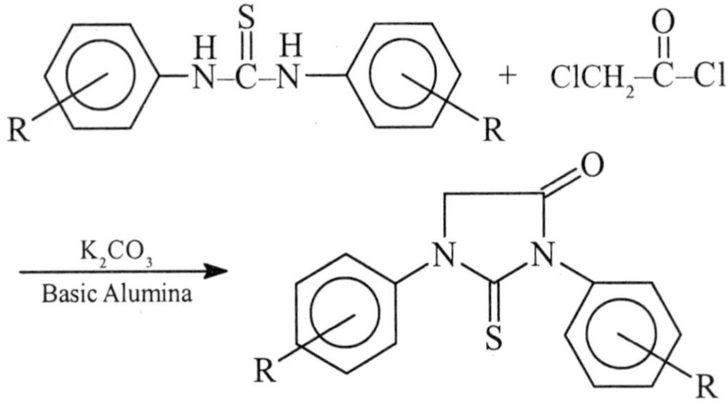

R = H, o-OMe, p-OMe, p-Cl, p-Br

Scheme 90

An interesting example is the solvent-free Michael addition reaction of nitromethane to chalcone in the presence of alumina under microwave irradiation condition that gives the adduct in 90% yield (Scheme 91).[87]

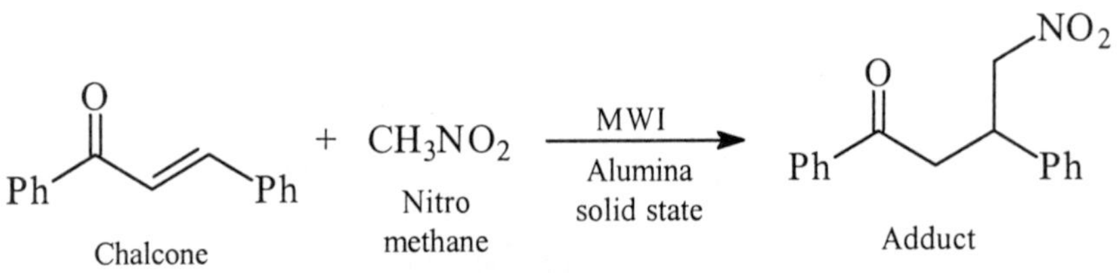

Scheme 91

The previous reaction (Scheme 91) takes about 15 days under conventional conditions.[88]

References

1. J.M. Thomas and S.S. Mori, J.P. Desvergne, *Adv. Phys. Org. Chem.*, 1977, **15**, 63.
2. J.M. Thomas, *Pure Appl. Chem.*, 1979, **51**, 1065.
3. D.Y. Curtin and I.C. Paul, *Chem. Rev.*, 1981, **81**, 525.
4. A. Gavezzoti and M. Simonetta, *Chem. Rev.*, 1982, **82**, 1.
5. V. Ramamurty, *Tetrahedron*, 1986, **42**, 5753.
6. V. Ramamurty and K. Venkatesan, *Chem. Rev.*, 1987, **87**, 433.
7. G.R. Desiraju, Organic Solid State Chemistry, Elsevier, 1987.
8. F. Toda, *Acc. Chem. Res.*, 1995, **28**, 480.
9. K. Tanaka and F. Toda, *Chem. Rev.*, 2000, **100**, 1025.
10. L. Claisen, *Ber.*, 1912, **45**, 3157; L. Claisen and E. Tietze, 1925, **58**, 275; D.S. Tabell, *Org. React.*, 1944, **2**, 1.
11. A. Schmiht, *Liebigs Ann. Chem.*, 1863, **127**, 319.
12. G. Kaupp and D. Matthies, *Chem. Ber.*, 1987, **120**, 1897.
13. K. Penzien and M.J. Schmidt, *Angew. Chem. Int. Ed. Engl.*, 1968, **8**, 608; D. Rabinovich and Z. Shakked, *Acta Crystallogr.*, 1974, **B30**, 2829.
14. Y. Tanaka, H. Sakuraba, Y. Oka and H. Nakanishi, *J. Incl. Phenom.*, 1984, **2**, 841.
15. B. Satish, K. Panneersel-Vam, D. Zacharids and G.R. Desivaju, *J. Chem. Soc. Perkin Trans.*, 1995, **2**, 325.
16. E. Diez-Barra, A. de la Hoz, S. Merino and P. Sanchez-Verdu, *Tetrahedron Lett.*, 1997, **38**, 2359.
17. F. Toda, K. Tanka and J. Sato, *Tetrahedron Asymmetry*, 1993, **4**, 1771.
17a. A. Loupy, J. Sansoulet, A. Zaparucha and C. Merinne, *Tetrahedron Lett.*, 1989, **30**, 333.
18. F. Toda, H. Takumi and M. Akehi, *J. Chem. Soc. Chem. Commun.*, 1990, 1270; F. Toda and K. Okuda, *J. Chem. Soc. Chem. Commun.*, 1991, 1212.
19. Y. Wef, R. Bakthavatechalam, *Tetrahedron Lett.*, 1991, **32**, 1535.
20. F. Toda, K. Tanaka and K. Hamai, *J. Chem. Soc., Perkin Trans. I*, 1990, 3207.
21. W. Dieckmann, *Ber.*, 1894, **27**, 102, 965; 1900, **33**, 2670; *Ann.*, 1901, **317**, 53, 93; C.R. Hauser and B.E. Hudson, *Organic Reactions*, 1942, **I**, 274.
22. F. Toda, T. Suzuki and S. Higa, *J. Chem. Soc. Perkin Trans. 1*, 1998, 3521.
23. F. Toda, H. Takumi and H. Yamaguchi, *Chem. Exp.*, 1980, **4**, 507.
24. H. Tanka, S. Kishigami and F. Toda, *J. Chem. Soc.*, 1991, **56**, 4333.
25. F. Toda and H. Akai, *J. Org. Chem.*, 1990, **55**, 3446.
26. K. Smit, A. Musson, G.A. De Boss, *J. Chem. Soc. Chem. Commun.*, 1996, 469.
27. M. Etter, G.M. Frankenbach and J. Bernstein, *Tetrahedron Lett.*, 1989, **30**, 3617.
28. F. Toda and T. Shigemasa, *J. Chem. Soc. Perkin Trans. I*, 1989, 209.
29. F. Toda, K. Tanaka, Y. Kagawa and Y. Sakaino, *Chem. Lett.*, 1990, 373.
30. I. Almena, A. Diaz-Ortiz, E. Diez-Barra, A. Hos and A. Loupy, *Chem. Lett.*, 199○, 333; S. Caddick, *Tetrahedron*, 1995, 10400.
31. F. Toda and H. Akai, *J. Org. Chem.*, 1990, **55**, 4973.
32. M. Dessolin and M. Goltier, *J. Chem. Soc. Chem. Commun.*, 1984, 38; M. Dessolin, O. Eisenstein, M. Golfier, T. Prange and P.J. Sautet, *J. Chem. Soc. Chem. Commun.*, 1992, 132.
33. M.L. Tosato and I. Soccorsi, *J. Chem. Soc. Perkin Trans. 2*, 1981, 1321.

34. K. Naruchi and M. Miura, *J. Chem. Soc. Perkin Trans.*, 1987, **2**, 113.
35. F. Akutsu, K. Aovagi, N. Nishimura, M. Kudon, Y. Kasashima, M. Inoki and K. Naruchi, *J. Chem. Soc. Perkin Trans.*, 1996, 889.
36. H. Tanaka, S. Kishigami and F. Toda, *J. Org. Chem.*, 1990, **55**, 2982.
37. F. Toda, K. Tanaka and S. Iwata, *J. Org. Chem.*, 1989, **54**, 3007; T. Higashizima, N. Sakai, K. Nozaki and H. Takaya, *Tetrahedron Lett.*, 1994, **35**, 2023.
38. F. Toda and Y. Tokumaru, *Chem. Lett.*, 1990, 987.
39. G.W. Wang, K. Komatsu, Y. Murata and M. Shiro, *Nature*, 1997, **387**, 583.
40. G.W. Wang, Y. Murata, K. Komatsu and T.S.M. Wan, *J. Chem. Soc. Perkin Trans. 1*, 1996, 2059.
41. N.S. Khruscheva, N.M. Loim, V.I. Sokolov and V.D. Makhaev, *J. Chem. Soc. Perkin Trans. 1*, 1997, 2425.
42. J. Schmiyers, F. Tuda, J. Boy and G.J. Kaupp, *J. Chem. Soc. Perkin Trans. 2*, 1998, 989.
43. F. Toda, H. Akai, *J. Org. Chem.*, 1990, **55**, 3446.
44. F. Toda and N. Imai, *J. Chem. Soc. Perkin Trans. 1*, 1994, 2673.
45. C.J. Abelt and J.M. Pleier, *J. Org. Chem.*, 1988, **53**, 2159.
46. A.R. Ramesha and S. Chandresekaran, *J. Chem. Soc. Perkin Trans. 1*, 1994, 767.
47. H. Hagiwara, S. Ohtsubo and M. Kato, *Mol. Cryst. Liq. Cryst.*, 1996, **279**, 291.
48. F. Toda, H. Takumi and M. Akehi, *J. Chem. Soc. Chem. Commun.*, 1990, 1270; F. Toda and K. Okuda, *J. Chem. Soc. Chem. Commun.*, 1991, 1212.
49. H. Kanazawa, *Polymer*, 1992, **32**, 2557.
50. A. Saoudi, J. Hamelin and H. Benhaoua, *J. Chem. Res. (s)*, 1996, 492.
50a. A.K. Bose, M. Jayaraman, A. Okawa, S.S. Bari, E.W. Robb and M.S. Manhas, *Tetrahedron Lett.*, 1996, **37(39)**, 6989.
51. F. Texier-Boullet, R. Latouche and J. Hamelin, *Tetrahedron Lett.*, 1993, **34**, 2123.
52. M. Kidwai, R. Venkataramanan and S. Kohli, *Synth. Commun.*, 2000, **30(6)**, 989.
53. M. Kidwai, B. Dave, K.R. Bhushan, P. Misra, R.K. Saxena, R. Gupta, R. Gulati and M. Singh, *Biocatalysis and Biotransformation*, 2002, **20(5)**, 377-379.
54. M. Kidwai, P. Sapra, K.R. Bhushan, P. Mishra, R.K. Saxena, R. Gupta and M. Singh, *Bioorg. Chemistry*, 2001, **29**, 380.
55. I.D. Flemeng, M.K. Turner and S.J. Brewer, *US 4075061*, 1978; *Chem. Abstr.*, 1978, **88**, 168488.
56. A. Petit, A. Loupy, P. Maillard and M. Momenteau, *Synth. Commun.*, 1992, **22**, 1137.
57. B.C. Ranu, A. Hajra and U. Jana, *Synlett.*, 2000, 75.
58. R.S. Varma, D. Kumar and P.J. Liesen, *J. Chem. Soc. Perkin Trans. 1*, 1998, 4093.
59. K. Bougrin, M. Soufiaoui, A. Loupy and P. Jacquault, *New J. Chem.*, 1995, **19**, 213.
60. D. Villemin, M. Hammadi and B. Martin, *Synth. Commun.*, 1996, **26**, 2895.
61. G. Kerneur, J.M. Lerestif, J.P. Bazureau and J. Hamelin, *Synthesis*, 1997, 287.
62. M. Kidwai, P. Sapra, K.R. Bhushan and P. Misra, *Synthesis*, 2001, **10**, 1509.
63. A.L. Marrero-Terrero and A. Loupy, *Synlett.*, 1996, 245.
64. A. Padwa, L. Fisera, K.F. Koehler, A. Rodriguez and G.S.K. Wong, *J. Org. Chem.*, 1984, **49**, 276.
65. M. Kidwai, R. Venkataramanan and B. Dave, *J. Heterocycl. Chem.*, 2002, **39**, 1-3.
66. B. Oussaid, L. Moeini, B. Martin, D. Villemin and B. Garrigues, *Synth. Commun.*, 1995, **25(10)**, 1451.
67. A.P. Combs, S. Saubern, M. Rafalski and P.Y.S. Lam, *Tetrahedron Lett.*, 1999, **40**, 1623.
68. M. Kidwai and P. Misra, *Oxidation Communications*, 2001, **24**, 287.
69. M. Kidwai and P. Sapra, *Org. Prep. Proced. Int.*, 2001, **33**, 381.
70. J.S. Yadav, B.V.S. Reddy and P.T. Reddy, *Synth. Commun.*, 2001, **31(3)**, 425.
71. R.S. Varma and R. Dahiya, *J. Org. Chem.*, 1998, **63**, 8038.
72. R.S. Varma, R.K. Saini and D. Kumar, *J. Chem. Res. (s)*, 1998, 348.

73. G. Sabitha, R.S. Babu, B.V.S. Reddy and J.S. Yadav, *Synth. Commun.*, 1999, **29(24)**, 4403.
74. R.S. Varma and R.K. Saini, *Synlett.*, 1997, 857.
75. M. Kidwai, K.R. Bhushan, P. Sapra, R.K. Saxena and R. Gupta, *Bioorg. Med. Chem.*, 2000, **8**, 69.
76. M. Kidwai, R. Venkataramanan, R.K. Garg and K.R. Bhushan, *J. Chem. Res. (S)*, 2000, 586.
77. M. Rahmouni, A. Derdour, J.P. Bazureau and J. Hamelin, *Tetrahedron Lett.*, 1994, **35**, 4563.
78. S. Balalaie, M.S. Hashtroudi and A. Sharifi, *J. Chem. Res. (S)*, 1999, 392.
79. M. Kidwai and P. Sapra, *Synth. Commun.*, 2002, **32(11)**, 1693.
80. M. Kidwai, P. Sapra, P. Misra, R.K. Saxena and M. Singh, *Bioorg. Med. Chemistry*, 2001, **9**, 217-220.
81. S. Kobayashi, I. Hachya, H. Ishrleni and Araki, *Synlett.*, 1993, 472; S. Kobayashi and S. Nagayama, *J. Am. Chem. Soc.*, 1998, **120**, 2985.
82. A. Kawada, S. Mitamura and S. Kobayashi, *Synlett.*, 1994, 545.
83. S. Kobayashi, M. Araki and M. Yasuda, *Tetrahedron Lett.*, 1995, **36**, 5773.
84. S. Kobayashi, M. Arcki and S. Hachiya, *Synlett.*, 1995, 233.
85. W.F. Hoelderich, *Stud. Surf. Sci. Catal.*, 1993, **75**, 127.
86. M. Kidwai, R. Venkataramanan and B. Dave, *Green Chemistry*, 2001, **3**, 278.
87. A. Boruah, M. Boruah, D. Prajapati and J.S. Sandhu, *Chem. Lett.*, 1996, 965.
88. M.C. Kloetzel, *J. Am. Chem. Soc.*, 1947, **69**, 2271.

14. Versatile Ionic Liquids as Green Solvents

14.1 Green Solvents

The commonly used solvents like benzene, toluene, methylene chloride etc. for organic synthesis, particularly in industrial production, are known to cause health and environmental problems. In view of this, the search for alternatives to the damaging solvent is of highest priority. This is particularly important as solvents are used in huge amounts (in industrial production) and these are mostly volatile liquids, which are difficult to contain.

The ionic liquids, comprising entirely of ions were and mainly of interest to the electrochemists.[1] It is possible, by careful choice of starting materials, to prepare ionic liquids that are liquid at and below room temperature. Different aspects of ionic liquids have been reviewed by a number of authors.[2] The first ionic liquid [EtNH$_3$] [NO$_3$] (m.p. 12 °C) was discovered in 1914.[3]

Broadly speaking, ionic liquids are of two types: simple salts (made up of a single anion and cation) and binary ionic liquids; the latter are salts where an equilibrium is involved. It is the binary ionic liquids that are used as green solvents. Some examples of simple salt are given (Scheme 1).

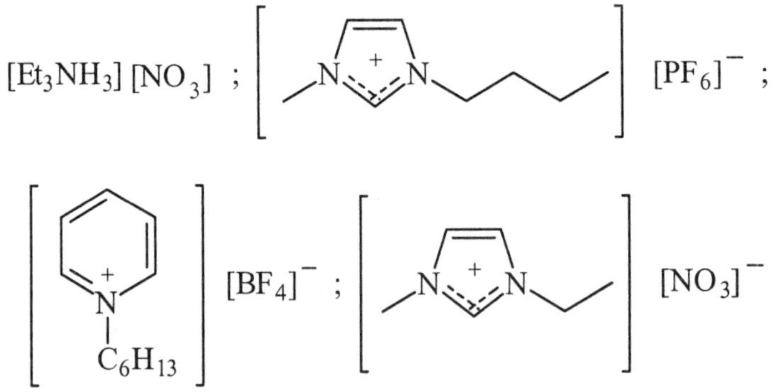

$$[Et_3NH_3] [NO_3] \; ; \qquad [PF_6]^- \; ;$$

$$[BF_4]^- \; ; \qquad [NO_3]^-$$

Scheme 1

An example of a binary ionic liquid system is a mixture of aluminium (III) chloride and 1,3-dialkylimidazolium chloride. It contains several different ionic species and their m.p. and properties depend upon the mole fractions of AlCl$_3$

and 1,3-dialkyl imidazolium chloride present. For the binary systems, the m.p. depends upon the composition[4] and is designated as [emin]Cl-AlCl$_3$, where [emin]$^+$ is 1-ethyl-3-methyl imidazolium.

The above binary system, [emin]Cl-AlCl$_3$ can be basic, acidic or neutral in nature. The composition of the binary ionic liquid is described by the apparent mole fraction of AlCl$_3$ [X(AlCl$_3$)] present. Ionic liquids with X(AlCl$_3$) < 0.5 contain an excess of Cl$^-$ ions over [Al$_2$Cl$_7$]$^-$ ions are called 'basic'. On the other hand, those with X(AlCl$_3$) > 0.5 contain an excess of (Al$_2$Cl$_7$)$^-$ ions over Cl$^-$ and are called 'acidic'. Mixtures with X(AlCl$_3$) = 0.5 are called 'neutral'. These ionic liquids (basic, acidic or neutral) are used in different types of reactions. Thus, the properties such as m.p., viscosity and hydrophobicity and misicibility with water can be varied by changing the structure and composition of the ions.

The most common salts in use are those with alkylammonium, alkylphosphonium, N-alkylpyridinium and N,N'-dialkyl imidazolium cation (Scheme 1).

The reactions in ionic liquids are easy to perform and need no special apparatus or methodologies. Also, the ionic liquids can be recycled and this leads to reduction of the costs of the processes.

14.2 Reactions in Acidic Ionic Liquids

The chloroaluminate (III) ionic liquids (where X(AlCl$_3$) > 0.50) behave like powerful Lewis acid and promote reactions that are usually promoted by AlCl$_3$. In fact, chloroaluminate ionic liquids are powerful solvents and can be prepared by mixing the appropriate organic halide salt with AlCl$_3$ and heating to form the ionic liquid. This synthesis should be performed in inert atmosphere.

The well known Friedel-Crafts reaction works very well with the chloroaluminate (III) ionic liquids.[5] Thus, using this methodology, traseolide (5-acyl-1,1,2,6-tetramethyl-3-isopropylindane) and tonalid (6-acetyl-1,1,2,4,4,7-heaxmethyltetralin have been synthesised in high yield in the ionic liquid [emin]Cl-AlCl$_3$ (X = 0.67) (Scheme 2).

Similarly Friedel-Crafts reaction of naphthalene gives 1-acetyl derivative as the major product (Scheme 3).

Another interesting application in the use of ionic liquids is in the hydrogenation of polycyclic aromatic hydrocarbons, which are soluble in chloroaluminate (III) ionic liquids to form highly coloured paramagnetic solutions[6], which on treatment with a reducing agent such as an electropositive metal and a proton source results in selective hydrogenation of the aromatic compound. Using this method, the anthracene can be reduced to perhydroanthracene at normal temperature and pressure to give the most stable isomer (the sequence of reduction of anthracene is depicted in Scheme 4).[7]

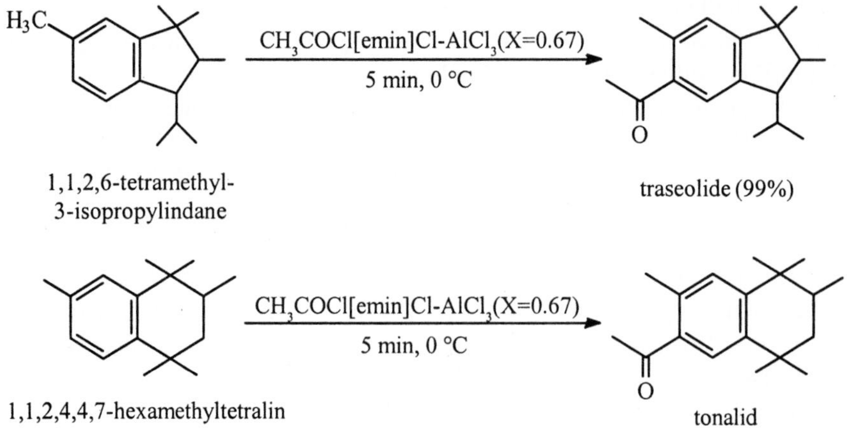

1,1,2,6-tetramethyl-
3-isopropylindane

traseolide (99%)

1,1,2,4,4,7-hexamethyltetralin

tonalid

Scheme 2

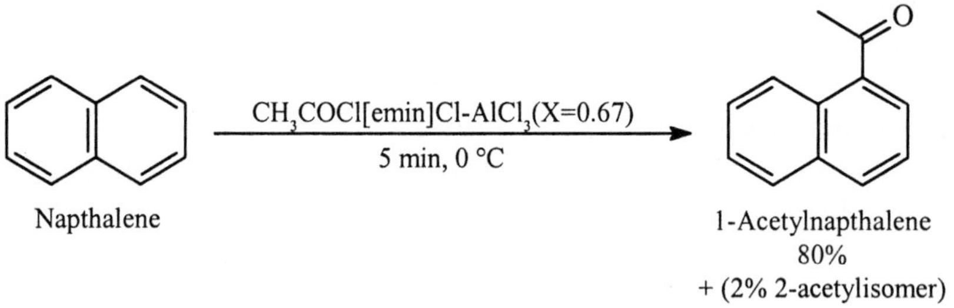

Napthalene

1-Acetylnapthalene
80%
+ (2% 2-acetylisomer)

Scheme 3

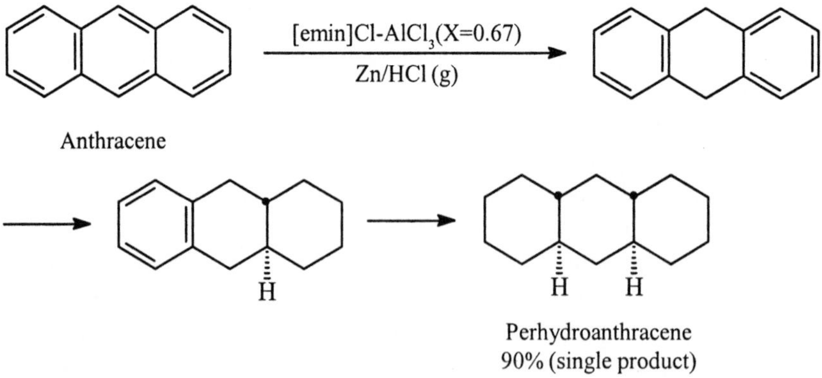

Anthracene

Perhydroanthracene
90% (single product)

Scheme 4

The above reduction is in contrast to catalytic hydrogenation reaction, which requires high temperature and pressure and expensive platinum oxide catalyst and gives a mixture of products.[8] In the above case (Scheme-4) the sequence of chemical reduction can be determined by careful monitoring of the reduction in ionic liquid.

Using the above procedure (Scheme-4) pyrene can be reduced to perhydro pyrene.

14.3 Reactions in Neutral Ionic Liquids

As already stated, chloroaluminate (III) ionic liquids are excellent solvents in many reactions. The main problem arises due to their moisture sensitivity and difficulty in separation of products (containing heteroatoms) from the ionic liquid. In view of this, water-stable ionic liquids have been developed. One example of this is the ionic liquid [bmin][PF_6] [(bmin)$^+$ = 1-butyl-3-methylimidazolium)].[9] The ionic liquid [bmin][PF_6] forms triphasic mixture with water and alkanes, which makes it useful for clean synthesis. Such ionic liquids can be used without any special conditions needed to exclude moisture and the isolation of the reaction products is convenient.

Some applications in the use of neutral ionic liquids are discussed here.

14.3.1 Hydrogenations

The most important advantage of using neutral ionic liquids is that the reaction products can be easily separated from the ionic liquids and the catalyst.[10] Using the neutral ionic liquids, cyclohexene can be reduced to cyclohexane.[11] Even benzene could be reduced to cyclohexane.[12] An interesting asymmetric hydrogenation using a chiral catalyst [$RuCl_2$-(S)-BINAP]$_2$,NEt_2 has enabled the synthesis of (S)-Naproxen (Scheme 5).[13]

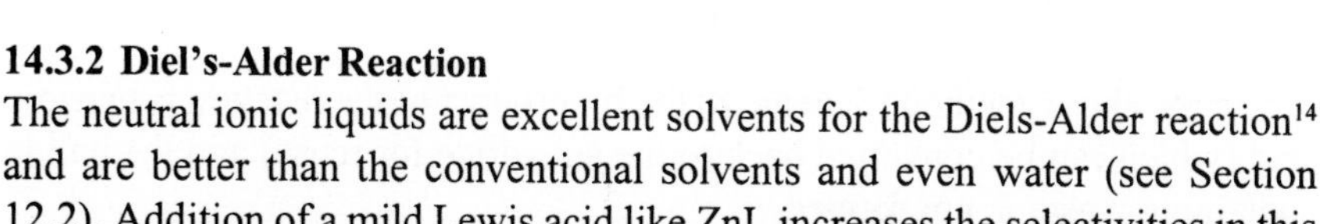

Scheme 5

14.3.2 Diel's-Alder Reaction

The neutral ionic liquids are excellent solvents for the Diels-Alder reaction[14] and are better than the conventional solvents and even water (see Section 12.2). Addition of a mild Lewis acid like ZnI_2 increases the selectivities in this reaction. A special advantage of this system is that the ionic liquid and catalyst can be recycled and reused after extraction or direct distillation of the product from the ionic liquid. A typical Diel's-Alder reaction is given (Scheme 6).

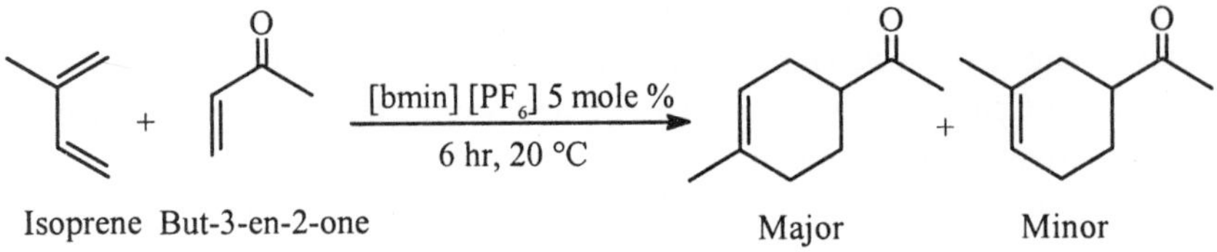

Isoprene But-3-en-2-one   Major   Minor

Scheme 6

14.3.3 Heck Reaction

Neutral ionic liquids are excellent solvents for the palladium catalysed coupling of alkyl halides with alkenes (Heck reaction). The special advantage of using neutral ionic liquids is that many palladium complexes are soluble in ionic liquids[15] and that the products or product of the reaction can be extracted with water or alkane solvents. So the expensive catalyst can be recycled compared to the routine Heck reaction in which the catalyst is lost at the end of the reaction (see Section 12.11). A typical Heck reaction is given (Scheme 7).

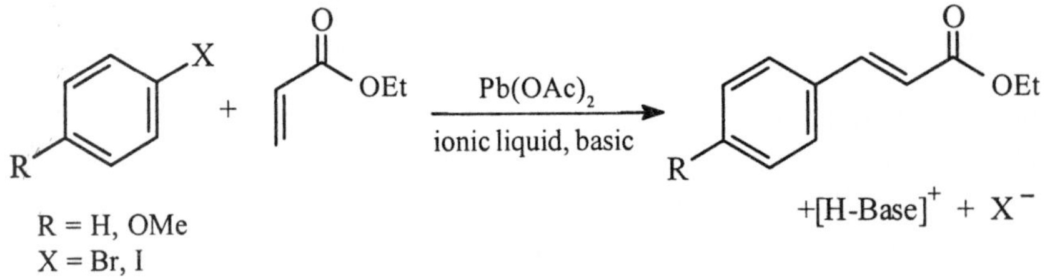

R = H, OMe
X = Br, I

$+[H\text{-}Base]^+ + X^-$

Scheme 7

An alternative Heck reaction uses aromatic anhydrides as a source of the aryl group (Scheme 8).

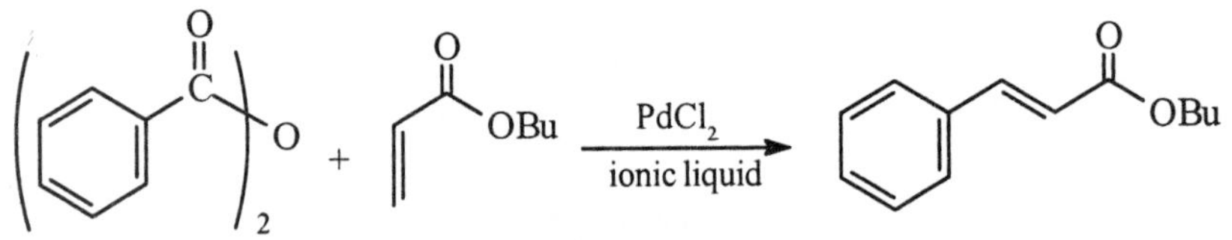

Scheme 8

In the above method (Scheme 8) the by-product of the reaction is benzoic acid (which can be converted back to the anhydride for reuse) and the halide containing waste is not formed.

14.3.4 O-Alkylation and N-alkylation

A common reaction in organic synthesis is the nucleophilic displacement reaction. Thus 2-naphthol undergoes alkylation to give O-alkyl ether on treatment with a haloalkane and base (NaOH or KOH) in [bmin][PF$_6$].[16] Similarly indol undergoes N-alkylation. Though both the above O- and N-alkylations occur with similar rates to those carried out in dipolar solvents (e.g. DMF or DMSO), but the advantage of ionic liquid process is that the reaction products can be extracted into an organic solvent such as toluene leaving behind the ionic liquid, which can be recycled after separation from sodium or potassium halide by extraction with water.

14.3.5 Methylene Insertion Reactions

Aldehydes and ketones are known to react with sulphur ylides to give epoxide (Scheme 9).

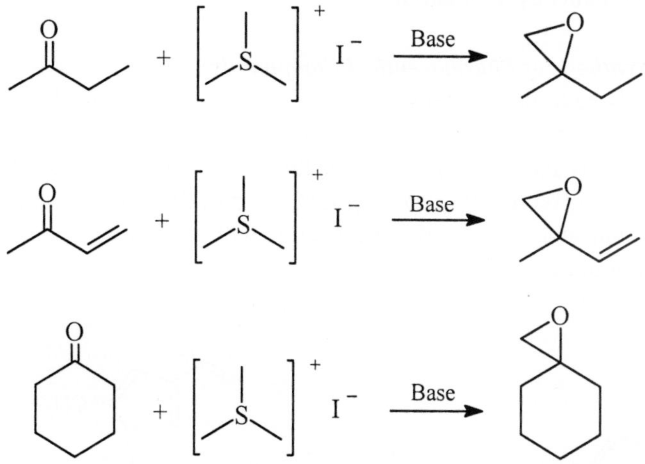

Scheme 9

The methylene insertion reactions have been found to proceed better in ionic liquids such as [bmin] [PF$_6$] or [bmin] [BF$_4$] on treatment with sulphur ylides. Sulphur ylides are obtained *in situ* by the reaction of alkyl halides with sulphides. The reaction works equally well with preformed sulphonium salts (Scheme 10).

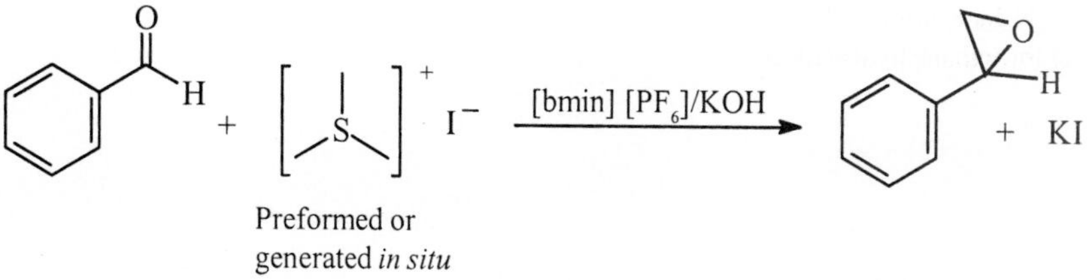

Preformed or
generated *in situ*

Scheme 10

It is also possible to carry out asymmetric methylene insertion reaction in good enantiomeric excess by using a chiral sulphide such as 2R,5R-tetrahydrothiophene (Scheme 11).

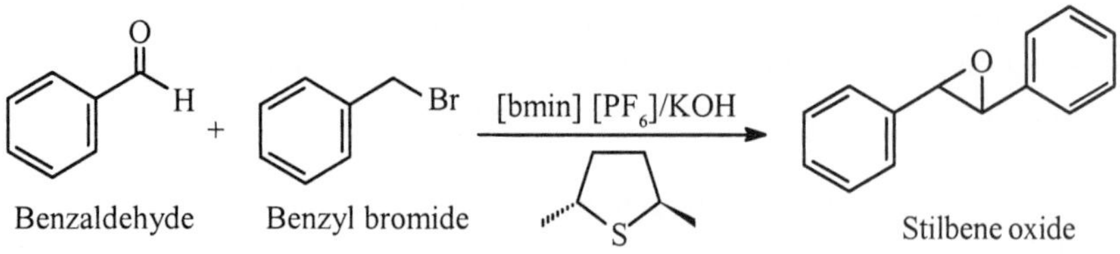

Benzaldehyde   Benzyl bromide   [bmin] [PF$_6$]/KOH   Stilbene oxide

Scheme 11

14.3.6 Miscellaneous Applications

14.3.6.1 Synthesis of Pharmaceutical Compounds
Ionic liquids have been extensively used in the synthesis of pharmaceutical compounds. A representative example is the synthesis of pravadoline.[17] The method consists the alkylation of 2-methylindole with 1-(N-morpholino)-2-chloroethane in [bmin] [PF$_6$] to give 95% yield of the corresponding N-alkyl derivative. Subsequent Friedel Craft reaction with p-methoxybenzoyl chloride in chloroaluminium (III) ionic liquid gives pravadoline (Scheme 12).

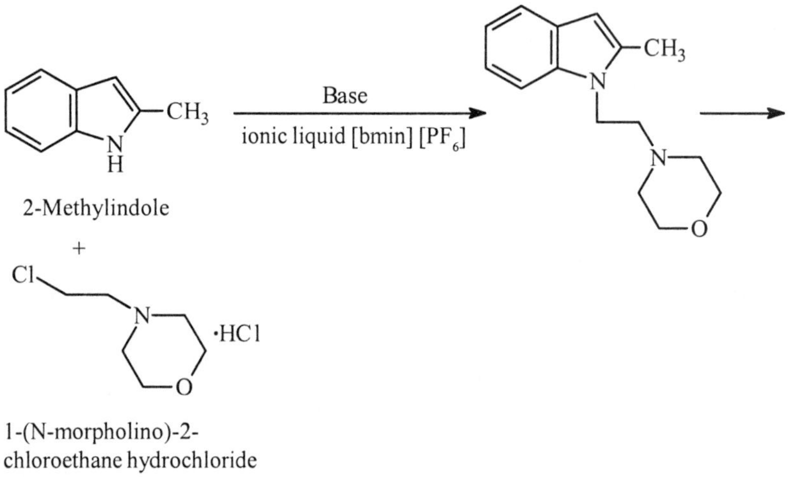

2-Methylindole

Base
ionic liquid [bmin] [PF$_6$]

+

1-(N-morpholino)-2-
chloroethane hydrochloride

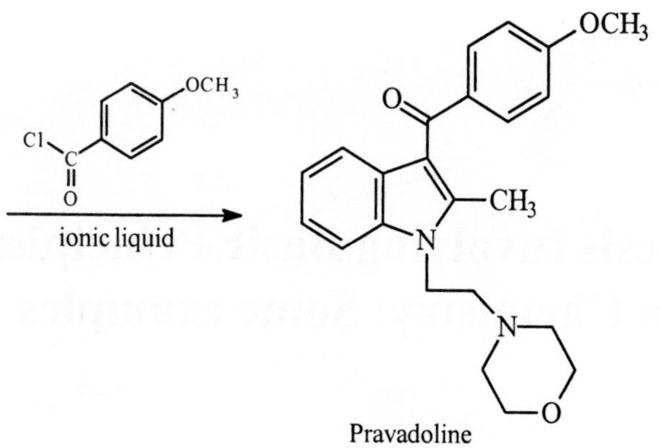

Pravadoline

Scheme 12

References

1. C.L. Hussey, *Adv. Molten Salt Chem.*, 1983, **5**, 185.
2. T. Welton, *Chem. Rev.*, 1999, 2071-2083; J. Holbrey and K.R. Seddon, *Clean Prod. Proc.*, 1991, **1**, 223-236; K.R. Seddon, *J. Chem. Tech. Biotech.*, 1997, **68**, 351-356; K. Sheldon, *Chem. Commun.*, 2001, 2399-2407; M.J. Earle and K.R. Seddon, *Pure Appl. Chem.*, 2000, **72**, 1391-1398; P. Wasserscheid and W. Keim, *Angew. Chem. Int. Ed. English*, 2002, 3773-3784.
3. P. Walden, *Bull. Acad. Imper. Sci.* (St. Petesburg), 1914, 1800.
4. C.L. Hussey, J.B. Scheffler, J.S. Wilkes and A.A. Fannin, Jr., *J. Electrochem. Soc.*, 1986, **133**, 1389.
5. J.A. Boon, J.A. Levosky, J.L. Pfung and J.S. Wikes, *J. Org. Chem.*, 1998, **51**, 480.
6. P. Tarakeshwar, J.Y. Lee and K.S. Kim, *J. Phys. Chem.*, 1998, **102A**, 2253.
7. C.J. Adams, M.J. Earle and K.R. Seddon, *Chem. Commun.*, 1999, 1043.
8. D.K. Dalling and D.M. Grant, *J. Am. Chem. Soc.*, 1974, **96**, 1827.
9. J.D. Huddlesten, H.D. Willauer, R.P. Swatloski, A.E. Visser and R.D. Rogers, *Chem. Commun.*, 1998, 1765.
10. T. Welton, *Chem. Rev.*, 1999, **99**, 2071-2083.
11. P.A.Z. Suarez, J.E.L. Dullius, S. Einloft, R.F. de Souza and J. Dupont, *Inorg. Chim. Acta.*, 1997, **225**, 207.
12. P.J. Dyson, D.J. Ellis, D.G. Parker and T. Welton, *Chem. Commun.*, 1999, 25.
13. A.L. Monterio, K.F. Zinn, R.F. de Souza and J. Dupont, *Tetrahedron Asymmetry*, 1997, **8**, 177.
14. M.J. Earle, P.B. McCormac and K.R. Seddon, *Green Chem.*, 1999, **1**, 23; T. Fisher, A. Sethi, T. Welton and J. Woolf, *Tetrahedron Lett.*, 1999, **40**, 793.
15. W.A. Herrmann and V.P.W. Bohn, *J. Organomet. Chem.*, 1999, **572**, 141.
16. M.J. Earle, P.B. McCormac and K.R. Seddon, *Chem. Commun.*, 1998, 2245.
17. M.J. Earle and K.R. Seddon, *Pure Appl. Chem.*, 2000, **7**, 1397.

15. Synthesis Involving Basic Principles of Green Chemistry: Some examples

15.1 Introduction

Since the beginning of the twenty-first century it was considered of paramount importance to reduce the impact on the environment caused basically by pollutants (including waste products) generated by the chemical industries. Thus there was urgent need to develop environmentally benign or green synthesis. We already know that a number of ways are available to reduce the impact on the environment of a large scale process. These include carrying out reactions in safer aqueous systems instead of hazardous organic solvents. The reactions, as far as possible, should be carried out at ambient temperature instead of using heat energy. If possible, the materials should be recycled. The pathways for synthesis be selected so as to avoid the generation of toxic materials. All these may reduce the impact of the process on the environment in terms of pollution or consumption of resources.

It is most advantageous to carry out reactions in aqueous media. The most important factor for industrial process is the economic value of the process. In view of this, water is the cheapest abundantly available solvent. Also when water is used as a solvent, the final product can be separated (or isolated easily) and there is least waste generation. Also the reactions carried out in aqueous medium are comparatively much more safer compared to reactions conducted in organic solvents. However, the most important thing is that reactions in aqueous medium are generally environmentally benign.

The objective of green chemistry is not only to design new green synthesis (environmentally benign synthesis) but also to devise green methods for the synthesis of already existing molecules (products) whose known synthesis are responsible for environmental pollution.

Following are some typical examples of green synthesis.

15.2 Synthesis of Styrene

The chemical industry consumes styrene, the monomer used for the manufacture of polystyrene in large quantities each year. The commonly used industrial method for making it converts benzene into styrene by a Friedal-Crafts alkylation followed by dehydrogenation (Scheme 1).

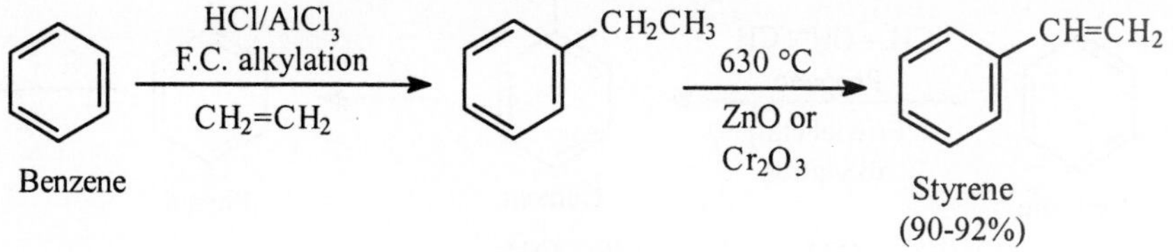

Scheme 1

A new synthesis developed by Chapman uses a single step to convert mixed xylenes (compounds that are noncarcinogenic) into styrene. This new method eliminates the use of ton quantities of benzene each year (details of this process are not available).

15.3 Synthesis of Adipic Acid, Catechol and 3-Dehydroshikimic Acid (a Potential Replacement for BHT)

Adipic acid is required in large quantities (about 1 billion kg a year) for the synthesis of nylon, plasticizers and lubricants. Conventionally, adipic acid is made from benzene (Scheme 2).

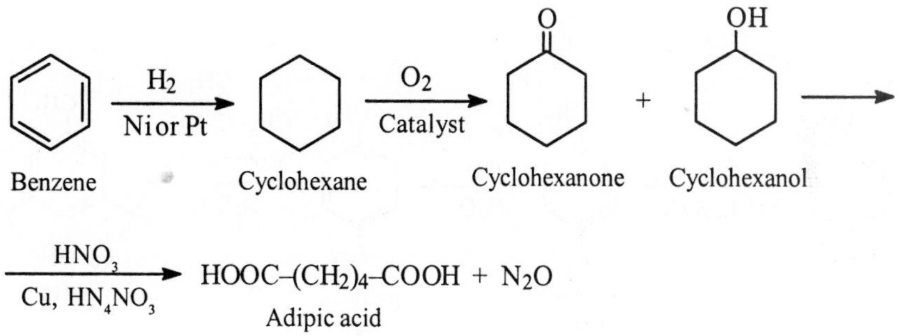

Scheme 2. Conventional synthesis of adipic acid

Like adipic acid, catechol is also manufactured using benzene as the starting material. The procedure is given (Scheme 3).

As seen, both adipic acid and catechol are obtained from benzene, which causes environmental and health problem. Also, benzene is produced from non-renewable source. In addition, in the synthesis of adipic acid, nitrous oxide is generated as a byproduct, which contributes to the greenhouse effect as well as destruction of the ozone layer.

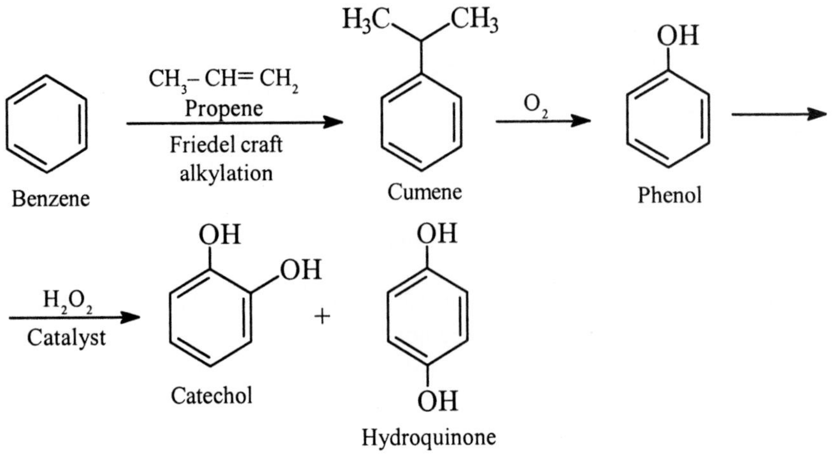

Scheme 3. Conventional synthesis of catechol

BHT is obtained from toluene (which unlike benzene is not carcinogenic but is toxic in nature) as shhown (Scheme 4).

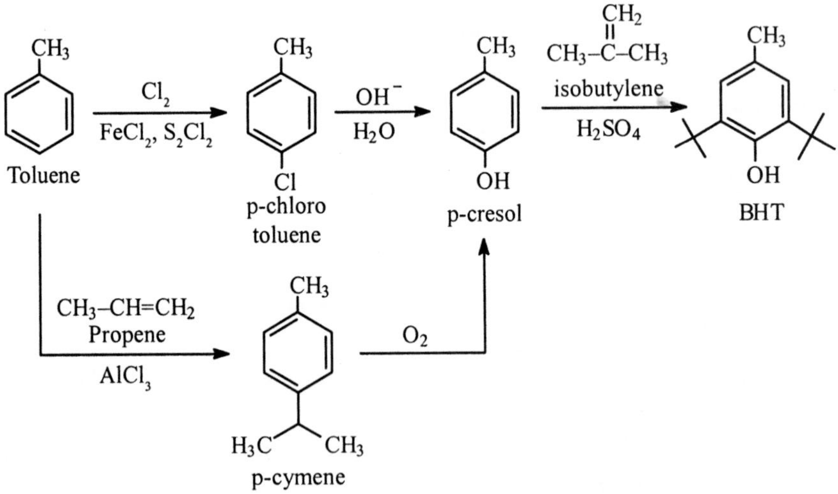

Scheme 4. Conventional synthesis of BHT

An environmentally benign (or green) synthesis of adipic acid, catechol and BHT (a potential replacement for BHT) has been developed by John W. Frost and Karen M. Draths starting with glucose and using a biocatalyst (genetically altered *E. coli* bacteria)[1,2] (Scheme 5).

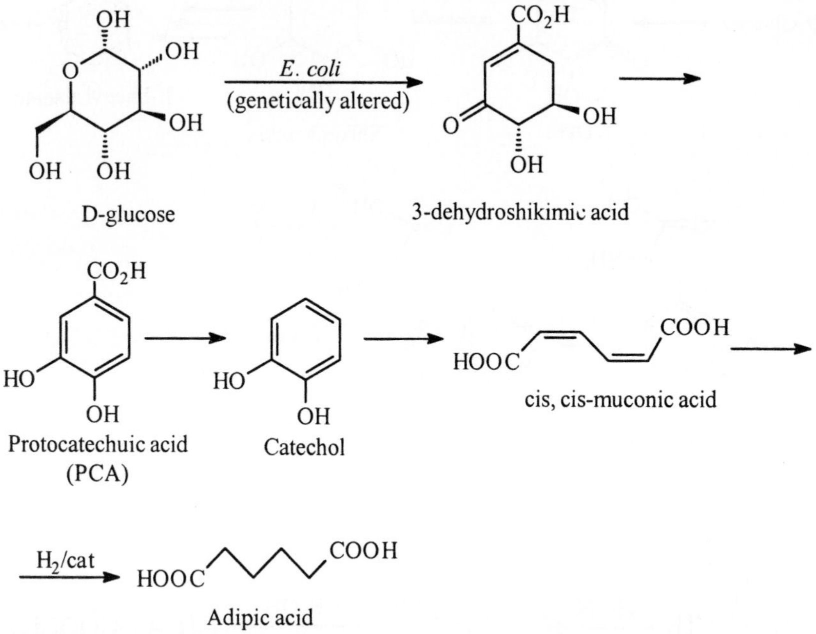

Scheme 5. Green synthesis of catechol and adipic acid

The above environmentally benign synthesis of catechol and adipic acid uses D-glucose (a non-toxic and a renuable resource) as the starting material. Also the synthesis is conducted in water instead of organic solvents. The reaction can also be used to stop at either catechol stage or at the adipic acid stage by using another genetically altered *E. coli*.[2]

It may be appropriate to say that in the above synthesis, use of unmodified *E. coli* gives the amino acids, L-phenylalanine, L-tyrosine and L-tryptophan via the formation of shikimic acid from dehydroshikimic acid (Scheme 6).

15.4 Synthesis of Methyl Methacrylate

Methyl methacrylate is used in large quantities for manufacture of polymers. It was earlier synthesised as shown in Scheme 7.

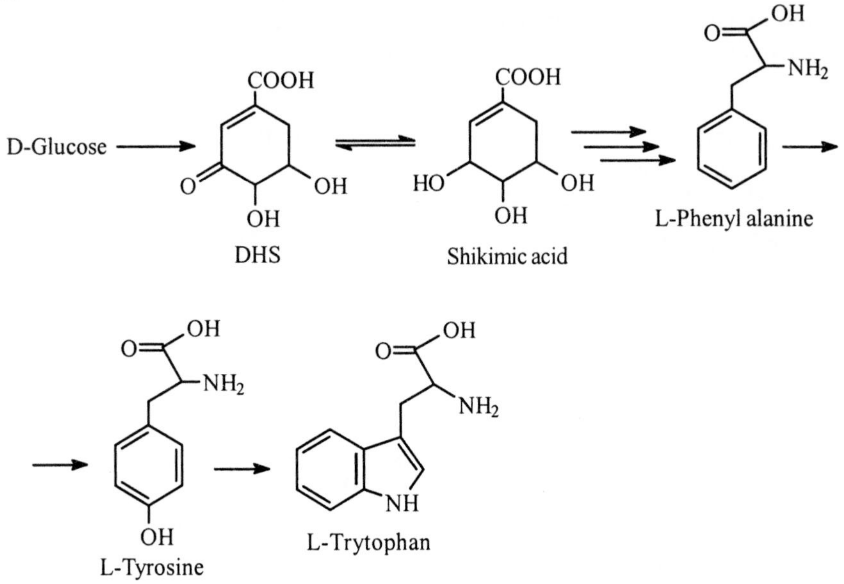

Scheme 6

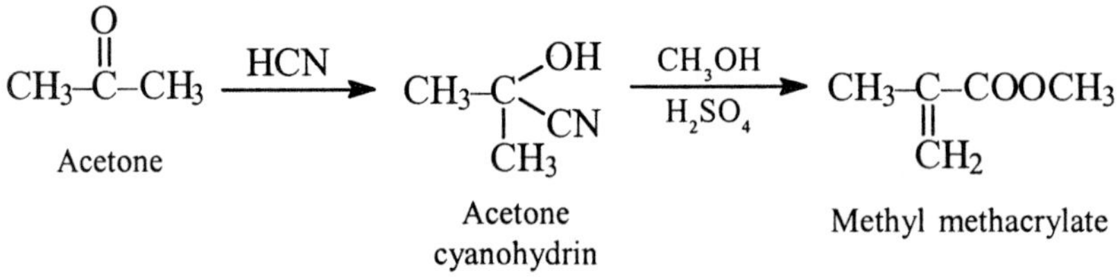

Scheme 7. Conventional synthesis of methyl methacrylate

A very convenient synthesis of methyl methacrylate, developed by the shell corporation is given (Scheme 8).

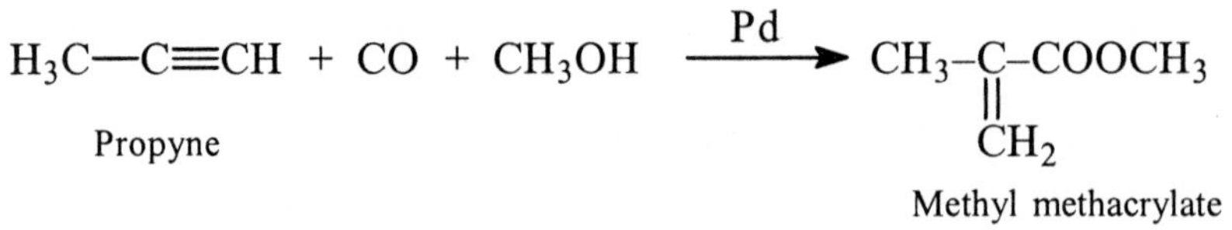

Scheme 8. New synthesis of methyl methacrylate

This new synthesis[3] employs a palladium catalyst and enjoys 100% atom economy compared to the conventional synthesis, which besides using an extremely poisonous HCN enjoys only 47% atom recovery.

15.5 Synthesis of Urethane

Urethane is required in large quantities for the manufacture of polyurethanes, a class of important polymers which are used for a number of commercial applications.

Urethane was synthesised earlier using phosgene, an extremely hazardous chemical (Scheme 9).

$$RNH_2 + COCl_2 \longrightarrow RNCO + 2HCl \xrightarrow{R'OH} RNHCO_2R'$$

Amine   Phosgene   Isocyanate   Urethane

Scheme 9. Synthesis of urethane using phosgene

Monsanto company has developed a method for the synthesis of urethanes, eliminating the use of phosgene[4] (Scheme 10).

$$RNH_2 + CO_2 \longrightarrow RNCO + H_2O \xrightarrow{R'OH} RNHCO_2R'$$

Amine   Carbon   Isocyanate   Urethane
        dioxide

Scheme 10. New synthesis of urethane without using phosgene

15.6 An Environmentally Benign Synthesis of Aromatic Amines

The usual synthesis of aromatic amines involve chlorination of benzene followed by nitration and nucleophilic displacement of the chlorine with a new substitution group. This process is illustrated by the following synthesis of 4-aminodiphenylamine (Scheme 11).

An environmentally benign synthesis of 4-aminodiphenylamine was developed by Monsanto.[5] In this process, nitrobenzene and aniline are heated in presence of tetramethyl ammonium hydroxide to give condensation products as tetramethyl ammonium salts, which on catalytic hydrogenation give 4-aminodiphenylamine while regenerating tetramethyl ammonium hydroxide (Scheme 12).

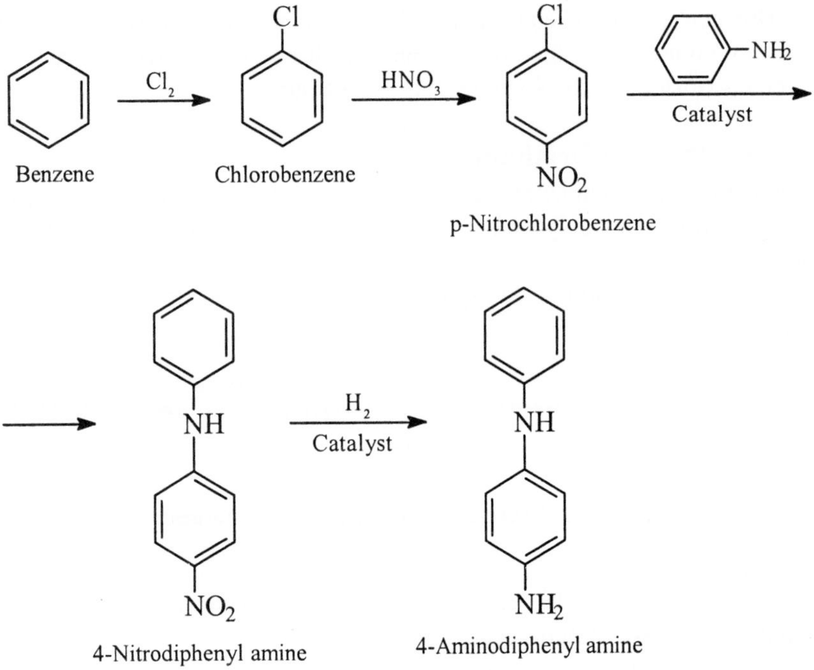

Scheme 11. Conventional synthesis of 4-Aminodiphenyl amine

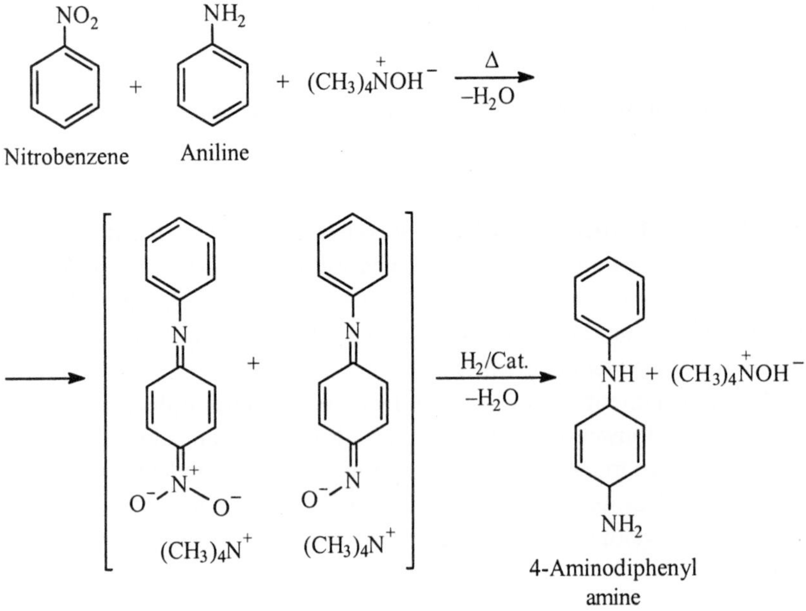

Scheme 12. Convenient synthesis of 4-Aminodiphenyl amine

The new process avoids the use of halogenated intermediates and also the use of nitric acid and is atom economical.

15.7 Selective Alkylation of Active Methylene Group

The conventional alkylation of active methylene group employs the reaction of alkyl halides using a base like sodium ethoxide. At times, this method results in multiple alkylations (Scheme 13).

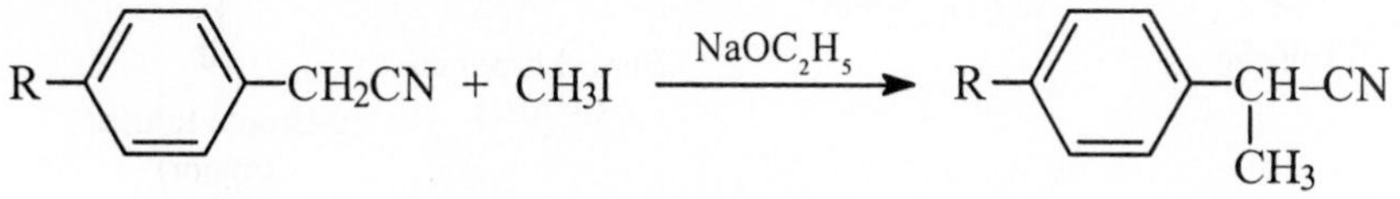

Scheme 13. Conventional alkylation of active methylene group

A convenient method developed by Tudo involves the use of dimethyl carbonate in presence of potassium carbonate[6] (Scheme 14).

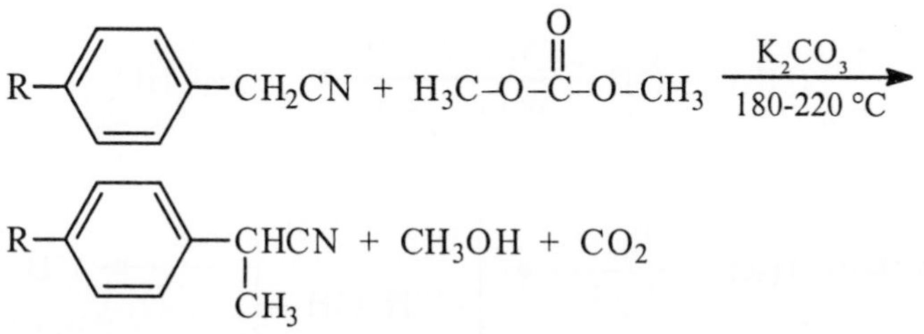

Scheme 14

15.8 Free Radical Bromination

The usual free radical bromination of toluene with N-bromosuccinimide gives benzyl bromide. This bromination is carried out in a solvent, e.g., CCl_4. It has been found[7] that free radical bromination of toluene with NBS in supercritical carbon dioxide gave 100% yield of benzyl bromide (Scheme 15).

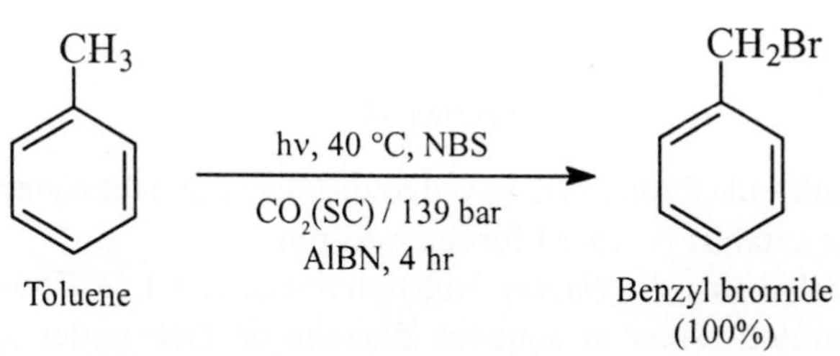

Scheme 15

However, bromination of toluene with bromine in supercritical CO_2 gave 70% benzyl bromide and minor amount of p-bromotoluene (Scheme 16).

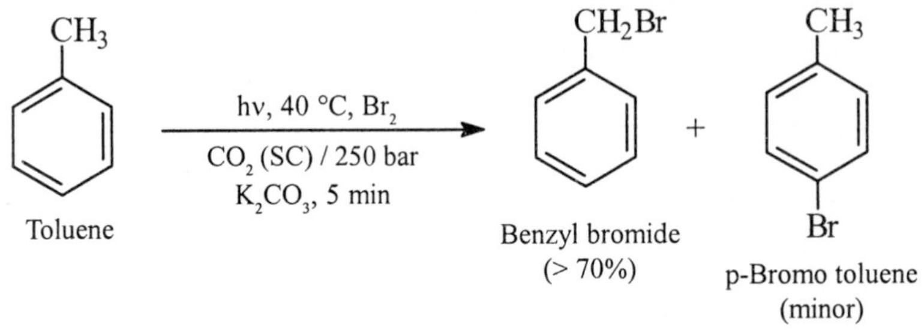

Scheme 16

15.9 Acetaldehyde

It was obtained commercially by catalytic oxidation of ethyl alcohol or by the hydration of acetylene (Scheme 17).

$$CH_3CH_2OH \xrightarrow[\text{or Cu /300 °C}]{\text{Ag/Air/250 °C}} CH_3CHO$$

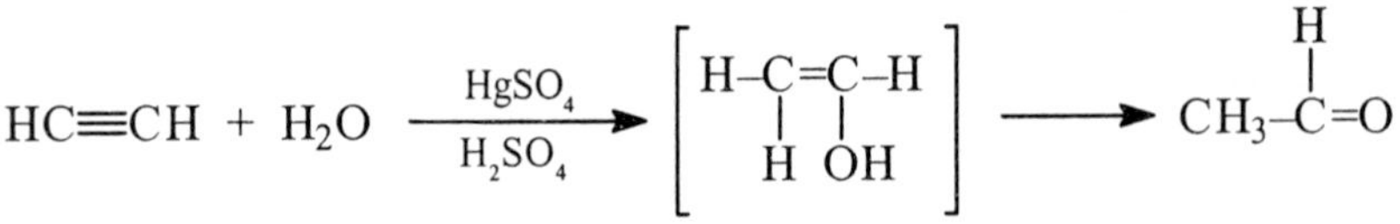

Scheme 17

It is most conveniently obtained by the Wacker-chemie's oxidation process. In this process ethylene is oxidised with oxygen in presence of the catalyst solution[8] (Scheme 18).

$$CH_2{=}CH_2 + O_2 \xrightarrow[\text{H}_2\text{O}]{\text{Pd(II)/Cu(II)}} CH_3CHO$$
$$90\%$$

Scheme 18

The acetaldehyde formed is removed by distillation and the aqueous solution containing the catalyst is reused for the next run.

For internal olefins, the Wacker oxidation is regioselective. Thus, oxidation of β,γ-unsaturated esters in aqueous dioxane or THF under appropriate condition gives γ-ketoester (Scheme 19).[9]

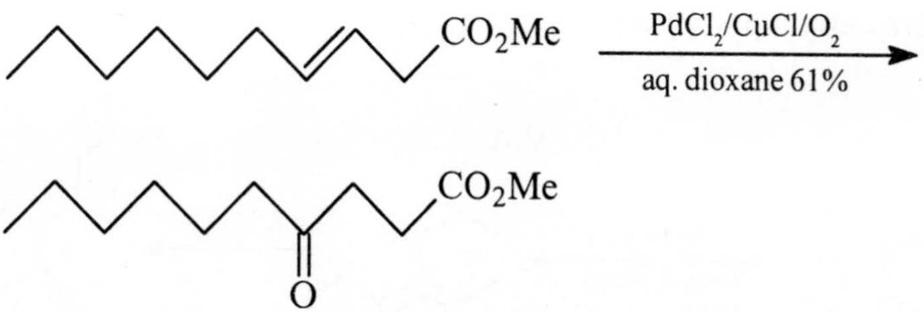

Scheme 19

15.10 Furfural from Biomass

Furfural is an industrial chemical and is used for the manufacture of furfural-phenol plastics. It is also used as solvent in refining of petroleum oils and in the preparation of pyromucic acid.

Furfural is prepared industrially from pentosans which are contained in cereal strains and brans. It is obtained in the laboratory from corncobs.[10]

In fact, cellulosic biomass has been converted into useful products.[11] The furfural formation is via a dehydration process. The mechanism from D-xylose is given (Scheme 20).

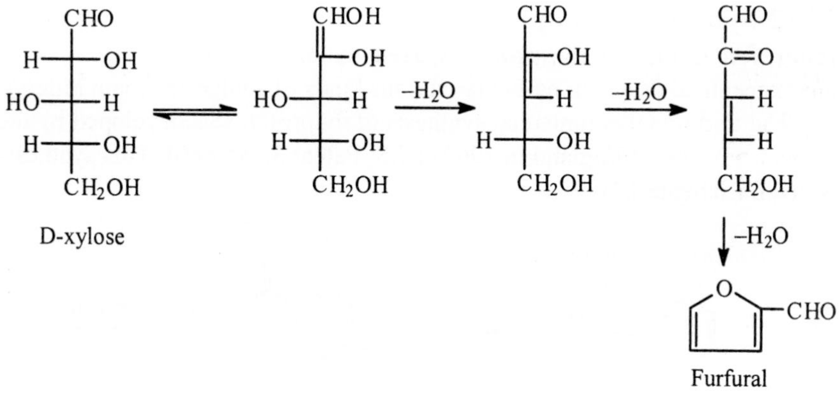

Scheme 20

15.11 Synthesis of (S)-metolachlor, an Optically Active Herbicide

In the field of pharmaceuticals and pesticides, the desired big activity of the molecule is in a pure enantiomer (chiral molecule). This has been made possible by using either enzymes or chiral metal complexes (asymmetric catalysts).[12] These chiral metal complexes, though expensive, are important for industrial application (i.e., have a high turnover frequency). A typical example is the synthesis of (S)-metolachlor, a herbicide. It is prepared by asymmetric hydrogenation of the prochiral imine catalysed by an iridium (I) complex of a chiral ferrocenyldiphosphine[13] (Scheme 21).

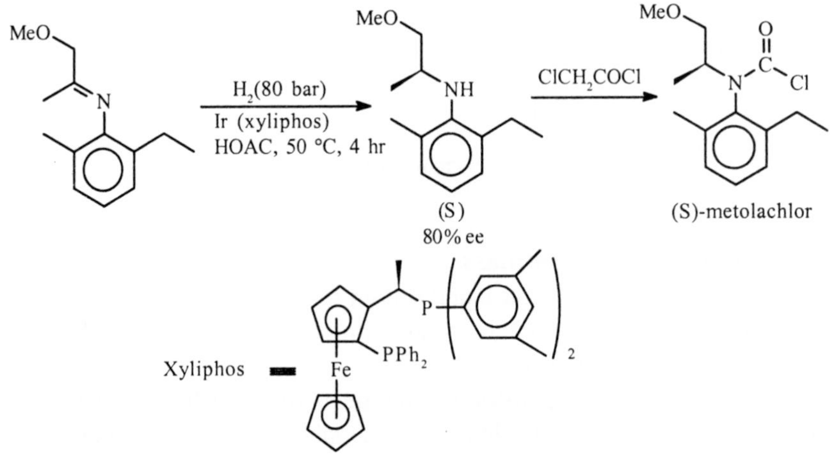

Scheme 21

15.12 Synthesis of Ibuprofen

Ibuprofen is one of the products used in large quantities for making pharmaceutical drugs, in particular various kinds of analgesics (pain killers).

The traditional commercial synthesis of ibuprofen was developed by the Boots Company of England in 1960s (U.S. Patent 3,385,886). This synthesis is given (Scheme 22).

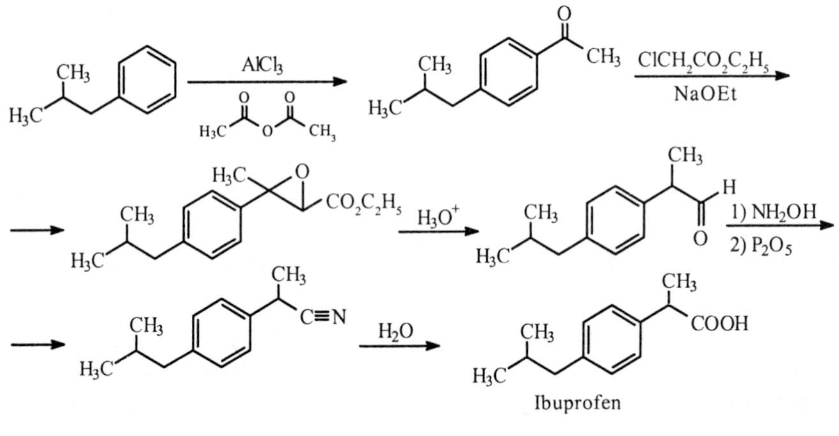

Scheme 22

The given synthesis (Scheme 22) is a six step process and results in large quantities of unwanted waste chemical byproducts that must be disposed of. There is 40% atom economy in this synthesis.

The BHC company developed a new greener commercial synthesis of ibuprofen that consists of only three steps[14] (Scheme 23).

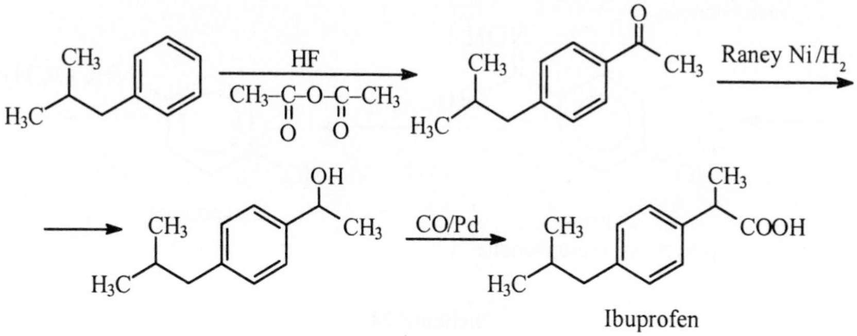

Scheme 23

The above synthesis results in only small amount of unwanted products and has very good atom economy (77%).

15.13 Synthesis of Paracetamol

Paracetamol is used in broad spectrum of arthritic and rheumatic conditions linked with musculoskeletal pain, headaches, neuralgias and dysmenorrhea. It is generally prepared from p-nitrophenol by reduction (Sn + HCl) followed by reaction with acetic anhydride-acetic acid mixture. Alternatively it is obtained by the Beckmann rearrangement of oxime of p-hydroxyacetophenone.

In the green synthesis of paracetamol[15] p-hydroxyacetophenone is reacted with ammonia and hydrogen peroxide in presence of titanium(IV)-silicate (TS-1) catalyst[16] to give the oxime of p-hydroxyacetophenone. In fact, in the above reaction, ammoximation proceeds via *in situ* formation of hydroxylamine by titanium-catalysed oxidation of ammonia with H_2O_2 in the micropores of the catalyst. Final reaction of hydroxylamine with the ketone occurs in the bulk solution. This methodology can be applied to any ketone or aldehyde for the formation of oximes. Typical is in preparation of cyclohexanone oxime, which is used in the manufacture of caprolactam via the Beckmann rearrangement of the oxime.

The oxime of p-hydroxyacetophenone (obtained above) on Beckmann rearrangement gives paracetamol (Scheme 24).

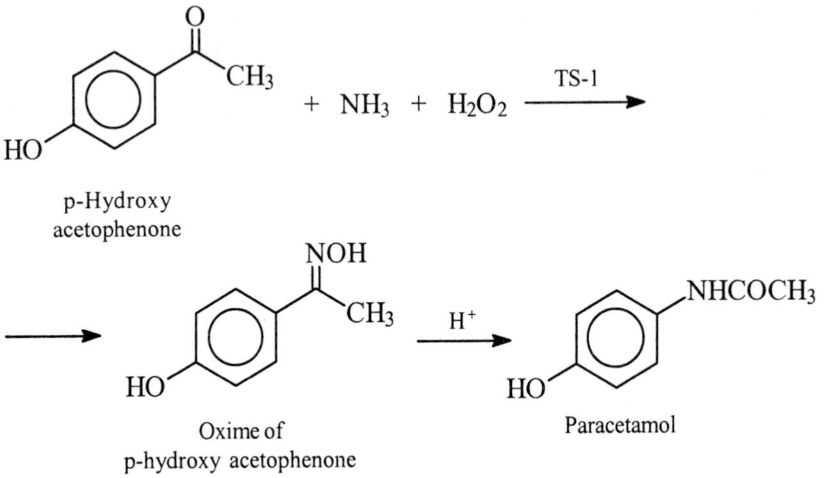

Scheme 24

15.14 Green Synthesis of 3-phenyl Catechol

The 3-substituted catechols are important building blocks for the chemical and pharmaceutical industries. Their chemical synthesis is cumbersome, requiring organometallic reagents, HBr etc. An industrial green synthesis of 3-phenyl catechol consist in the transformation of 2-phenyl phenol (a man made compound that has been widely used as a food protecting agent and as a germicide) into 3-phenylcatechol by a 2-hydroxybiphenyl 3-monooxygenase[17] (Scheme 25).

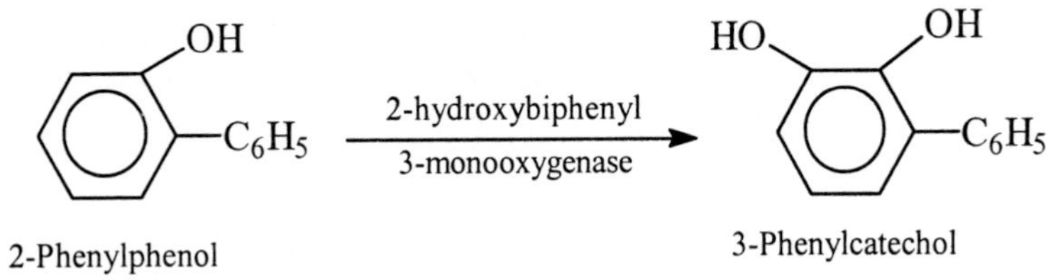

Scheme 25

This enzyme has the potential for the synthesis of 3-substituted catechols. However, this strain could not be used for the synthesis of catechol.

15.15 Synthesis of Epoxystyrene

Styrene epoxide is a valuable building block and is used in the production of the antihelmintic drug Levamisole.[18] The styrene epoxide for the synthesis of drugs must be enantiopure. There are some chemical asymmetric synthetic routes[19] but the yields are only in the range 45-50%.

A very convenient green synthesis of epoxystyrene has been developed. It consist of the reaction of styrene with xylene monooxygenase, which introduces an epoxide in the vinyl double bond of styrene[20] (Scheme 26).

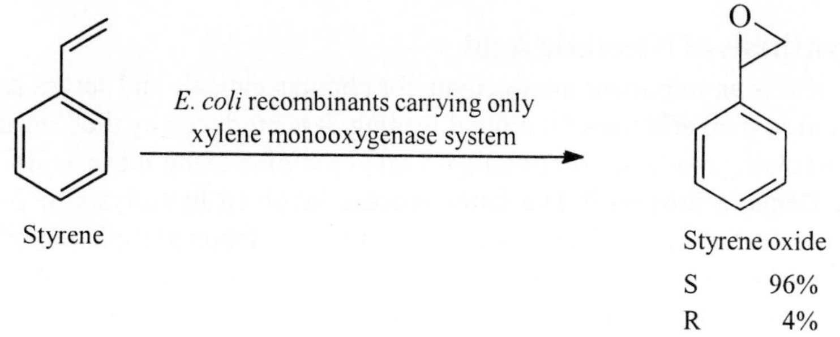

E. coli recombinants carrying only
xylene monooxygenase system

Styrene	Styrene oxide
	S 96%
	R 4%

Scheme 26

15.16 Synthesis of Citral

Citral is a valuable intermediate for the production of fragrances (like α- and β-ionones) and also for building units for carotenoids and vitamin A. The conventional industrial method using β-pinene as the starting material requires chlorine as oxidatant and involves five steps and the yields are low.

In the new BASF route, the cheap starting materials isobutene and formaldehyde are made to react to form isoprenol (3-methyl-but-3-en-1-ol). A part of this is isomerized and the other part is oxidised to the corresponding aldehyde[21] (Scheme 27).

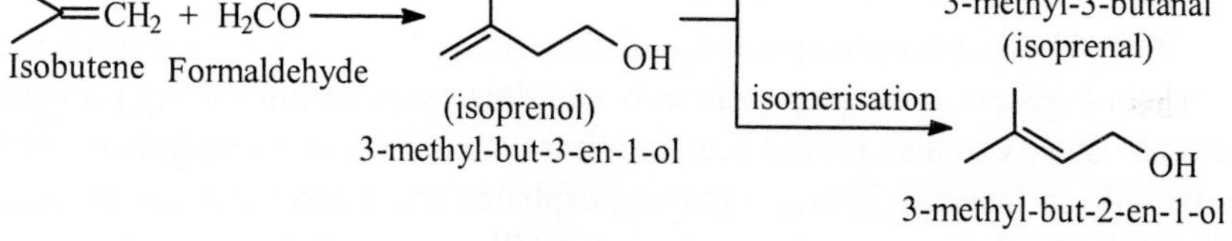

[O]

CH_2 + H_2CO ⟶

Isobutene Formaldehyde

(isoprenol)
3-methyl-but-3-en-1-ol

isomerisation

CHO

3-methyl-3-butanal
(isoprenal)

OH

3-methyl-but-2-en-1-ol

Scheme 27

The 3-methyl-3-butanal and 3-methyl-but-2-en-1-ol thus formed, react to form citral in 95% overall yield (Scheme 28).

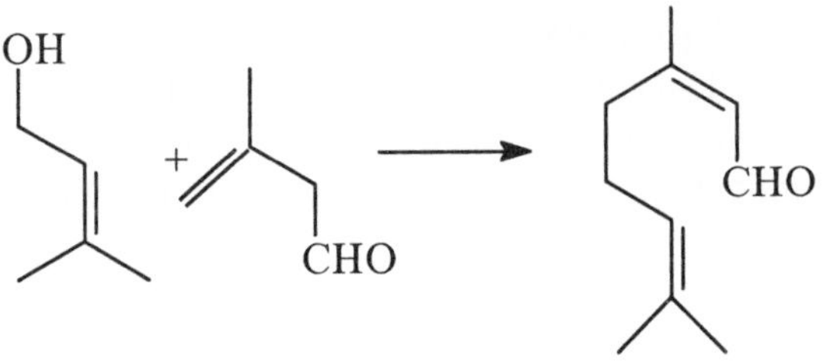

citral

Scheme 28

15.17 Synthesis of Nicotinic Acid

Nicotinic acid is an important intermediate for pharmaceuticals and serves as a provitamin in food additives for animal feeding. It is produced by the Lonza process (involving oxidation of 2-methyl-5-ethyl pyridine using nitric acid)[22] or by the Degussa process.[23] The latter process involved hydrolysis of β-cyanopyridine, which in turn was produced by amminoxidation of β-picoline.[24] A third process involving selective vapour phase oxidation of β-picoline catalysed by vanadium titanium oxide catalyst has also been described.[25]

A convenient process involving direct oxidation of β-picoline has been developed for the synthesis of nicotinic acid.[26]

15.18 Use of Molting Accelerators to Replace More Toxic and Harmful Insecticides

A number of insecticides have been used to control crop-destroying insects. Some of the common class of insecticides in use are organochlorines, organophosphates and carbamates. The well known DDT, which was in use for the maximum period of time, is now known to be toxic to humans and other non-target organisms.[27] The organophosphates and carbamates are less persistent in the environment and are not readily incorporated into the food chain as organochlorine insecticides (like aldrin, dieldrin and DDT). The organophosphate and carbamates are not bioaccumulated (like DDT etc.), but they readily decompose in the environment and tend to be more toxic to humans and other non-target organisms. However, these are readily decomposed to nontoxic products by reaction with water.

Both, the organophosphates and carbamates, work by inhibiting the acetylcholinesterase enzyme in insects and thus cause failure of the nervous system. These can also inhibit acetylcholinesterase enzyme in mammals (just as they do in insects). Hence, organophosphates and carbamates are toxic to humans and other mammals, as they also kill non-target insects, which are beneficial.

A new class of insecticides, viz. diacetylhydrazines have been developed by Rohm and Haas. This company uses tebufenozide as the active ingredient, which control caterpillars. Two other members in this series, halofenozide and methoxyfenozide have also been developed (Scheme 29).

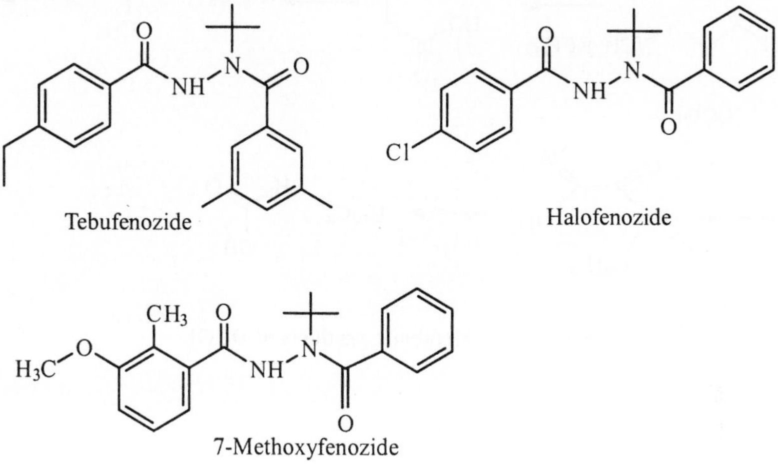

Tebufenozide

Halofenozide

7-Methoxyfenozide

Scheme 29

For more details refer to real-world cases in green chemistry.[28]

15.19 An Environmentally Safe Marine Antifoulant

The antifouling agents are used on boat hulls to reduce the build up of marine organisms (such as barnacles, algae, plants and diatoms). A build up of these organism causes additional costs involved in increased fuel consumption and cleaning time. Tributyltin (TBT) compounds were earlier used as antifoulant agents. One of the main drawbacks was their persistence in the environment and bioaccumulation in various nontargent marine organisms.[29]

Rohm and Haas has developed the use of 4,5-dichloro-2-n-octyl-4-isothiazolin-3-one (DCOI) as an antifouling agent.[28]

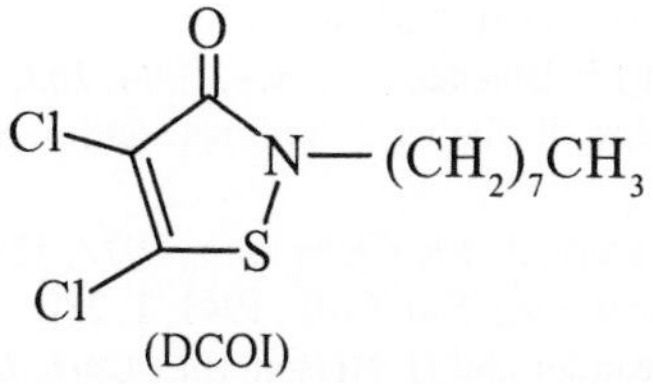

(DCOI)

Unlike tributyltinoxide (TBTO), the DCOI is far less persistent in marine environments. Also in the case of DCOI, the product of the metabolism

(Scheme 30) are non-toxic.[28]

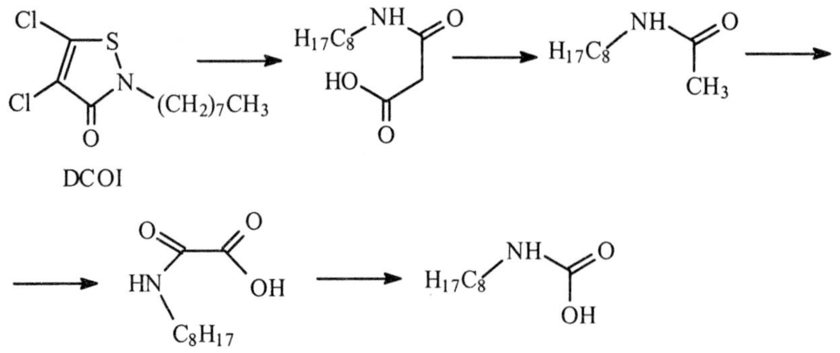

Scheme 30. Metabolic products of DCOI

References

1. M. Karen Draths, W. John Frost, Environmentally Compatible Synthesis of Adipic Acid from D-Glucose, *J. Am. Chem. Soc.*, 1994, **116**, 399-400.
2. T. Paul Anastas, C. Tracy Williamson, Eds., Green Chemistry: Frontiers in Benign Chemical Synthesis and Processes, Oxford University Press, New York, 1998.
3. A. Roger Sheldon, Catalyst and Pollution Prevention, *Chem. Ind.*, 1997, 12-15.
4. W.D. McGhee and D.P. Riley, Organometallics II, 1993, 900-907; W.D. McGhee, D.P. Riley, M.E. Chrost and K.M. Christ, *Organometallics*, 1993, **12**, 1429-1433; D.P. Riley, Y. Pan and D.D. Riley, *J. Chem. Soc. Chem. Commun.*, 1994, 699-700; T.E. Waldman, D.P. Riley, *J. Chem. Soc. Chem. Commun.*, 1994, 957-958.
5. M.K. Stern, F.D. Hileman and J.K. Bashkin, *J. Am. Chem. Soc.*, 1992, **114**, 9237-8; M.K. Stern and B.K. Cheng, *J. Org. Chem.*, 1993, **58**, 6883-8; M.K. Stern, B.K. Cheny, F.D. Mileman and J.M. Allman, *J. Org. Chem.*, 1994, **59**, 5627-32.
6. P. Tundo, F. Trotta and G. Moraglio, *J. Org. Chem.*, 1987, **52**, 1300; P. Tundo, F. Trotta, G. Moraglio and F. Ligorati, *Ind. Eng. Chem. Res.*, 1998, **27**, 1565; P. Tundo, F. Trotta and G. Moraglio, *J. Chem. Soc. Perkin Trans. 1*, 1989, 1070; P. Tundo and M. Selven, *Chemtech.*, 1995, 31-35.
7. J.M. Tanko, R.H. Mas and N.K. Suleman, *J. Am. Chem. Soc.*, 1990, **112**, 5557; J.M. Tanko, N.K. Suleman, G.A. Hulvey, A. Park and J.E. Powers, *J. Am. Chem. Soc.*, 1993, **115**, 4520-26; J.M. Tanko and N.K. Suleman, *J. Am. Chem. Soc.*, 1994, **116**, 5112-6; J.M. Tanko and J.F. Blackert, *Science*, 1994, **263**, 203-5.
8. J. Smidt, W. Hafner, R. Jira, R. Sieber, J. Sedlmeer and A. Sabel, *Angew. Chem. Int. Ed. Engl.*, 1962, **1**, 80.
9. J.K. Stille and R. Divakaruni, *J. Am. Chem. Soc.*, 1978, **100**, 1303.
10. R. Adams and V. Voorhees, *Org. Syn. Coll.*, 1941, **1**, 280.
11. For a review, see O. Theander and D. Nelson, *Adv. Carbohydr. Chem. Biochem.*, 1988, **46**, 273; F. Hoppe-Seyler, *Ber.*, 1871, **4**, 15; M.S. Feather, *Tetrahedron Lett.*, 1970, 4143; H.S. Ibell, *J. Res. Natt. Bur. Stand.*, 1994, **32**, 45.
12. R.A. Sheldon, Chriotechnology: The industrial synthesis of optically active compounds, Marcel Dekker, New York, 1993.

13. H.U. Blaser and F. Spindler, *Topics in Catalysis*, 1998, **5**, 275.
14. U.S. Patents, 4,981,995 and 5,068,448; 1991.
15. J. Le Bars, J. Dakka and R.A. Sheldon, *Appl. Catal. A. General*, 1966, **136**, 69.
16. B. Notari, *Stud. Surt. Sci. Catal.*, 1988, **37**, 413.
17. W.A. Suske, M. Held, A. Schmid, T. Fluschmann, M.G. Wubbolts and H.P.E. Kohler, *J. Biol. Chem.*, 1997, **227**, 24257; A. Schmid, H.P.E. Kohler and K.H. Engesser, *J. Mol. Catalysis*, 1998, **5**, 311.
18. J. Hasegawa and T. Ohashi, *Chemistry Today*, 1996, 44.
19. M. Tokunaga, J.F. Larrow, F. Kakuichi and E.N. Jacobsen, *Science*, 1997, **277**, 936; M.G. Wubbolts, J. Hoven, B. Melgert and B. Witholt, *Enzyme and Microbial Technology*, 1994, **16**, 887.
20. M.G. Wubbolts, P. Reuvekamp and B. Witholt, *Enzyme Microbial Technology*, 1994, **16**, 608.
21. W.F. Hoelderich, *Stud. Surf. Sci. Catal.*, 1993, **75**, 127; W. Aquita, H. Fuchs, O. Wörz, W. Ruppel and K. Halbritter, *U.S. 6013 843*, 2000.
22. A. Stocker, O. Marti, T. Pfammatter, G. Schreiner and S. Brander, *DE 2 04 6 566*, 1970.
23. H. Beschke, H. Friedrich, K.P. Müller and G. Schreyer, *DE 2517 054*, 1975.
24. T. Lüssling, H. Schaefer and W. Weigert, *DE 1948 715*, 1969.
25. S. Jaras and S.T. Lundin, *J. Appl. Chem. Biotechnol.*, 1977, **27**, 499; R. Yokoyama and S. Sawada, *U.S. 3803 156*, 1969; M. Gasier, J. Haber and T. Macheg, *Appl. Catal.*, 1987, **33**, 1.
26. W.F. Foelderich and F. Kollmer, *Pure Appl. Chem.*, 2000, **72**, 1273 and the references cited therein.
27. Colin Baird, *Environmental Chemistry*, 2nd edn., W.H. Freeman, New York, 1999.
28. For more details refer to real world cases in Green Chemistry, Michael C. Cann and More E. Cannelly, American Chemical Society, 2000 (ISBN 0-8412-37336).
29. A. Rouhi, *Chem. Eng. News*, 1998, 41-42.

Suggested Readings

1. Green Chemistry, Theory and Practice, Paul T. Anastas and John C. Warner, Oxford University Press, 1998, New York, USA.
2. Real-World Cases in Green Chemistry, Michael C. Cann, Mare E. Connelly, American Chemical Society, 2000.
3. Green Chemical Synthesis and Processes, Paul T. Anastas, Luren G. Heine and Tracy C. Williamson (Editors), ACS Publication 2000.
4. Green Chemistry in India, M. Kidwai, Pure and Applied Chmistry, Vol. 73, No. 8, 1261-1263, 2001.
5. Pure and Applied Chemistry, Special Topic Issued on Green Chemistry, IUPAC, Vol. 72, No.7, July 2000.
6. Organic Synthesis: Speical Techniques, V. K. Ahluwalia and Renu Aggarwal, Narosa Publishing House, 2001, New Delhi.
7. Green Chemistry: An Introductory Text, Mike Lancaster, Green Chemistry Network, University of York, RSC, 2002.
8. Green Chemistry in Indian Context: Challenges Mandates and Chances of Success, Upasana Bora, Mihir K. Chaudhri and Sanjay K. Dehury, Current Science, Vol. 82, 1427, 2002.
9. Organic Synthesis in Water, Paul A. Grieco (Editor), Blackie Academic and Professional, London, UK.
10. Organic Reactions in Aqueous Media, CHAU-JUN Li and TAK-Hang Chan, John Wiley & Sons Inc., New York.

Index